COLSTON RESEARCH SOCIETY

COLSTON PAPERS

General Editor
R. L. GREGORY

This book is Volume XXXII of the Colston Papers. Previous volumes are:

Vol.   I.   1948   *Cosmic Radiation*—Editor, F. C. Frank (out of print)
Vol.   II. 1949   *Engineering Structures*—Editor, A. G. Pugsley (out of print)
Vol.   III. 1950   *Colonial Administration*—Editor, C. M. MacInnes (out of print)
Vol.   IV. 1951   *The Universities and the Theatre*—Editor, D. G. James (out of print)
Vol.   V.   1952   *The Suprarenal Cortex*—Editor, J. M. Yoffrey (out of print)
Vol.   VI. 1953   *Insecticides and Colonial Agricultural Development*—Editors, T. Wallace and J. T. Martin (out of print)
Vol.   VII. 1954   *Recent Developments in Cell Physiology*—Editor, J. A. Kitching (out of print)
Vol.   VIII. 1956 *The Neurohypophysis*—Editor, H. Heller (out of print)
Vol.   IX. 1957   *Observation and Interpretation*—Editor, S. Körner (out of print)
Vol.   X.   1958   *The Structures and Properties of Porous Materials*—Editors, D. H. Everett and F. S. Stone (out of print)
Vol.   XI. 1959   *Hypersonic Flow*—Editors, A. R. Collar and J. Tinkler
Vol.   XII. 1960   *Metaphor and Symbol*—Editors, L. C. Knights and Basil Cottle (out of print)
Vol.   XIII. 1961 *Animal Health and Production*—Editors, C. S. Grunsell and A. I. Wright
Vol.   XIV. 1962   *Music and Education*—Editor, Willis Grant (out of print)
Vol.   XV. 1963   *Reality and Creative Vision in German Lyrical Poetry*—Editor, A. Closs
Vol.   XVI. 1964   *Economic Analysis for National Economic Planning*—Editors, P. E. Hart, G. Mills and J. K. Whitaker (out of print)
Vol.   XVII. 1965 *Submarine Geology and Geophysics*—Editors, W. F. Whittard and R. Bradshaw
Vol.   XVIII. 1966 *The Fungus Spore*—Editor, M. F. Madelin
Vol.   XIX. 1967   *The Liver*—Editor, A. E. Read
Vol.   XX. 1968   *Towards a Policy for the Education of Teachers*—Editor, William Taylor (out of print)
Vol.   XXI. 1969   *Communication and Energy in Changing Urban Environments*—Editor, D. Jones
Vol.   XXII. 1970 *Regional Forecasting*—Editors, M. Chisholm, A. E. Frey and P. Haggett (out of print)
Vol.   XXIII. 1971 *Marine Archaeology*—Editor, D. J. Blackman
Vol.   XXIV. 1972 *Bone: Certain Aspects of Neoplasia*—Editors, C. H. G. Price and F. G. M. Ross
Vol.   XXV. 1973   *Foreign Relations of African States*—Editor, K. Ingham
Vol.   XXVI. 1974 *Structure of Fibrous Polymers*—Editors, E. D. T. Atkins and A. Keller
Vol. XXVII.   1975   *The Eruption and Occlusion of Teeth*—Editors, D. F. G. Poole and M. V. Stack
Vol. XXVIII.   1976   *Remote Sensing of the Terrestrial Environment*—Editors, R. F. Peel, L. F. Curtis and E. C. Barrett
Vol.   XXIX. 1977 *Ions in Macromolecular and Biological Systems*—Editors, D. H. Everett and B. Vincent
Vol.   XXX. 1978   *Tidal Power and Estuary Management*—Editors, R. T. Severn, D. L. Dineley and L. E. Hawker
Vol.   XXXI. 1979 *Income Distribution: the Limits to Redistribution*—Editors, D. Collard, R. Lecomber and M. Slater

Biological Clocks in Seasonal Reproductive Cycles

Editors:
B. K. Follett and D. E. Follett

Proceedings of the
Thirty-second Symposium of the Colston Research Society

held in the
University of Bristol   March–April 1980

Bristol: Scientechnica
1981

Published by
John Wright & Sons Ltd,
42–44 Triangle West, Clifton,
Bristol, BS8 1EX. 1981

British Library Cataloguing in Publication Data
Biological clocks in seasonal reproductive
   cycles. — (Colston papers; 32)
   1. Biological rhythms—Congresses
   2. Estrus—Congresses
   I. Follett, B. K.   II. Follett, D. E.
   591.2'1'882   QH527

ISBN 0 85608 032 2

Text set in 10/11 pt Linotron Times,
printed and bound in Great Britain
at The Pitman Press, Bath

FOREWORD

For a number of years the Colston Society, in association with the University of Bristol, has supported an Annual Symposium on some subject of particular scientific or cultural significance. It was my pleasure to be President of the Society in 1980 and to be involved at first hand in further promoting the interests of the Society.

The subject of the 1980 Symposium—Biological Clocks in Seasonal Reproductive Cycles—was of great interest to the expert participants and also to lay members of the Society, who particularly enjoyed Professor Aschoff's brilliantly judged Public Lecture.

It will be in keeping with the spirit of the Colston Society if the proceedings of the 1980 Symposium promote greater understanding of the complex nature of biological rhythms. It would be exciting if they were to play a part in stimulating the commercially valuable application of present knowledge to agriculture and horticulture.

Sir John Wills

CONTENTS

Page

	Preface	ix
	Participants	xi
C. S. Pittendrigh	Circadian organization and the photoperiodic phenomena	1
W. Engelmann	Photoreception and the clock	37
W. W. Schwabe	Flowering hormones and inhibitors in photoperiodism	45
J. K. Gaunt and E. S. Plumpton	Photoperiodic control of tocopherol oxidase in *Xanthium*: an *in vitro* model	57
D. S. Saunders	Insect photoperiodism: entrainment within the circadian system as a basis for time measurement	67
B. Dumortier and J. Brunnarius	Involvement of the circadian system in photoperiodism and thermoperiodism in *Pieris brassicae* (Lepidoptera)	83
S. Masaki and S. Kikukawa	The diapause clock in a moth: response to temperature signals	101
T. L. Page	Localization of circadian pacemakers in insects	113
A. D. Lees and J. Hardie	The photoperiodic control of polymorphism in aphids: neuroendocrine and endocrine components	125
H. Underwood	Circadian clocks in lizards: photoreception, physiology and photoperiodic time measurement	137
E. Gwinner	Circannual rhythms: their dependence on the circadian system	153
M. Menaker, D. J. Hudson and J. S. Takahashi	Neural and endocrine components of circadian clocks in birds	171
B. K. Follett, J. E. Robinson, S. M. Simpson and C. R. Harlow	Photoperiodic time measurement and gonadotrophin secretion in quail	185
J. A. Elliott	Circadian rhythms, entrainment and photoperiodism in the Syrian hamster	203
B. Rusak	Mammals—general discussion	219
R. L. Goodman and F. J. Karsch	The hypothalamic pulse generator: a key determinant of reproductive cycles in sheep	223
K. Hoffmann	The role of the pineal gland in the photoperiodic control of seasonal cycles in hamsters	237
F. W. Turek and G. B. Ellis	Steroid-dependent and steroid-independent aspects of the photoperiodic control of seasonal reproductive cycles in male hamsters	251
J. Herbert	The pineal gland and photoperiodic control of the ferret's reproductive cycle	261
J. Aschoff	Twenty Years On: the Annual Colston Lecture	277
	Species Index	289
	General Index	291

PREFACE

Seasonal reproduction is an important feature in the life of most animals and plants, and should an individual breed at any time of the year other than the ideal it is likely to pass few if any genes to the next generation. Therefore, while organisms have adopted a bewildering array of tricks and adaptations to ensure reproductive success one feature is common to all those which practice seasonal breeding: they must have a clock telling them when the reproductive season is imminent. In this symposium an attempt has been made to consider this biological clock in a wide range of organisms and to this end sessions were devoted to plants, insects and to a variety of vertebrates. Many of the papers are devoted to photoperiodism and to the ways in which daylength is measured and then transduced into the production of hormones for flowering, the induction of diapause, or reproduction.

I was extremely fortunate in being able to persuade Professor Colin Pittendrigh to give the Introductory Lecture. This *tour de force* on the Sunday evening provided the stimulus for the rest of the papers and Pittendrigh's perceptive and intellectually challenging article opens this volume. In it he provides a persuasive theoretical case for believing that photoperiodic time measurement in many organisms is accomplished by some kind of seasonal rearrangement of the internal circadian rhythms and this point emerges in a number of the later papers. I was also fortunate in that Professor Jürgen Aschoff was willing to give the annual Colston Lecture. This highly entertaining lecture, which also included some intriguing new scientific ideas, appears at the end of the volume. Since Professor Aschoff retires this year as Director of the rhythm group of the Max-Planck-Institut für Verhaltensphysiologie it is surely only appropriate to dedicate this volume to him.

Seasonal breeding and photoperiodism in particular is not simply a topic of great academic interest but is also one of substantial commercial importance. Already, many farmers and horticulturalists use quite sophisticated changes in daylength to enhance production and many of these treatments have their origins in work on the seasonal clock. I hope that one product of this symposium will be to stimulate scientists in agricultural research to develop new and better ways of increasing food production.

Earlier symposia on biological rhythms (Cold Spring Harbor, 1960; Fedalfing, 1964; Friday Harbor, 1969; Dahlem, 1975; Naito Foundation Symposium in Tokyo, 1978) were not simply highly successful gatherings but played a valuable role in shaping subsequent research in the field. I would like to imagine that this Colston Symposium—the thirty-second of the series sponsored jointly by the University and the City of Bristol—will fit naturally with the previous meetings. The programme was devised with the advice of an internationally based committee and it owes everything to the enthusiastic contributions made by the speakers and those who took part in the many discussions. Some of these discussions have been included after the individual lectures and I hope they convey some feeling of the meeting itself. I am indebted not only to the speakers but also to the colleagues who led the various formal discussions: Professors E. Bünning, D. S. Farner, C. S. Pittendrigh, B. Rusak and R. V. Short.

Thanks for financial support are due to a number of organizations. The Colston Society itself provided many of the funds for travel, local accommodation and the annual dinner and has underwritten the production costs of this volume. In addition, grants for travel were received from The Royal Society, The British Council, British Imperial Tobacco Ltd, Hoechst GMBH, The Naito Foundation and the Journal of Reproduction & Fertility Limited. I should also like to thank the members of my Agricultural Research Council Research Group in Bristol who carried much of the responsibility for the local arrangements. Without their efforts the symposium would have been far less successful. Finally, I would like to express my appreciation to the Chairman of the University Committee, Professor D. H. Everett, and to the Administrative Secretary of the Colston Society, Mr J. H. M Parry, for their support and advice.

B.K.F.

PARTICIPANTS

Dr K. J. Adams	Department of Biology, North East London Polytechnic
Mr O. X. Almeida	MRC Reproductive Biology Unit, Edinburgh
Dr J. Arendt	Department of Biochemistry, University of Surrey
Professor J. Aschoff	Max-Planck-Institut für Verhaltensphysiologie, Erling Andechs
Dr E. Bittman	Department of Pathology, University of Michigan
Dr J. Brady	Imperial College Field Station, Ascot, Berks
Dr K. Brinkmann	Botanisches Institut der Universität, Bonn
Dr N. Bromage	Department of Biological Sciences, Aston University
Dr K. Brown-Grant	Department of Anatomy, University of Newfoundland
Dr J. Brunnarius	Station de Zoologie, INRA, Versailles
Professor E. Bünning	Institut für Biologie, University of Tübingen
Dr H. M. Charlton	Department of Human Anatomy, University of Oxford
Dr J. Clarke	Department of Agricultural Sciences, University of Oxford
Dr K. E. Cockshull	Glasshouse Crops Research Institute, Littlehampton, Sussex
Dr C. Coen	Department of Human Anatomy, University of Oxford
Dr S. Daan	Department of Zoology, University of Groningen
Dr D. T. Davies	Institute of Terrestrial Ecology, Abbots Ripton, Cambs
Dr S. Dobson	Institute of Terrestrial Ecology, Abbots Ripton, Cambs
Professor B. T. Donovan	Institute of Psychiatry, University of London
Dr B. Dumortier	Station de Zoologie, INRA, Versailles
Dr. J. A. Elliott	Department of Biological Sciences, Stanford University
Mr D. Ellis	Department of Zoology, University of Bristol
Dr W. Engelmann	Institut für Biologie, University of Tübingen
Professor D. S. Farner	Department of Zoology, University of Washington
Professor B. K. Follett	Department of Zoology, University of Bristol
Mr R. G. Foster	Department of Zoology, University of Bristol
Dr J. K. Gaunt	Department of Biochemistry and Soil Science, University College of North Wales
Professor J. Glover	Department of Biochemistry, University of Liverpool
Dr B. Goldman	Department of Zoology, University of Connecticut
Dr A. R. Goldsmith	Department of Zoology, University of Bristol
Dr C. A. Grocock	Department of Human Anatomy, University of Oxford
Dr E. Gwinner	Max-Planck-Institut für Verhaltensphysiologie, Erling Andechs
Mr C. R. Harlow	Department of Zoology, University of Bristol

Dr S. Hansen	Department of Psychology, University of Göteborg
Dr J. Herbert	Department of Anatomy, University of Cambridge
Dr K. Hoffmann	Max-Planck-Institut für Verhaltensphysiologie, Erling Andechs
Miss J. Holmes	Department of Zoology, University of Bristol
Dr D. J. Hudson	Department of Biology, University of Oregon
Dr J. B. Hutchison	MRC Unit on Behaviour, University of Cambridge
Dr J. W. Jacklet	Neurobiology Research Centre, University of New York
Dr M. D. R. Jones	Department of Biology, University of Sussex
Dr F. J. Karsch	Department of Pathology, University of Michigan
Dr H. Kawamura	Mitsubishi-Kasei Institute of Life Sciences, Tokyo
Dr M. Klinowska	Department of Anatomy, University of Cambridge
Dr C. Laud	Department of Biochemistry, University of Surrey
Professor A. D. Lees	Imperial College Field Station, Ascot, Berks
Dr P. C. B. MacKinnon	Department of Human Anatomy, University of Oxford
Dr S. Masaki	Department of Entomology, Hirosaki University
Professor M. Menaker	Department of Biology, University of Oregon
Dr T. J. Nicholls	Department of Zoology, Queen Mary College
Dr.T. L. Page	Department of Biological Sciences, Stanford University
Dr E. Peterson	Department of Biology, University of Sussex
Dr P. Pevet	Netherlands Institute of Brain Research, Amsterdam
Professor C. S. Pittendrigh	Department of Biological Sciences, Stanford University
Mrs J. M. Prosser	Department of Zoology, University of Bristol
Dr J. E. Robinson	Department of Zoology, University of Bristol
Dr B. Rusak	Department of Psychology, Dalhousie University
Dr D. S. Saunders	Department of Zoology, University of Edinburgh
Professor W. W. Schwabe	Wye College, University of London
Dr P. J. Sharp	Poultry Research Centre, Edinburgh
Professor R. V. Short	MRC Reproductive Biology Unit, Edinburgh
Dr S. M. Simpson	Department of Zoology, University of Bristol
Dr S. Skopik	Department of Biology, University of Delaware
Dr P. Södersten	Department of Psychology, University of Göteborg
Ms E. Steel	MRC Unit of Behaviour, University of Cambridge
Dr M. Takeda	Department of Biology, University of Delaware
Dr F. W. Turek	Department of Biological Sciences, Northwestern University
Dr H. Underwood	Department of Zoology, North Carolina State University
Mr H. F. Urbanski	Department of Zoology, University of Bristol
Dr C. Whitehead	Department of Biological Sciences, Aston University
Dr H. L. Williams	Department of Animal Husbandry, Royal Veterinary College
Dr S. Wilson	Department of Physiology & Biochemistry, University of Reading

CIRCADIAN ORGANIZATION AND THE PHOTOPERIODIC PHENOMENA*

Colin S. Pittendrigh

Hopkins Marine Station of Stanford University,
Pacific Grove, California, USA

The great majority of plants and animals, invertebrate and vertebrate, have evolved, as part of their innate organization, oscillating systems whose periods are a close match to one, or more, of the major physical cycles in the environment. These oscillations with circa*dian*, circa*tidal*, circa*lunar* or circa*annual* periods are clock-like in several respects: their period is remarkably stable and homeostatically conserved in the face of environmental change;[1] and they are readily entrained by some environmental cycle (especially the light/dark cycle in the circadian case) that not only imposes (as zeitgeber) its own frequency on the internal clock-oscillator but phases it appropriately to local time.[2]

Circadian oscillators serve as clocks in several different ways: (*a*) they underlie the remarkable phenomena of zeitgedachtnis,[3] first studied in honeybees which, having found a nectar source at, say 3 pm, return to that source at the same clock time on following days; (*b*) they constitute the chronometer used by many arthropods and vertebrates in time-compensated sun-orientation;[4] (*c*) more generally, they provide the framework for a temporal programming of biological activity to appropriate times of day; and (*d*) as Bünning correctly suggested 45 years ago,[5] they are involved in the time measurement underlying the photoperiodic phenomena. A particular 'daylength' or 'photoperiod' serves as a cue to the time of year appropriate for flowering, for gonadal growth or for diapause etc. There is a time measurement implicit in the organism's discrimination between long and short photoperiods and, in the great majority of species, this timing again involves circadian organization.

My concern here is with the last two categories: the nature of the innate temporal programme in a circadian system and its potentiality for photoperiodic time measurement.

A mouse retained in an aperiodic environment (food, temperature and light held constant) goes through a succession of undamped circadian cycles, not only of the more obvious functions of activity and rest (including sleep) but of physiological change in many (all?) of its organs, central and peripheral. The pineal synthesizes and secretes melatonin only during the animal's 'subjective night' when it is active, but growth hormone is released by day when the mouse sleeps. The spontaneous electrical activity of most brain centres is maximal during the night but it peaks during the day[6] in the suprachiasmatic nuclei (SCN). The chemistry of blood and

* This paper is dedicated to my friend Professor Jürgen Aschoff on the occasion of his 68th birthday and retirement from the Max-Planck Gesellschaft. Professor Aschoff and his institute at Erling Andechs have provided leadership and inspiration in the development of circadian physiology for nearly 25 years.

Biological Blocks in Seasonal Reproductive Cycles 1–35 (1981) (ed. B. K. and D. E. Follett: Bristol, Wright).

urine varies systematically through the cycle owing to circadian timing of adrenal and kidney function; and the enzymatic content of intestinal wall and liver shows major surges in anticipation of the animal's regularly programmed feeding behaviour. One of the central questions in circadian physiology is how this elaborate programme is 'read'; and one of the most important results in recent years has been the demonstration of its dependence on an anatomically localized pacemaker.

In mammals the SCN is such a pacemaker; its ablation leads to aperiodicity of all other functions that are circadian in intact animals. Its isolation, by Halász knife surgery,[6] from the rest of the brain has the same effect—rhythmicity elsewhere is lost but persists within the hypothalamic island that contains the SCN pacemaker. In sparrows removal of the pineal has similar consequences[7] and its reimplantation into the anterior chamber of the eye promptly restores rhythmicity of the whole animal. The central question is now sharpened: how does a discrete pacemaker oscillating in a small anatomical centre of the central nervous system assure proper mutual timing of events scattered spatially throughout peripheral organ systems and temporally throughout the entire day?

One of the general propositions to be pursued here is that the pacemaker is not the only autonomous oscillatory component in a circadian system. The arrhythmicity of the whole animal which follows pacemaker removal is essentially a loss of synchrony among a host of slave oscillations in other tissues and organs; their entrainment by the system pacemaker is what assures not only synchrony but proper mutual phasing (and hence a temporal programme) of the slave oscillations in the intact animal.

The multi-oscillator nature of circadian systems is not, itself, a novel proposition: it was a central theme in my discussion of circadian organization at Cold Spring Harbor in 1960;[8] in the last decade Aschoff's laboratory[9] has produced the most compelling evidence of separable autonomous oscillators controlling the circadian rhythms of body temperature and activity/rest in man; and in reporting the system arrhythmicity that follows pacemaker removal in sparrows and mammals respectively, both Menaker and co-workers[10] and Rusak[11] have explicitly noted evidence of residual oscillations whose loss of synchrony is what renders the animal as a whole aperiodic. There are comparable findings from crickets[12] and roaches.[13] Moreover it has seemed almost self-evident for some time[8,14-16] that the normal temporal structure must derive from a mixture of hierarchical and mutual couplings within the multi-oscillator system. There has, however, been little progress in developing more explicit models with promise of explaining why, for instance, major phase-shifts should impair the system; why there are penalties in driving the system at other than 24-hour periods; and how, even in principle, circadian organization might sense—as it does—the seasonal change in daylength.

This paper has three related goals. The first is to report some of the now compelling evidence that the circadian system gating emergence time in *Drosophila* involves a distinct slave oscillator as well as a pacemaker. Some of the principles involved in this analysis lead to the second theme, which concerns the strategic advantages and consequences (in an evolutionary sense) of a pacemaker/slave architecture for circadian organization. The third aim is to show that the temporal programme of such an organization is, almost inevitably, subject to seasonal change as its pacemaker's waveform is altered by the lengthening or shortening of the daily photoperiod.

PACEMAKER AND SLAVE OSCILLATIONS IN *DROSOPHILA*

In a population of developmentally asynchronous *Drosophila* the insects become pharate (fully adult) at random times through the day but the population's emergence activity is limited to a narrow window, about 6 h wide, once per day. The emergence of each insect, no matter when it became pharate, awaits the next signal from a circadian gating system. The synchrony of individual gating systems, effected by the daily light/dark cycle, is responsible for the population's rhythm of emergence activity. We are concerned here with the structure of the gating system. It comprises two distinct oscillatory components: (*a*) the pacemaker and (*b*) a slave oscillator responsible for the gating signal itself.

Pacemaker of the gating system

The system's pacemaker is a self-sustaining oscillator that is entrainable by temperature cycles but has a temperature-compensated free-running period.[20] This clock-like feature is characteristic of all circadian pacemakers. It is remarkably responsive to light and hence readily entrainable by light/dark cycles to which it is 'directly' coupled by an extra-ocular pathway. The time course of the pacemaker's steady-state motion can be assayed experimentally as a succession of phases (ϕ_n) each of which is characterized by a unique phase shift ($\Delta\phi_n$) response to a standard light pulse. A plot of this relation [$\Delta\phi_n(\phi_n)$] yields a so-called phase response curve or PRC.

Figure 1 gives the PRC for *D. pseudoobscura* pupae (or larvae) subjected to 15 min pulses of white light ($\sim$50 lux). Its general pattern is common to the PRCs of all circadian pacemakers: $\Delta\phi$ responses are small during the so-called 'subjective day' which is the half-cycle (CT 0 < CT 12) that coincides with light in the entrained steady-state; the bigger responses in the 'subjective night' are clearly phase delays to begin with but in the late subjective night can be interpreted as either advances or delays. For example, the phase shift following a pulse at CT 20 is either an 18 h delay or a 6 h advance. Figure 1 gives two alternative forms of the PRC: one treats the shifts in the late subjective night as advances, the other treats *all* shifts as delays, yielding a smooth monotonic curve.

The monotonic PRC is a useful tool in several ways. In figure 2 it tracks the pacemaker's time course through three full cycles of a free run at 20 °C in constant darkness following release from entrainment by a light/dark cycle (LD 12 : 12). The phase (CT 18·5) responding to a pulse by a 12 h delay is taken as the phase reference point for the pacemaker. It occurs 6·5 h (then modulo $\tau = 24$ h) after entry into darkness: the free run began from CT 12 at the end of the last seen light. The median of the eclosion peak is taken as phase reference for the more easily assayed rhythm itself. It occurs 15·3 h (then modulo τ thereafter) after entry into darkness and, hence, coincides with the pacemaker phase designated CT 3·3. The phase relation (ψ_{EP}) of eclosion peak (E) to pacemaker (P) is thus −8·8 h.

Comparable data in the upper panels of figure 3 show the phase relation of the pacemaker's PRC to the last seen light/dark transition for several photoperiod durations. For all photoperiods longer than $\sim$9 h the interval between light-off and CT 18·5 is constant at −6·5 h, but it is longer after shorter photoperiods. The lower panel in figure 3 plots the reference phases for both pacemaker (CT 18·5) and eclosion peak (median) for essentially all possible photoperiods in a 24-hour cycle. Although the phase relation to dawn of both pacemaker (ψ_{PL}) and eclosion peak

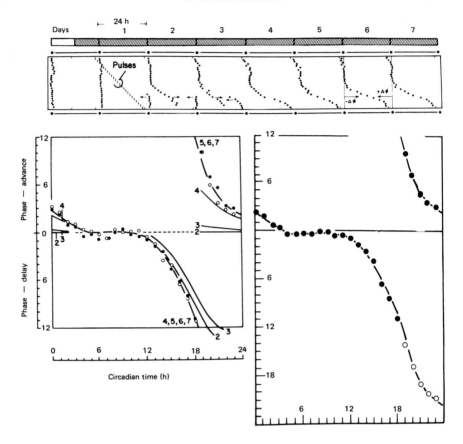

Figure 1. Phase response curve: *Drosophila pseudoobscura* eclosion rhythm—15 min pulses of white light (50 lux) at 20 °C. *Upper panel:* Twenty-five pupal populations are released from LD 12 : 12 into DD. The history of each population is given as a horizontal line of points that mark the successive eclosion peaks. The control population is shown separately above and (repeated) below the box. Each population within the box receives a light pulse (vertical bar, marked Pulses) at a different phase of the free-running cycle. The phase shifts caused by the light are characteristic of the phase pulsed and reach steady-state by the fourth or fifth day after the pulse. Some populations delay to new phase $(-\Delta\phi)$, some advance $(+\Delta\phi)$; in some populations part of the population delays and part advances. *Bottom left:* PRCs based on the data of the upper panel (*see text*). Curves based on the transients (peaks on days 2, 3 and 4) are not the same as those based on the steady-state. The advances proceed more slowly than the delays. *Bottom right:* The steady-state PRC. When the advances are displaced 360° one obtains a monotonic form of the PRC: all phase shifts are treated as delays.

(ψ_{EL}) becomes more negative as the photoperiod increases, the phase relation (ψ_{EP}) of eclosion peak to pacemaker is never the same. This is especially clear in photoperiods of longer than 16 h, as ψ_{EL} becomes more positive. *Clearly the gating signal that times the flies' eclosion behaviour is not associated with any fixed phase point in the pacemaker's cycle.*

   The same result emerges from analysing the effects of temperature on the gating system. First, the PRCs measured at 10°, 20° and 26° show that the detailed time course of the pacemaker (PRC shape) in each cycle is, like its frequency, temperature-compensated (figure 4). The phase relation (ψ_{PL}) of pacemaker to light cycle is accordingly temperature-compensated, but that is not true of the

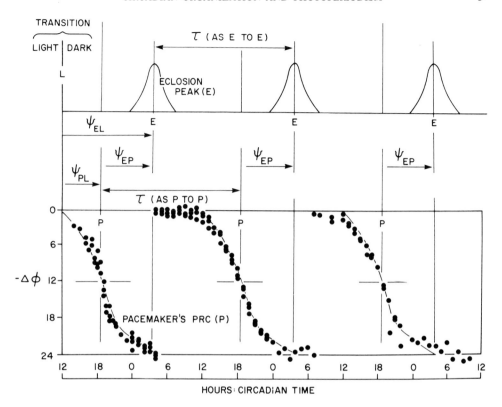

Figure 2. The phase relationship (ψ_{EP}) of eclosion peak (E) to its pacemaker (P) in *Drosophila*. The time course of the pacemaker through 72 h of a free run (after release from a 12 : 12 light/dark cycle) is measured as a succession of $-\Delta\phi$ responses. The pacemaker phase (CT 18·5) where $-\Delta\phi$ is 12 h is taken as phase reference for the pacemaker (P). The peak midpoint (E) is taken as phase reference for eclosion. The light/dark transition (L) is taken here as phase reference for the previous light cycle. ψ_{PL} is −6·5 h; ψ_{EL} is −15·3 h; ψ_{EP} is −8·8 h. The period (τ) of the gating system can be estimated as the interval from E to E or from P to P.

phase relation (ψ_{EL}) of eclosion peak to light cycle. The peak occurs progressively later as the temperature is lowered (figure 4).

The timing of eclosion relative to the pacemaker is subject to genetic as well as phenotypic variation. At temperatures near 20 °C ψ_{EP} is more negative in males than females. Moreover, laboratory selection of parents emerging at the beginning or at the end of the daily distribution of emergence activity has produced two strains (*Early* and *Late*) of D. *pseudoobscura* in which ψ_{EP} differs by about 4 h (figure 5).

The lability of ψ_{EP} again emerges as a conspicuous feature of the gating system's resetting behaviour following exposure to a single light pulse. The PRC used in earlier paragraphs was based on steady-state phase shifts several days after the light pulse was seen. In reaching the new steady-state the system goes through several 'transient' cycles, each of which yields a different PRC (figure 1). The important conclusion that these transients do not accurately reflect the pacemaker's resetting behaviour is based on experiments exemplified by figure 6. Many pupal populations synchronized by a prior light régime are released into DD with their free-running motion beginning from CT 12 as figure 2 showed. All 25 populations

Hours Since Light/Dark Transition

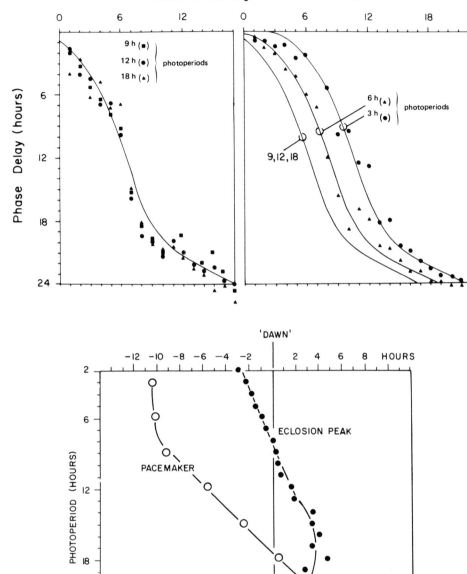

Figure 3. The dependence of ψ_{PL} and ψ_{EL} on photoperiod in *Drosophila*. *Upper panels:* The phase relation (ψ_{PL}) of pacemaker (P) to light cycle (L) is assayed by measuring the phase of the PRC relative to lights-out following entrainment by LD 3 : 21, 6 : 18, 9 : 15, 12 : 12 and 18 : 6. For all photoperiods longer than ~9 h the phase of the PRC is the same relative to the onset of darkness. CT 18 : 5 (where $\Delta\phi$ = 12 h) falls 6·5 h after lights-out; pacemaker motion at dark onset begins from CT 12. After 6 : 18 it begins from ~CT 10 and after 3 : 21 from CT 7·5. *Lower panel:* The phase where a light pulse causes $-\Delta\phi$ of 12 h is used as phase reference for the pacemaker; the median of the daily peak is taken as phase reference for the associated eclosion. The phase relation of both (pacemaker and eclosion) relative to dawn changes as photoperiod increases; so does their phase relation (ψ_{EP}) to each other.

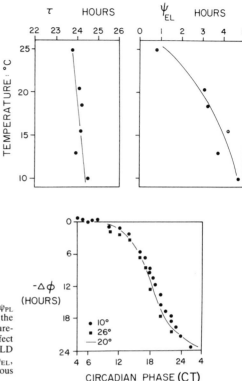

Figure 4. The differential effect of temperature on τ, ψ_{PL} and ψ_{EL} in *Drosophila*. *Top left:* The period (τ) of the free-running rhythm is only very slightly temperature-dependent. *Bottom:* Temperature has no measurable effect on PRC shape, nor on its phase (ψ_{PL}) to the previous LD cycle. *Top right:* Temperature has a marked effect on ψ_{EL}, the phase of eclosion peaks in DD relative to the previous LD cycle, and, hence, on ψ_{EP}.

in the lower panel receive a 'first' pulse 8 h later when the pacemaker is at phase CT 20. That pulse is known to cause a 6 h phase advance (or 18 h delay) of the eventual steady-state which develops only gradually over four or five cycles if measured by the time of the eclosion peak (figures 1 and 12). One of the 25 populations is left as a control; each of the other 24 receives a 'second' or 'tester' pulse at successively later hours after the first. If the 'first' pulse were to cause an instantaneous 6 h phase advance, the pacemaker's subsequent time course would be described by an abrupt 6 h advance of the steady-state PRC; if, however, the pacemaker's phase shift behaviour were accurately reflected by the timing of subsequent (transient) eclosion peaks, it would not, because the first two peaks after the pulse show no or very little shift. As figure 6 shows, an instantaneous shift of the (steady-state) PRC does accurately predict the phase shifts caused by subsequent 'tester' pulses. The upper panel in figure 6 gives a comparable analysis of a 6 h phase delay. Clearly the *gradual* resetting behaviour of the rhythm of eclosion peaks (figure 1, upper panel; figure 12, upper panel) does not accurately reflect the pacemaker's resetting kinetics; and during the transients intervening between old and new steady-states there is a profound disruption of the steady-state value of ψ_{EP}.

The discrete resetting behaviour of the pacemaker makes it easy to analyse its entrainment by cycles of light pulses. An entrained steady-state is reached when the

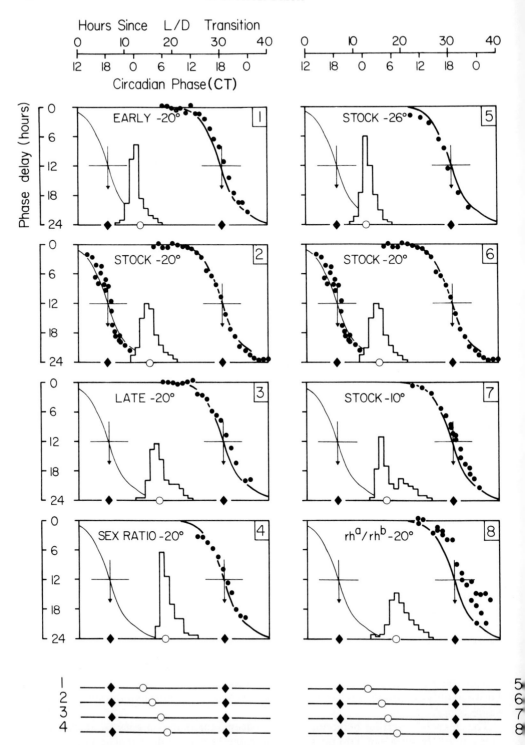

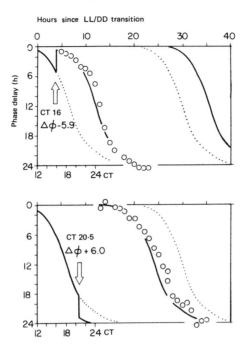

Figure 6. The nearly 'instantaneous' resetting behaviour of the pacemaker. *Upper panel:* Nineteen pupal populations are synchronized by an LL/DD transition. All 19 receive a 'first' pulse (at CT 16) 4 h after entry into DD; it causes an eventual 5·9 h phase delay of the rhythm. An instantaneous 5·9 h delay of the steady-state PRC at the time of the 'first' pulse accurately predicts the phase shifts caused by second, 'tester', pulses given (to separate replicates) 1,2,3 . . . 18 h after the 'first' pulse. The dotted PRC marks the time course of the pacemaker, had it not been exposed to the first pulse. *Lower panel:* A similar experiment in which the tester pulses track the time course of the pacemaker phase advance caused by the 'first' pulse falling at CT 20.

pulse in each cycle (with a period of T hours) falls at the same pacemaker phase (ϕ_n) such that

$$\Delta\phi_n = \tau - T \qquad (1)$$

Equation (1) allows us to predict the steady-state phase relation (ψ_{PL}) of pacemaker to light cycle for any value of T. The solid curve at the left of figure 7 (left panel) plots the times—predicted in this way—at which the pacemaker's reference phase (CT 18·5) should occur for all light cycles from $T = 18$ to $T = 30$ h. This prediction of ψ_{PL} has been tested experimentally by measuring the phase relation of the pacemaker's PRC (relative to the last seen entraining light pulse) following release into DD from cycles of $T = 21$ and $T = 27$ h. As the observed values (open circles) for $T = 21$ and $T = 27$ show, the prediction of ψ_{PL} from equation (1) is remarkably accurate (*see* Pittendrigh[15]). In a DD free run ψ_{EP} is −8·8 h (figure 2); if it were insensitive to pacemaker period the eclosion peaks in

Figure 5 (opposite). The lability of ψ_{EP} in *Drosophila.* Circadian time 18·5, where a light pulse causes −$\Delta\phi$ of 12 h is again taken as phase reference point (◆) for the pacemaker and the median of the peak as phase reference (○) for eclosion activity. Both are measured relative to the L/D transition preceding the free run in which both PRC and eclosion activity are assayed. The solid curve describing the PRC is the same in all eight panels and is the best fit to the data (●) for *Stock* at 20 °C (panels 2 and 6). The observed PRC data (●) in other panels show that the pacemaker's phase relation (ψ_{PL}) to the last seen light is the same for *Stock* at all temperatures and in *Early, Stock* and *Late* at 20 °C. The phase relation (ψ_{EL}) of eclosion to the light is, however, temperature-dependent and different in *Early, Stock* and *Late* at the same temperature; so, therefore, is ψ_{EP} (*see* summary at bottom of figure).

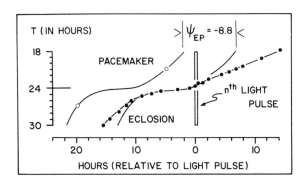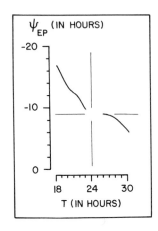

Figure 7. The dependence of ψ_{EP} on T (light cycle period) in *Drosophila pseudoobscura*. (Based on the data given in figure 19.) *Left:* The phase relations of pacemaker (ψ_{PL}) and eclosion peak (ψ_{EL}) to light cycles with periods ranging from $T = 18$ to $T = 30$ h (one 15 min pulse/cycle). The solid curve to the left is the computed time at which the pacemaker's reference phase (CT 18·5) occurs relative to the light pulse; the open circles (○) at $T = 21$ and $T = 27$ are observed values confirming prediction. The solid curve to the right indicates where the associated eclosion peaks would occur if the value of ψ_{EP} found in a DD free run (figure 2) was realized in all cycles independent of T. The solid points are medians of the observed eclosion peaks. Clearly ψ_{EP} varies with T, becoming more negative as T shortens (*right*).

cycles ranging from $18 < 30$ h would therefore fall on the ψ_{PL} curve displaced 8·8 h to the right. As figure 7 shows, however, they do not: ψ_{EP} is only $-8\cdot8$ at or near $T = 24$; it becomes more negative when entrainment shortens pacemaker period and more positive when it is lengthened.

The gating oscillation (slave to the pacemaker)

The dependence of ψ_{EP} on the ratio τ/T is promptly explained by the proposition, introduced many years ago by Victor Bruce, Peter Kaus and myself,[17] that the signal releasing eclosion behaviour is given by a second (B-) oscillator in the gating system that is coupled to and driven by the pacemaker (A-oscillator). Recent computer simulations with an explicit version of this model have provided a remarkably satisfactory explanation of the eclosion rhythm in *D. pseudoobscura* and leave no doubts about its essential validity. Its utility is illustrated in the following sections with a few of its most easily reported applications.*

* I am indebted to my friend Professor T. Pavlidis for the main Fortran program we have used in simulating pacemaker/slave relationships. The system is defined by the following equations:

Pacemaker (A oscillator):
$$dR/dt = R + (0\cdot5 - 0\cdot8S) - 0\cdot3S^2 - L, R \geqslant 0 \tag{1}$$
$$dS/dt = R - 0\cdot5S \tag{2}$$
Slave (B oscillator):
$$dX/dt = Y \tag{3}$$
$$dY/dt = C(S - 1\cdot13) - \varepsilon(X^2 - 1) \cdot Y - \omega_B^2 \cdot X \tag{4}$$

Equations (1) and (2) are Pavlidis' model[18] of the *D. pseudoobscura* pacemaker. Equations (3) and (4) are the van der Pol oscillator which we have used as the slave. L is the light intensity; C is a coefficient coupling the pacemaker to the slave; ε is a stiffness parameter (a damping coefficient in some sense); ω_B, or $(1/\tau_B)$ is the frequency of the B oscillator. Ms Rita Kan has written the extensive library of special programs that permit one to explore the model's responses to a wide range of simulated light and temperature régimes. A fuller account of the model is in preparation[19] and addresses several aspects of the eclosion rhythm that are ignored here. These include (*a*) the large rhythm (as distinct from pacemaker) transients generated by temperature steps that have been described,[8,20] (*b*) the dependence of peak width on, among other things, the period (T) of the light cycle driving the pacemaker; and (*c*) other complexities (*see* figure 18) in the system's response to light cycles with periods ranging from 12 to 36 h.

Our starting concern is with the lability of the phase relation (ψ_{EP}) between eclosion event (E) and pacemaker (P). The central proposition in the model is that E is determined by a particular phase in the slave or B-oscillator's motion; and that the observed lability of ψ_{EP} is explained by lability in ψ_{BA}, the phase relation between slave (gating) oscillator and its pacemaker. There are three parameters in the model which affect ψ_{BA} (figure 8).

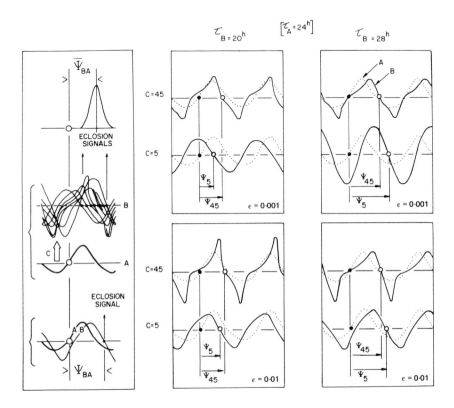

Figure 8. Pacemaker (A) and slave (B) oscillations in the gating system. *Left panel, bottom:* Time course of the pacemaker (A) and the slave oscillation (B) it drives. The upward crossing of zero is taken as phase reference for A; in this figure the downward crossing of zero is taken as phase reference for B and is assumed to be when the eclosion signal is given. ψ_{BA} is the interval between phase references for B and A. *Middle:* Eight slave oscillations differing in τ_B, ε and C are driven by the same pacemaker. Because ψ_{BA} is different for each slave, each gives an eclosion signal at different times relative to the pacemaker. *Top:* The two-tailed distribution of eclosion events in each daily peak is attributed to a two-tailed distribution of ψ_{BA} values among the flies involved. The inter-fly differences in this figure are caused by differences in τ_B, ε and C, not in τ_A; small differences between flies in τ_A do, however, exist. *Middle and right panels:* The waveform and phase relation (ψ) of the eight different slaves (solid curves) to a common pacemaker (dotted curves). ψ_{BA} always increases as τ_B is lengthened, e.g. from 20 to 28 h. When $\varrho < 1.0$ (e.g. $\tau_B = 20$ h) an increase of C or ε increases ψ, but when $\varrho > 1.0$ (e.g. $\tau_B = 28$ h) an increase of those parameters reduces ψ, ψ_5 and ψ_{45} are the ψ_{BA}s developed when C is 5 and 45 respectively.

The first is $\varrho = \tau_B/\tau_A$, where τ_B is the free-running period of the slave and τ_A that of the pacemaker.

The second is ε, which is essentially a damping coefficient in the van der Pol equation used as slave oscillator.

The third is C, the strength with which the slave is coupled to the pacemaker.

When ϱ is increased ψ_{BA} always becomes more negative. When $\varrho < 1\cdot0$ an increase of either ε or C makes ψ_{BA} more negative, but when $\varrho > 1\cdot0$ ψ_{BA} becomes less negative as C or ε is increased. These aspects of ψ_{BA} variation are summarized by the families of curves in figure 9. Their precise shape is dependent on the choice

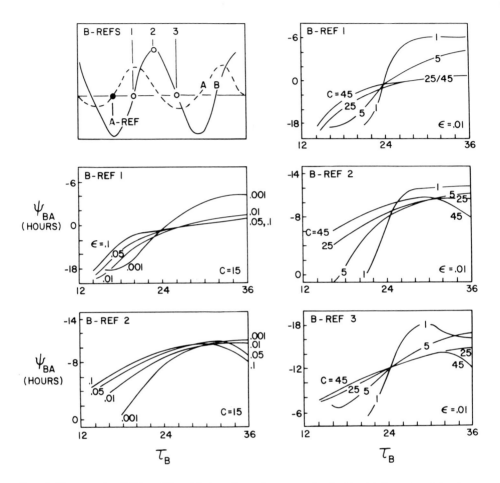

Figure 9. The pacemaker (A)/slave (B) model: dependence of ψ_{BA} on ϱ, C and ε. *Top left:* Pacemaker (A) and slave (B) in steady-state. The upward zero crossing is taken as phase reference for A; three different possible phase references are indicated for the slave. *Right:* Three families of curves describe the dependence of ψ_{BA} on ϱ and C. Since τ_A is held constant at 24 h, ϱ variation reduces to τ_B variation. ε is held constant at 0·01. The curves are somewhat different depending on the B-reference chosen. In all cases, however, an increase of C makes ψ_{BA} larger (more negative) when $\varrho < 1\cdot0$; and when $\varrho > 1\cdot0$ an increase of C makes ψ_{BA} more positive. *Lower left:* Comparable curves showing the effect of ε variation. C is constant at 15.

of a phase reference point for the slave but the major features are detail-independent; indeed, they are found again when a damped version of the van der Pol equation is used for the slave oscillation. For present purposes the principal features are as follows:

1. The range of ψ_{BA} variation realizable by the model is limited to a little more than $180°$ (<12 h).

2. The phase relation between any two slaves (differing in C or ε) is inverted when their free-running periods (τ_B) are changed from less than τ_A to more than τ_A.

3. Neither C nor ε is a major source of ψ_{BA} variation when ϱ is close to $1\cdot0$.

The daily distribution of eclosion events

In an eclosion rhythm free running in darkness, the width of successive peaks grows at an almost negligible rate. It follows that individual flies in the population have very similar pacemaker periods, and what little variance there is on τ_A cannot account for the large differences (<8 h) between flies in precisely when they emerge. Previous discussion of ψ_{EP} lability has focused on variation of $\overline{\psi}_{EP}$, the phase relation of peak midpoints to pacemaker at different temperatures and in different strains. Here we are evidently confronted with substantial ψ_{EP} variation between flies within a peak. The model interprets this as inter-fly variation in ψ_{BA}: flies emerging later in the peak have slave oscillators whose parameters (τ_B, ε and C) make ψ_{BA} more negative than those emerging earlier in the peak. Figure 10

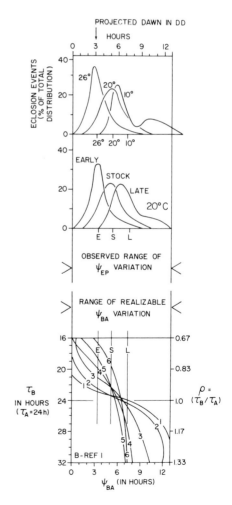

Figure 10. Interpretation of the 'daily' eclosion peak in *Drosophila* in terms of inter-fly variation in ψ_{BA}. *Upper panels* give the observed range of eclosion events (relative to pacemakers synchronized by a common previous LD cycle) for *Stock* at three temperatures, and for *Early*, *Stock* and *Late* at 20 °C. The observed range spans about 14 h which is a good match to the theoretical range (~12 h) of ψ_{BA} variations realizable by the model. *Lower panel:* The range of realizable ψ_{BA} values is illustrated by the dependence of ψ_{BA} for 6 slave oscillations (differing in C and ε) on change in their free-running period. Slave 1 has $\varepsilon = 0\cdot01$ and $C = 1$; 2 has $\varepsilon = 0\cdot001$ and $C = 5$; 3 has $\varepsilon = 0\cdot001$ and $C = 15$; 4 has $\varepsilon = 0\cdot001$ and $C = 35$; 5 has $\varepsilon = 0\cdot01$ and $C = 5$; and 6 has $\varepsilon = 0\cdot01$ and $C = 45$.

pursues this interpretation. The upper panels give normalized emergence peaks for three strains (*Stock*, *Early* and *Late*) of *D. pseudoobscura* at 20 °C and for *Stock* at three temperatures (26°, 20° and 10 °C). The lower panel plots ψ_{BA} as a function of τ_B/τ_A for a sample of gating (slave) oscillators differing in C and ε; τ_A is held constant at 24 h. The family of curves has been rotated 90° (relative to their orientation in figure 9) so that the ψ_{BA} axis now parallels the time axis on which the daily distribution of eclosion events is plotted. Several conclusions emerge from the details of figure 10.

1. The observed range of variation in emergence times within the species corresponds well with the maximum range of ψ_{BA} variation realizable by the model. The figure 'synchronizes' the earliest emergence events (smallest ψ_{EP} values) observed in the species with the smallest ψ_{BA} values realizable in the model.

2. The full (~14 h) range of observed emergence times (ψ_{EP} values) can only be matched by the model (ψ_{BA} values) if there are major differences between flies (within and between strains at different temperatures) in τ_B. If τ_B is invariant neither C nor ε can effect more than ~90° of ψ_{BA} variation (~6 h). Moreover, the necessary τ_B variation must be associated with weak C and small ε in order to change ψ_{BA} by 12 h.

3. The variation of eclosion times within a peak that is attributable to differences in ψ_{BA} can be caused by any combination of variations in τ_B, ε or C. An increase in τ_B always lengthens ψ_{BA} (later eclosion); but while an increase in C or ε similarly lengthens ψ_{BA} if $\tau_B < \tau_A$, it shortens ψ_{BA} (earlier eclosion) if $\tau_B > \tau_A$.

4. The peak midpoint ($\bar{\psi}_{EP}$) for *Stock* at 20 °C corresponds with a ψ_{BA} value that can only be realized in the model if $\bar{\tau}_B$ is less than τ_A. It follows that earlier flies in the peak (ψ_{BA} smaller) will have τ_B even shorter or C and ε smaller (or some combination of these three differences) than flies emerging later (ψ_{BA} larger). This, in turn, has strong implications concerning the effect of temperature on $\bar{\psi}_{BA}$ and the origin of the *Early* and *Late* strains.

5. Since $\bar{\psi}_{BA}$ must be shortened to explain the smaller $\bar{\psi}_{EP}$ found when *Stock* is exposed to higher temperatures (26 °C), we must conclude that if a rise in temperature affects τ_B it must be to shorten it; if it effects C and ε, it must be to reduce them. Conversely, lowering the temperature from 20 to 10 °C must lengthen τ_B and increase C and ε.

Similarly, the parameter changes underlying the origin of *Early* and *Late* (selection at 20 °C) must have been as follows: *Early* arose from some combination of reducing τ_B, C and ε while *Late* arose from a combination of lengthening τ_B and increasing C and/or ε.

All these implications from the model's interpretation of the daily distribution of emergence events are subject to experimental test.

ψ-reversals as ϱ passes through 1·0

The most compelling test involves the major delay ($\bar{\psi}_{BA}$ more negative) of *Stock's* peak at 10 °C, as compared to its phase at, say 27 °C. The peaks at these two temperatures span the entire 180° range of realizable ψ_{BA}. τ_B, unlike τ_A, must be temperature-dependent; and in lengthening from its average value at 27 °C to that at 10 °C $\bar{\tau}_B$ must pass through 24 h, i.e. ϱ must change from less to more than 1·0. Figures 10 and 11 show that if any inter-fly variation in $\bar{\psi}_{BA}$ traces to differences in C or ε, the passage of ϱ through 1·0 caused by the temperature effect on τ_B will

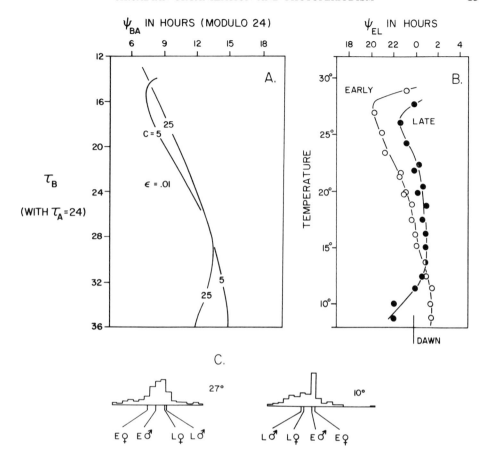

Figure 11. Phase reversals within the eclosion peak of *Drosophila* as τ_B lengthens at lower temperatures. A, ψ_{BA} for two slave oscillators with the same damping coefficient ($\varepsilon = 0.01$) but differing in C. As τ_B is lengthened and ϱ passes from <1 to>1 their phase relation is inverted. B, The effect of temperature on the phase relation of *Early* and *Late* to the light pulse (15 min) in a 24 h LD cycle. At higher temperatures *Early* is earlier than *Late*, but as temperature is lowered their phase relation is inverted. This is predicted if τ_B is lengthened at lower temperatures. C, The eclosion peaks (at 27 and 10 °C) of the synthetic population created by mixing *Early* and *Late*. At 27 °C the phase sequence is: *Early* females, *Early* males, *Late* females, *Late* males. At 10 °C the entire sequence is inverted.

necessarily entail extensive ψ-reversals within the emergence peak. Thus, if the sequence of flies within the peak at 27 °C is 1,2,3 . . . *n*, those same flies should emerge at 10 °C in the sequence *n* . . . 3,2,1. This wholly unexpected demand of the model is testable given the known differences in $\bar{\psi}_{EP}$ (hence $\bar{\psi}_{BA}$) between the *Early* and *Late* strains and between the sexes within each strain. By mixing pupae from *Early* and *Late* we establish a synthetic fly population with four markers of phase difference within the emergence peak (figure 11). At 27 °C the observed sequence is *Early* females, *Early* males, *Late* females and *Late* males. At 10 °C the model's prediction is met: the sequence is *Late* males, *Late* females, *Early* males and *Early* females. The curves in figure 11B give the medians for *Early* and *Late* populations (sexes mixed) on a temperature gradient from 27° to 8°C: the ψ-reversal of the strains occurs near 13 °C.

Rhythm transients

The phase-shifting behaviour of the slave oscillator in the model system has been explored for all values of τ_B from 12 to <36 h (holding τ_A at 24 h) and the full range of possible C and ε values. Several regularities emerge from these simulations of slave phase shifts (figures 12 and 15).

1. The instantaneous phase shift $(-\Delta\phi_A)$ of the pacemaker by a light pulse changes its phase relation to the slave from the steady-state value (ψ_{BA}) to ψ_{BA}^*. Unless C and ε are large, several (transient) cycles are necessary before ψ_{BA}^* relaxes to ψ_{BA} and the new steady-state is reached. The weaker the coupling (C small) and the smaller the slave's damping coefficient (ε) the longer (more cycles) the transients last.

2. The 'direction' of the slave's resetting behaviour (delay or advance) depends on the magnitude of the pacemaker's phase shift $(-D\phi_A)$. For small pacemaker delays the slave also delays but there is a critical ('watershed') value of $-\Delta\phi_A$ beyond which the slave advances from ψ_{BA}^* to ψ_{BA}. This watershed value of $-\Delta\phi_A$ depends on ψ_{BA} of the steady-state: the more negative ψ_{BA} the larger the $-\Delta\phi_A$ before the slave advances to new steady-state (figures 12–14).

3. There is often marked asymmetry in the rate at which advances and delays (of comparable magnitude) are effected. When $\varrho < 1.0$ and C and ε are small the advances go less rapidly than delays; when $\varrho > 1.0$ the converse is true (figure 15).

4. As C and ε get larger the slave resets very rapidly whether it delays or advances, and the asymmetry is lost.

All of these model properties bear on the observed behaviour of the gating system's transients.

The watershed value of $-\Delta\phi_A$ ('mixed transients')

The dependence of the slave's resetting behaviour (advance or delay) on a critical value of $-\Delta\phi_A$, combined with our interpretation of the daily peak as a frequency distribution of ψ_{BA} values among individual flies, immediately explains one of the strongest and otherwise perplexing features of rhythm transients. In *Stock* at 20 °C large pacemaker phase shifts reset the gating system of all flies in the population by the same amount, but the transient behaviour of individuals differs qualitatively: part of the population delays and the rest advances (figure 13; *see also* figure 1, upper panel). The pacemaker phase shift reaches or exceeds the watershed for earlier flies in the peak (they advance to new steady-state) but the later flies delay. The larger $-\Delta\phi_A$ the greater the fraction of the population that advances (figures 13 and 14); eventually, at CT 22, the $-\Delta\phi_A$ caused by the light pulse exceeds the watershed value for all individuals, all of which advance.

This explanation of 'mixed transients' requires that *Early* and *Late* will differ from *Stock*, and they do: the $-\Delta\phi_A$ value that causes 50 per cent of *Stock* to advance causes 75 per cent of *Early* but almost none of *Late* to advance. The same expectation is observed when the whole daily distribution is made earlier or later by temperature, e.g. after a 12 h $\Delta\phi_A$ the higher the temperature, the greater the fraction of the population that advances to the new steady-state.

Transient asymmetry

In accommodating the entire range of observed emergence times for the species (combining all genetic and temperature-induced variation) to the theoretical limits

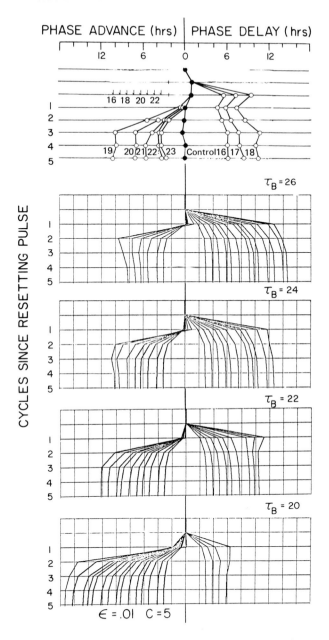

Figure 12. Transients of the eclosion rhythm in *Drosophila* following light pulses that (instantaneously) reset the pacemaker. *Upper panel:* Observed transients for *Stock* at 20 °C following pulses at CT 16, 17 . . . 23. The transients following pulses at CT 16, 17 and 18 are delays and proceed rapidly; the pulse at CT 19 (and later times) causes advances that proceed more slowly. *Lower panels:* Computer simulations of the slave oscillator's ($\varepsilon = 0\cdot01$, $C = 5$) resetting behaviour following light pulses causing a wide range of pacemaker phase shifts. When τ_B is 20 h the 'watershed' between delaying and advancing transients coincides with a pacemaker shift of ~6 h. As τ_B gets longer (and, hence ψ_{BA} more negative) the watershed is associated with progressively larger pacemaker phase shifts. (NB: The figure makes no attempt to illustrate the steady-state differences in ψ_{BA} associated with change in τ_B). The 'asymmetry' of the transients (delays reach steady-state faster than advances) observed in the flies is also seen in the model's behaviour when C and ε are small, as in the case illustrated by this figure.

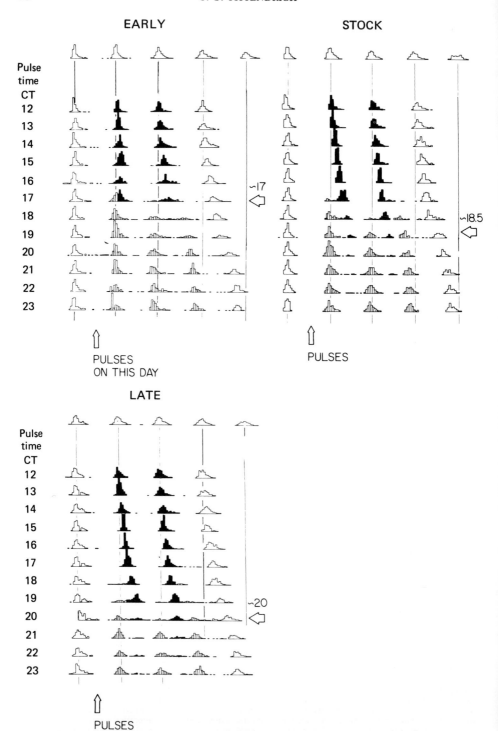

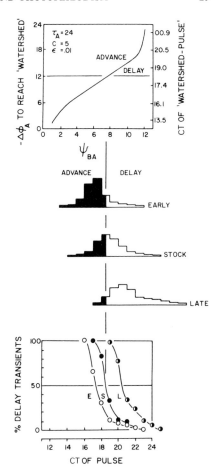

Figure 14. 'Mixed transients'. *Upper panel:* The 'watershed' value of $-\Delta\phi_A$ (pacemaker phase delay) separating advance and delay transients of the slave (C constant at 5, ε at 0·01) is computed for increasing values of τ_B and hence ψ_{BA}. The greater ψ_{BA} (corresponding with flies later in the peak) the greater is the 'watershed' value of $-\Delta\phi_A$. A pulse at ~CT 18 in the model causes a $-\Delta\phi_A$ that splits the pupal population of *Stock* into ~50 per cent delay/50 per cent advance. In *Early* a comparable value of $-\Delta\phi_A$ causes approximately 75 per cent of the pupae to advance but almost none of *Late* to advance.

of ψ_{BA} variation, we earlier concluded from the $\bar{\psi}_{EP}$ of *Stock* at 20 °C that its $\bar{\tau}_B$ was less than 24 h. That inference was then tested and confirmed by finding that ψ-reversal of the strains (*Late* emerging earlier than *Early*) occurred at a lower temperature. We have also had to assume that $\bar{C}$ and/or $\bar{\varepsilon}$ are small to account for the full 12 h range of realized ψ_{BA} values. When we set $\varrho < 1·0$ ($= \tau_B < \tau_A$) with C and ε small in the model system, there is a marked asymmetry of the slave's advance and delay transients. The advances require more cycles to reach steady-state, and the delays often show a small overshoot in the first cycle (figure 14). As figure 11 shows, these are both prominent features of the observed transients of *Stock* at 20 °C.

Figure 13 (*opposite*). 'Mixed transients': raw data. The records for *Early* (top left), *Stock* (top right) and *Late* (bottom) (20 °C) all have the same format. Each shows (top line) an unpulsed control if DD, and 12 experimental populations receiving a light pulse (15 min, 50 lux) at successively later circadian times (CT 12, 13 . . . 23). Populations *delaying* to new phases are shown in solid black; those *advancing* to new phases have vertical hatches. In *Stock* the transients consist of a mixture of delays and advances when the pulses fall between CT 17 and CT 22; the later the pulse the greater the pacemaker phase shift (expressed as $-\Delta\phi$) and the larger is the fraction of the population following an advance route to new steady-state. A pulse at ~CT 18·5 would split the population into 50 per cent delays, 50 per cent advances. In *Early*, the siwtch (50 per cent advances) requires a smaller pacemaker phase shift (pulse at ~CT 17) and in *Late* it requires a larger $-\Delta\phi_A$ (pulse at ~CT 20).

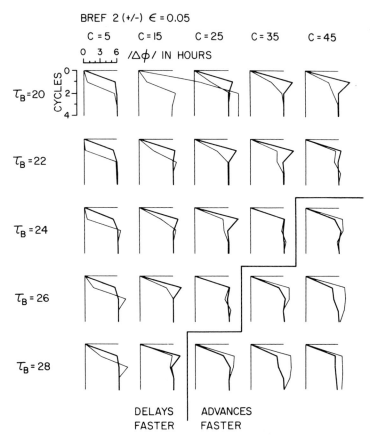

Figure 15. Asymmetry of advance and delay transients. Each panel in the figure uses the coordinates defined in the top left case: the heavy curve shows the growth of a 6 h delay phase through four successive cycles; the light curve traces the growth of a 6 h advance. In the case ($\tau_B = 20$, $C = 5$) the delay proceeds more rapidly than the advance. Bottom right, the advance goes faster than the delay. In general advances proceed more slowly than delays when $\tau_B < 24$ and C is small.

Transient kinetics: temperature effects and strain differences

Our starting interpretation of the daily distribution of emergence events (figure 10) involved the assumption that if, to delay the daily peak of *Stock*, low temperature may, even in part, affect C or ε it must do so by increasing these parameters. This leads, in turn, to the strong counterintuitive prediction that the rhythm should reset faster at lower temperatures. In the model the slave regains its steady-state phase relation to the pacemaker more rapidly when C or ε is increased. As figure 16 shows, this seemingly unlikely effect of temperature is indeed observed: the rhythm resets much more slowly at 26 ° than at 20 °C; and at 10 °C it resets essentially immediately. This temperature-dependence of the transients is, of course, in marked contrast to the invariance with temperature of the eventual steady-state phase shift expressed in the PRCs of figure 4.

By a similar argument we conclude that if genetic change in C or ε is involved in the (20 °C) origin of *Early* and *Late*, the slave oscillators in *Early* will be more

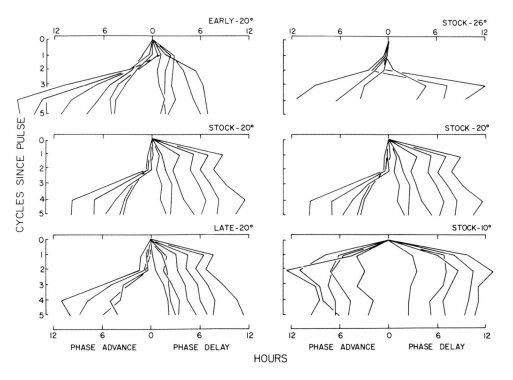

Figure 16. The duration of transients (rate of resetting) in *Drosophila*. Effects of temperature and strain differences. Each panel traces the time course (over five cycles) of the observed phase shift of eclosion peaks following light pulses that cause a wide range of pacemaker phase shifts. In *Stock* at 20° the delays proceed more rapidly than advances. In *Late* the pattern is the same. In *Early* all phase shifts proceed more slowly. In *Stock* at 26° the transients are even slower. At 10° the steady-state phase shift is almost complete in one cycle.

weakly coupled and more heavily damped than in *Stock* and, *á fortiori*, in *Late*. The transient kinetics of *Late* are not measurably different from those of *Stock*, but *Early* does indeed reset more slowly than *Stock* or *Late* as one would expect if C or ε were weakened in making ψ_{BA} more positive.

The origin of *Early* and *Late*

The phase relation of slave to pacemaker responds to change in ϱ or ε.* Since $\varrho = \tau_B/\tau_A$, the difference in $\bar{\psi}_{BA}$ caused by selection for *Early* and *Late* could therefore trace to change in either the slave oscillation (τ_B and ε) or its pacemaker (τ_A), and change in all three seems to be involved.

Change in τ_A: Differences in pacemaker period among the three strains have been measured by five different techniques which yield the same result. One of them is illustrated by figure 17. On the fifth day of a free run in DD at 17 °C the time course of *Early*'s pacemaker (measured by its PRC) is ~5 h behind that of *Late* and the difference grows in the sixth and again in the seventh cycle. The data

---

* In all the simulations with the model so far, an increase in C (strength of coupling between the oscillators) has had essentially the same effect as an increase in ε, the slave's damping coefficient. The discussion here is simplified by referring only to ε; but change in C would have the same effect.

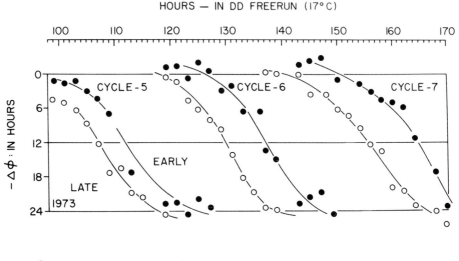

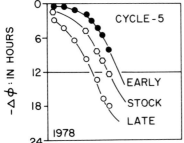

Figure 17. The pacemaker's period (τ_A) has been changed in the evolution of both *Early* and *Late. Upper panel:* The phase reference (-12 h phase shift response) of *Early's* pacemaker occurs ~5 h later than that of *Late* in the fifth cycle of a DD free run at 17 °C. The difference between the pacemakers grows in the sixth and seventh cycles. *Lower panel:* A comparable assay shows that τ_A in *Late* is shorter than that of *Stock* while τ_A in *Early* is longer.

in the lower panel show, moreover, that change in pacemaker period was involved in the evolution of both strains: τ_A in *Late* is shorter than it is in *Stock*, and in *Early* it is longer than in *Stock*.

There are several apparent riddles associated with this result. First, were pacemaker period the sole determinant of the rhythm's phase relation to the light cycle, *Early* (with τ_A longer) would clearly be later each day than *Late* (with τ_A short). Certainly the pacemaker of *Early* must be later (relative to the light cycle) than that of *Late* (figure 18). The riddle here is promptly resolved by reference to the pacemaker/slave structure of the gating system and the postulate that the slave, not the pacemaker, is the immediate determinant of emergence time. The slave's phase relative to the pacemaker (ψ_{BA}) responds to change in ϱ and hence to change in either τ_A or τ_B. When, as other data have implied, ε is small and ϱ is close to 1·0, a small lengthening of τ_A (e.g. 0·25 h) will have a greatly amplified impact on ψ_{BA}, shortening it by, for example, 2 h. Thus although the small lengthening of τ_A in *Early* surely makes ψ_{PL} slightly more negative than it is in *Stock*, the changed $\tau_B : \tau_A$ ratio not only offsets this but makes the phase relation (ψ_{BA}) of eclosion to pacemaker even more positive (figure 18) than in *Stock*.

The second riddle concerns the plausibility of a change in pacemaker period being attributed to selection that was exercised on pupal populations entrained to a light cycle (LD 12 : 12). Since all the pacemakers assume the same 24-hour period

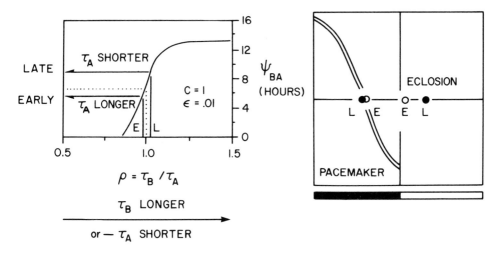

Figure 18. The effect of change in τ_A on ψ_{BA}. *Left:* The phase relation of slave to pacemaker (ψ_{BA}) as a function of ϱ ($=\tau_B/\tau_A$), with both C and ε small. When τ_A is lengthened ϱ is decreased and ψ_{BA} becomes smaller, as in *Early*. When τ_A is shortened ψ_{BA} is increased (as in *Late*). The resulting difference in ψ_{BA} is much greater than the difference in τ_A. *Right:* When the two stocks are entrained to a common light/dark cycle (LD 12 : 12) *Late*'s pacemaker phase-leads that of *Early* by a very small amount, but its slave oscillator (gating eclosion) phase-lags that of *Early*.

(T) as their zeitgeber, how can selection detect differences in their free-running periods? Simulations with the model again resolve the issue. When the slave oscillator in two gating systems is identical but the pacemakers have different periods (e.g. 23 and 25 h), ψ_{BA} is of course longer (more negative) in the free-running system with $\tau_A = 23$ h. When both systems are then entrained by LD 12 : 12 ($T = 24$ h) the pacemakers now have identical periods, but a residual difference in their waveform continues to cause a difference in ψ_{BA}, albeit smaller than in the free run (table 1).

Table 1. The effect of τ_A on ψ_{BA} in free-running and entrained steady-states of two oscillators (1) and (2)

	Pacemaker	Slave					
	τ_A	τ_B	C	ε	ϱ	ψ_{BA}	$\Delta\psi_{BA}$
Free-running							
(1)	23	24	5	0·01	1·04	−7·95	−0·66
(2)	25	24	5	0·01	0·96	−7·29	
*Entrained**							
(1)	24	24	5	0·01	1·0	−7·24	−0·29
(2)	24	24	5	0·01	1·0	−6·96	

* $T = 24$; LD 12 : 12.

Change in τ_B: Change in the slave's period (τ_B) also appears to be involved in the differentiation of both strains from *Stock*. That is, at present, the most likely explanation of the observations summarized in figure 19, which shows daily emergence peaks in pupal populations entrained to light cycles with periods (T) ranging from 16 to 34 h (one 15 min pulse/cycle). Using equation (1) (p. 9), we

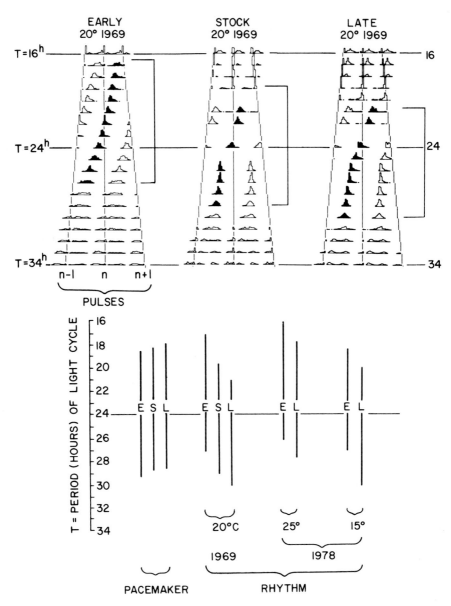

Figure 19. The limits of frequency-following behaviour in *Early*, *Stock* and *Late* strains of *Drosophila*. *Upper panels: Early*, *Stock* and *Late* (20 °C). The entrained steady-states effected by light cycles (one 15 min pulse/cycle) with periods (*T*) ranging from 16–34 h. Three pulses (*n* − 1, *n*, *n* + 1) are shown for each cycle, with pulse *n* synchronized for comparison of the different *T*s. In *Stock*, peaks of 'normal' width (blackened) are realized for all *T*s between 19 to 29 which is the range of the pacemaker's frequency following (FF) behaviour. Beyond those limits the peaks either broaden (*T* > 29) or become bimodal (*T* < 19) with a marked discontinuity of phase relative to that expressed in the FF range. In *Early* the corresponding limit of the eclosion rhythm's apparent FF behaviour is shifted to shorter *T* values, and in *Late* it is shifted to longer *T*s. *Lower panels (left):* The model predicts small differences between *Early*, *Stock* and *Late* in the pacemaker's FF range (*left*). The differences are in the opposite direction from the observed shifts in the rhythm's behaviour (*right*). Temperature also affects the rhythm's range of FF behaviour: it is shifted (in both *Early* and *Late*) to longer *T*s when the temperature is lowered.

(Selection on the *Early* and *Late* strains was discontinued in 1969 (after 60 generations) and the strain difference has decayed slightly following the relaxation of selection in subsequent years.)

find that the pacemaker will only entrain, in a frequency following (FF) mode, to cycles within the range $T = 18 < 29$. Within this FF range the steady-state is characterized by the light pulse falling at the same pacemaker phase (ϕ_i; such that $\phi_i = \tau - T$) in each successive cycle. Beyond the FF limits (T less than 18 or more than 29 h) the pacemaker also entrains but in a frequency dividing (FD) mode in which the light pulse falls at the same ϕ_i in every second cycle; it falls at ϕ_j in the intervening cycle; and $\phi_i + \phi_j = \tau - T$. The transition from FF to FD entrainment is gradual: in cycles with T only slightly less than 18 or greater than 29, ϕ_i and ϕ_j are so similar that what is, strictly, FD behaviour is, in a sense, only 'noisy' FF behaviour. The 'limits' of frequency following are therefore somewhat ill defined.

The data in figure 19 show that in *Stock* at 20 °C the emergence peak is essentially 'normal' (in width) in all cycles from $T = 19 < T = 29$; and as figure 7 showed earlier, their phase relation to the pacemaker (ψ_{EP}) is systematically more negative as T is shortened. That is, of course, predicted by the model—as τ_A is shortened by the light cycle ϱ increases and ψ_{BA} becomes negative, and as τ_A is lengthened ϱ becomes smaller and ψ_{BA} more positive. Beyond these FF 'limits' the emergence pattern becomes 'abnormal' (either clearly bimodal or much dispersed) as the slave oscillators respond to the complex waveform of the pacemaker in its FD mode. Figure 19 shows, moreover, that *Early* and *Late* differ from each other, and from *Stock*, in the range of T values within which 'normal' emergence peaks are found. *Early*'s range is displaced to shorter and *Late*'s to longer T values. The slave oscillators in the two selected strains evidently differ in their ability to follow the pacemaker to *its* limits of FF behaviour and this is most readily explained if their own free-running periods (τ_B) are different. Thus, if selection had shortened τ_B in *Early*, that would not only contribute to making ψ_{BA} more positive in any steady-state but improve the slave's ability to follow the pacemaker entraining to the shortest values of T, and on the other hand impair its ability to follow to the longest values of T. Clearly a lengthening of τ_B in *Late* would have the converse consequences.

If the range of light cycle periods to which the slave can entrain is indeed limited by its own period, then lowering the temperature (known from figure 11 to lengthen τ_B) should shift both lower and upper limits of that range to longer Ts. Figure 19 shows that this is indeed the case for both *Early* and *Late*. This temperature effect cannot, of course, be attributed to the pacemaker whose period (τ_A) is temperature-compensated.

Change in ε: The much slower time course of *Early*'s transients (figure 16) compared with those of both *Stock* and *Late*, leaves no doubt that change in ε (or C) has contributed to the origin of that strain. A smaller damping coefficient, like reduced coupling, not only makes ψ_{BA} more positive (earlier) but reduces the rate at which the slave regains its steady-state phase relation to the reset pacemaker.

Laboratory selection maintained for 60 generations at 20 °C has changed ψ_{BA} of *Stock* by essentially the same amount in *Early* and *Late*. Selection has clearly exploited genetic variation of ϱ in the evolution of both strains; and the $\Delta\tau_A$ in *Early* and *Late* appears comparable in magnitude (but with opposite sign). Change in ε, however, is evidently limited to the evolution of *Early*. Both observations are readily explained in terms of the curves describing ψ_{BA} as a function of ϱ and ε given in figure 9. When ϱ is close to 1·0, as it evidently is in *Stock* at 20 °C (*see* figure 10), ψ_{BA} is readily lengthened (*Late*) or shortened (*Early*) by change in ϱ. But with ϱ close to 1·0, change in ε has little impact on ψ_{BA}. Moreover, as the evolution of *Late* based on $\Delta\varrho$ continues, change in ε becomes even less effective as a basis for lengthening ψ_{BA}. On the other hand, as the initial evolution of *Early* based on $\Delta\varrho$ proceeds, $\Delta\varepsilon$ becomes progressively more effective as the basis for the further shortening of ψ_{BA}. It is then to be expected that genetic variation in ε should play a larger role in the origin of *Early* than of *Late*.

CIRCADIAN ORGANIZATION: THE TEMPORAL PROGRAMME

Dissection of the *Drosophila* gating system into separable pacemaker and slave oscillations provides some hints about the structure of circadian programmes in general. We began by noting the now clear evidence that the temporal programme of a circadian system promptly decays following removal of a central pacemaker. How can a localized central pacemaker ensure the proper sequential timing of events elsewhere in the system? The analysis of *Early* and *Late* suggests an answer: a central pacemaker can readily assure such temporal order if the subsystems it controls are themselves oscillatory. Slave oscillations in peripheral tissues, differing in τ_B, ε or C, will necessarily develop different ψ_{BA} values in the entrained steady-state imposed by the pacemaker: the system's programme then reduces to the array of divergent ψ_{BA} values among the subsystems. The genetic lability of ψ_{BA} for the individual slave oscillators is a central feature of this paradigm: it provides for the evolutionary adjustment of one component in the temporal programme without disturbance of the rest. It is recalled that in the moth *Pectinophora gossypiella* we found[21] that when *Early* and *Late* strains were established by selection on the emergence time of pupae there was a correlated response in the phase of the egg-hatch rhythm. The same slave oscillation (releasing eclosion hormone) is evidently involved in timing both events. However, the adult moth's oviposition activity, which is subject to equally strong circadian control, was unchanged in *Early* and *Late*—one component (slave oscillator) of the temporal programme had been changed without impact on another.

While the independent phase adjustment of one component in the programme can indeed be effected solely by change in the parameters (τ_B and ε) of the individual slave involved, the history of *Early* and *Late* shows that selection will inevitably influence the pacemaker's period. In nature, concurrent pressures to stabilize the phase of earlier and later components in a programme presumably cancel each other's influence on τ_A, leaving change in slave parameters as the more likely route for adaptive adjustment in the phase of any single component. On the other hand, if the direction of selection were to phase advance or phase delay the entire programme (or even most of it), as in the differentiation of day- and night-active species within some group, pacemaker period would surely be changed. It may well be that this intrinsic property of a pacemaker/slave basis for circadian programming is responsible for the well-known empirical rule ('Aschoff's') that pacemaker periods are longer in day-active species (c.f. *Early*) than in night-active species (c.f. *Late*).

Interactions among the constituent oscillators in a circadian organization are surely more complex than this simple picture of hierarchical control by a single pacemaker, which on the one hand imposes temporal order among slave oscillations and, on the other, by virtue of its own entrainment to the light/dark cycles, assures proper phasing of the entire programme to the external day. A conspicuous feature of the data already available is that some of the residual oscillators left after pacemaker ablation (in sparrows and hamsters) can themselves be entrained by the LD cycle: the control of these slaves' phase in the intact animal is clearly more complex than our simple model suggests. We also know that mutual as well as hierarchical couplings play a significant role in establishing the normal programme. Bilaterally redundant pacemakers are mutually coupled in roaches;[22,23] separate, non-redundant, pacemakers control the rhythms of body temperature and sleep/wakefulness in man[9] and in the majority (over 80 per cent) of people are mutually coupled, sharing the same frequency. 'The pacemaker' of circadian locomotor rhythms in rodents[15,24,25] and lizards[26] is equally clearly a complex of two, non-redundant, mutually coupled oscillators that appear to respond differentially to evening and morning light.

The overall qualification before proceeding with our model is that real circadian organization, in higher metazoa at least, involves more than the simple two-level hierarchy of a single 'pacemaker' entraining a single tier of slaves. Nevertheless, there are properties of this clearly too-simple model that have a useful bearing on the physiology of real circadian systems; and the poorly defined complexities it ignores can only enrich, not detract from, its biological interest.

Figure 20 lays out the temporal programme established by nine slave oscillators coupled to the same pacemaker: interslave differences in τ_B, ε and C are responsible for the differences in ψ_{BA} of each slave. The free-running programme is

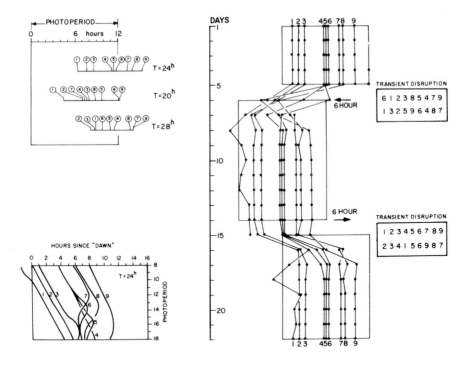

Figure 20. A temporal (circadian) programme established by nine slave oscillations driven by a common pacemaker. The nine slaves develop different phase relations (ψ_{BA}) from the pacemaker ($\tau_A = 24$ h) because of differences in τ_B, C and ε. Top left: The temporal sequence established by entrainment to three light cycles, each with a 12 h photoperiod and $T = 24$, 20 and 28 h. Right: Transient disruption of the programme caused by a 6 h phase advance and a 6 h phase delay of the light cycle. Bottom left: Change in the temporal programme caused by change of photoperiod in a 24 h LD cycle.

only slightly perturbed when the system is entrained to an LD 12 : 12 ($T = 24$ h) zeitgeber. When, however, the period (T) of the light cycle is changed, the programme is significantly disrupted; the sequence of some components is inverted. Similar programme disruptions are entailed when the pacemaker is subjected to an abrupt phase shift of the entraining cycle. Both abnormal lighting régimes (non-24-hour periods and repeated major phase shifts) are known to impair physiological performance (such as growth rate in plants and human performance) and entail significant penalties such as reduced longevity.[27-30] Disruption of the normal temporal programme, which is responsible for these penalties, is inevitable given the pacemaker/slave architecture on which it rests.

For present purposes the most interesting and unexpected property of the system is its response to change in 'photoperiod' (duration of the light pulse) in the normal 24-hour day. As photoperiod increases, the phase of all nine slaves (relative to dawn) tends to shift to the right, but several of them do so at different rates. Thus when photoperiod changes as it does in nature, seasonal change in the temporal programme is an inexorable consequence (figure 20).

THE PHOTOPERIODIC PHENOMENA

Forty-five years ago Erwin Bünning[5] published his remarkable intuitive insight that what we now call circadian rhythmicity was causally involved in the photoperiodic phenomena which had, then, only recently (1928) been discovered. 'Bünning's hypothesis' has been thoroughly validated in the great majority of species studied in the past 20 years. One of the most used, and most powerful, protocols that test the general proposition was introduced by Nanda and Hamner.[31] An 8 h photoperiod in an 18 h cycle fails to induce flowering in Biloxi soybeans, but when it is combined with lengthened dark periods creating light cycles whose periods (T) range from 24 to 72 h a striking result emerges—the same photoperiod that is ineffective in an 18 h cycle is effective when $T = 24$ h, ineffective again at $T = 36$ h, effective at $T = 48$ h and ineffective again at $T = 60$ h (figure 21). Success or failure depends on the length of the light cycle (T), not solely on the duration (photoperiod) of the light. All the effective cycles have a period that is modulo-24 h. As figure 21 shows, similar results have been found using plants, insects, birds and mammals. Circadian periodicity is clearly involved in these cases of photoperiodic induction. Failure to find a positive 'Nanda–Hamner result' in several species of insects and a lizard is currently the principal basis for concluding that in this minority of cases the photoperiodic time measurement is executed by an 'hour-glass' not an oscillator. We return later to the force of such negative evidence.

When it is found, the positive Nanda–Hamner result surely implicates circadian organization, but it says nothing about *how* it is involved in sensing the change of daylength. A common theme runs through both classes of model ('external coincidence', 'internal coincidence') that address the 'how' question. Both envisage some 'coincidence device' effecting the time measurement. Bünning[5] himself introduced the model, now called 'external coincidence',[32] which sees induction depending on the coincidence in time of a particular phase in the circadian programme with a particular phase (light) in the external cycle. Some phases of the programme that occur in darkness at one season coincide with light when daylength is increased. The coincidence of these phases is essential for a photochemical step that initiates induction.

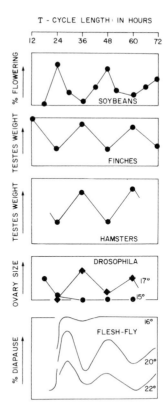

Figure 21. A sample of Nanda–Hamner transects. Each panel plots a changing photoperiodic response (percentage flowering, testis weight etc.) as a function of the length, in hours (*T*), of the entraining light/dark cycle. The duration of the photoperiod (light pulse) is held constant in each panel. The response rises and falls (modulo ~24 h) as a function of *T*. Biloxi soybeans from the data of Hamner;[31] finches (*Carpodacus*) redrawn from the data of Hamner;[41] hamsters (*Mesocricetus auratus*) from the data of Elliott et al.;[42] *Drosophila aurariae*;[40] flesh-fly (*Sarcophaga*) from the data of Saunders.[39]

A different model was introduced 25 years later,[38] and subsequently[32] distinguished from Bünning's under the label 'internal coincidence'. This model was prompted by my inference in 1960 that multicellular circadian systems were, almost necessarily, multi-oscillator systems and the implications, then vaguely understood, of the *Drosophila* system comprising A- and B-oscillators. The wholly intuitive guess at that time was that temporal order *within* the circadian system itself would be responsive to change in photoperiod: 'on one side of the critical photoperiod phase relations among constituent oscillator elements allow a particular reaction sequence to proceed; on the other side of the [photo]period widely different [phase] relations keep this metabolic pathway closed'.[38] The pacemaker–slave paradigm of circadian organization outlined here provides a more substantial foundation for this 'internal coincidence' approach. When daylength changes, the phase relation (ψ_{BA}) of each slave oscillation to its pacemaker changes, and some slaves differing in τ_B, C or ε, will change their phase relation to each other. Indeed the behaviour of the model multi-slave programme in figure 20 suggests that the problem for natural selection has not been so much to find ways in which circadian organization can sense seasonal change, but rather to evolve a programme that is sufficiently insensitive to changing photoperiod to be useful throughout the annual cycle.

The dependence on photoperiod of ψ_{BA} and $\psi_{B_1B_2}$ (the phase relation of any two slave oscillations to each other), exemplified in figures 20 and 24, promises explanation of several other, related aspects of photoperiodic induction that are difficult to reconcile with an 'external coincidence' model. One of these is the meaning of the 'circadian surfaces' in figures 22 and 23.

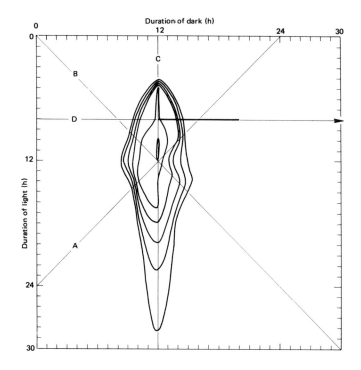

Figure 22. A circadian surface based on the photoperiodic induction of diapause in the European cornborer (*Ostrinia nubilalis*). (From Pittendrigh[33] based on the data of Beck.[34]) Each line is an iso-induction (or iso-diapause) contour defining all the cycles (differing in duration of light and dark) that will induce the same percentage of diapause. The surface defined by the contours peaks near $T = 24$, 12 L + 12 D. The lowest contour is 50 per cent, the highest 100 per cent. *See* text for discussion of the A, B and D transects across the surface.

Some years ago I detected[33] some circadian features in Beck's data[34] on the photoperiodic induction of diapause in the moth *Ostrinia*. Beck himself had concluded that the clock involved was an hour-glass. The circadian aspect of the responses emerges if one plots *iso-induction contours* on a surface whose x and y coordinates are the duration of the light pulse and the duration of darkness in each LD cycle to which the system has been exposed. An iso-induction contour defines all those LD cycles that elicit a given (per cent) photoperiodic response. Per cent diapause is then the z-axis of the surface that emerges when iso-induction contours are plotted for increasing levels of induction. The surface peaks at a value of $T (= L + D)$ near 24 h. Three transects across the peak are of interest: (*a*) transect A (with T constant at 24 h) yields the photoperiodic response curve for *Ostrinia*; (*b*) transect B, in which T is systematically lengthened and the photoperiod is $T/2$, corresponds with the protocols used by Went,[27] Pittendrigh and Minis[28] and St Paul and Aschoff,[29] showing that physiological performance is impaired when T is either longer or shorter and 24 h; and (*c*) transect D, in which the photoperiod is held constant and the dark interval lengthened to increase T, and which is the beginning

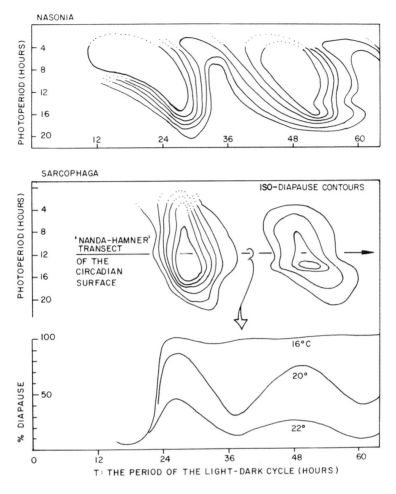

Figure 23. Extended circadian surfaces based on the photoperiodic induction of diapause in a wasp (*Nasonia*) and a flesh-fly (*Sarcophaga*). *See* text. The surface changes with temperature as the Nanda–Hamner results at 16 °, 20 ° and 22 °C show. (Redrawn from Saunders.[39]).

of a Nanda–Hamner experiment. The *Ostrinia* data then available did not involve light cycles long enough to test the inference[32] that the Nanda–Hamner curves found in other species were similar transects across an *'extended circadian surface'* that would include a succession of peaks. Saunders,[35,36] working in my laboratory at that time, tested the inference and found the remarkable surfaces for *Nasonia* and *Sarcophaga* given in figure 23. The meaning of these intriguing surfaces has never been clear, although the results associated with transect B suggested from the beginning[32] that maintenance or disruption of 'normal' circadian organization was involved. Physiological function is evidently impaired when the system is driven at periods other than 24 h. One of the most useful features of the pacemaker–slave model of circadian programmes is that it readily yields an explanation of extended circadian surfaces. The phase relation (ψ_{BA}) of slave to pacemaker not only varies with change in photoperiod but also change in cycle length (T); and this variation in

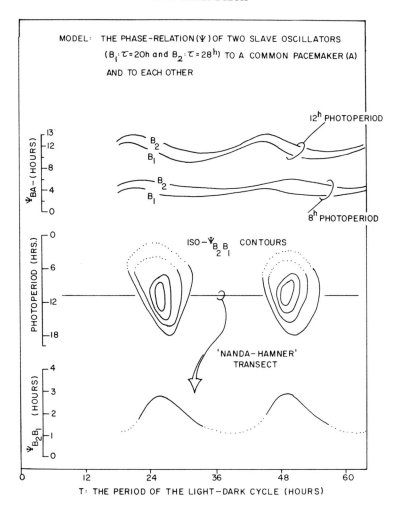

Figure 24. A model circadian surface. *Upper panel:* The phase relation (ψ_{BA}) of 2 slave oscillators, differing in τ_B (B_1, $\tau = 20$ h and B_2, $\tau = 28$ h), to a common pacemaker as a function of T (length of light cycle), for two photoperiods (8 and 12 h). The phase relation ($\psi_{B_1B_2}$) changes with change in both T and photoperiod. When T and photoperiod are systematically varied the contours defining iso-$\psi_{B_1B_2}$ values yield a circadian surface.

ψ_{BA} is likely not to be the same in two slaves differing in τ_B, ε and C. Figure 24 gives an example. If we take $\psi_{B_1B_2}$ (the phase relation between two such slaves) as a measure of the system's state, it is found to vary with cycle length and photoperiod, and an extended circadian surface can be constructed plotting iso-$\psi_{B_1B_2}$ contours. Watrus[37] (and personal communication) has independently shown that the model can generate an equally plausible circadian surface based on iso-ψ_{BA} contours.

An additional merit of this interpretation of circadian surfaces is its clear potential for explaining the perplexing findings, emphasized by Danilevskii,[38] that in many insects the critical daylength for photoperiodic induction is temperature-dependent. This immediately precludes interpretation of the time measurement exclusively in terms of a circadian pacemaker as such: *its* period and detailed time

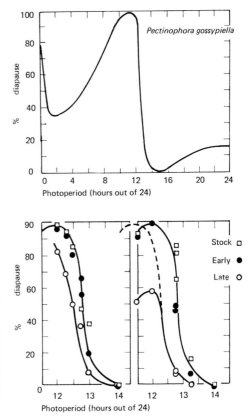

Figure 25. The effect of changing ψ_{BA} on the photoperiodic response curve of the moth *Pectinophora gossypiella.* (From Pittendrigh and Minis.[21]) *Upper panel:* The photoperiodic response curve (percentage diapause as a function of photoperiod length with $T = 24$ h) of the unselected strain. *Lower panels:* The response curve is depressed in the *Late* strain but not in *Early*. *Left*, assayed after 4 and 5 generations of selection; *right*, after 6 and 8 generations of selection. Depression of the entire curve causes a change in the 'critical day length', which is the photoperiod that induces a 50 per cent response.

course (as reflected in PRC shape) are both invariant with change in temperature. Saunders[39] has offered an explanation of the temperature-dependence in terms of a 'photoperiodic counter' that is some device additional to the 'photoperiodic clock' and not part of the circadian system. There are, however, alternative explanations based on the strong temperature-dependence of ψ_{BA}, that refer both the time measurement and its temperature-dependence to circadian organization among slave oscillations in the system. Even an 'external coincidence' version of circadian time measurement can be made temperature-dependent if the 'photo-inducible phase' in the circadian system is part of a slave's cycle, as distinct from the pacemaker's. On the other hand, the internal coincidence version which sees recognition of day length as a function of phase relations within the system (slave to slave, or slave to pacemaker), just as readily accommodates the temperature effects, again because ψ_{BA} is temperature-dependent. As the individual ψ_{BA} values of two slaves (B_1, B_2) differing in τ_B, C or ε change with temperature, so also, in the majority of cases, will $\psi_{B_1B_2}$. We should then expect the height and possibly the shape of circadian surfaces to be temperature-dependent, and although we still lack whole surfaces measured at two different temperatures, there are now two cases, one from Saunders[39] and one from our laboratory,[40] where the amplitude of a Nanda–Hamner transect has been found to change with temperature (figures 21 and 23).

A significant feature of the temperature effects in both the *Sarcophaga* and *Drosophila* cases is that one and the same species gives a positive Nanda–Hamner result (implicating circadian organization) at one temperature but a clear negative Nanda–Hamner result at another. Had *Sarcophaga* been analysed only at 16 °C and *Drosophila* only at 15 °C those species would have joined the small class of cases where an hour-glass, not circadian periodicity, is thought to execute the time measurement. Clearly the case for an hour-glass remains unconvincing until the Nanda–Hamner and other similar protocols have been carried out over a range of temperatures.

The Danilevskii school[38] in particular has drawn attention to adaptive change in the photoperiodic mechanism of latitudinal races: in northerly strains the critical daylength is commonly longer than in southern strains. Here again the genetic lability of ψ_{BA} and, hence, the amplitude of circadian surfaces, promises an explanation. It is recalled that selection for *Early* and *Late* strains in the moth *Pectinophora*[21] had a significant impact on the photoperiodic timing of that insect. Those experiments were done to test a version of 'external coincidence' in which the photoinducible phase was part of a slave oscillator's cycle. We found that the responses of *Early* were not distinguishable from those of *Stock*, but in *Late* the critical daylength was indeed shortened. One conspicuous feature of the data, however, precluded interpretation of this expected result in terms of 'external coincidence': the change in critical photoperiod was not effected by a displacement of the response curve on the daylength axis—it was effected by a *depression of the entire curve*. This is also true for Danilevskii's geographic races: like the temperature-dependence of Nanda–Hamner curves (figure 25), depression of the entire photoperiodic response curve suggests that what is involved is depression (or distortion) of the entire circadian surface, and that this is not only compatible with selection on ψ_{BA}, it is expected in terms of the pacemaker–slave model of circadian organization.

The detailed mechanism of photoperiodic induction, even in a formal sense, will almost surely differ from species to species. Given the very rich possibilities for time measurement in a multi-oscillator system it would be surprising if that were not the case. Nevertheless, even the little we now understand about circadian organization makes the general idea of 'internal coincidence' attractive: temporal programmes based on the phase relations of constituent oscillators in the system cannot escape seasonal change as the pacemaker's waveform is modified by changing photoperiod.

REFERENCES

1. C. S. Pittendrigh and P. C. Caldarola. *Proc. Natl Acad. Sci. USA*, **70**, 2697–701 (1973).
2. C. S. Pittendrigh. In: *Handbook of Behavioral Neurobiology*, Vol. V, Chap. 7. New York, Plenum Press (in the press) (1981).
3. J. Beling. *Z. Vergl. Physiol.* **9**, 259–338 (1929).
4. K. Hoffman. *Cold Spring Harbor Symp. Quant. Biol.* **XXV**, 379–88 (1960).
5. E. Bünning. *Berichte Dtsch. Bot. Gesellschaft.* **54**, 590–607 (1936).
6. S. T. Inouyé and H. Kawamura. *Proc. Natl Acad. Sci. USA* **76**, 5962–6 (1980).
7. N. H. Zimmerman and M. Menaker. *Proc. Natl Acad. Sci. USA* **76**, 999–1003 (1979).
8. C. S. Pittendrigh. *Cold Spring Harbor Symp. Quant. Biol.* **XXV**, 159–82 (1960).
9. J. Aschoff. *Aerospace Med.* **40**, 844–9 (1969).
10. S. Gaston and M. Menaker. *Science* **160**, 1125–7 (1968).
11. B. Rusak. *J. Comp. Physiol.* **118**, 145 (1977).
12. B. Rence and W. Loher. *Science* **190**, 385–7 (1975).
13. T. Page. This volume pp. 113–24.

14. C. S. Pittendrigh and V. G. Bruce. In: R. B. Winthrow (ed.) *Photoperiodism*. American Association for the Advancement of Science, Washington, DC, pp. 475–506 (1959).
15. C. S. Pittendrigh. In: F. O. Schmitt and F. G. Worden (ed.) *Neurosciences: Third Study Program*. MIT Press, pp. 437–58 (1974).
16. M. Moore-Ede. *Fed. Proc.* **35**, 2333–8 (1976).
17. C. S. Pittendrigh, V. G. Bruce and P. Kaus. *Proc. Natl Acad. Sci. USA* **44**, 965–73 (1958).
18. T. Pavlidis. *Biological Oscillators: their Mathematical Analysis*. New York, Academic Press (1973).
19. C. S. Pittendrigh. *Pacemaker and Slave Oscillations in a Circadian System.* (In preparation.)
20. C. S. Pittendrigh. *Proc. Natl Acad. Sci. USA* **40**, 1018–29 (1954).
21. C. S. Pittendrigh and D. H. Minis. In: M. Menaker (ed.) *Biochronometry*. Washington, DC, National Academy of Sciences, pp. 212–50 (1971).
22. T. Page, P. Caldarola and C. S. Pittendrigh. *Proc. Natl Acad. Sci. USA* **74**, 1277–81 (1977).
23. T. Page. *J. Comp. Physiol.* **124**, 225–36 (1978).
24. C. S. Pittendrigh and S. Daan. *J. Comp. Physiol.* **106**, 333–55 (1976).
25. S. Daan and C. Berde. *J. Theor. Biol.* **70**, 297–313 (1978).
26. H. Underwood. *Science* **195**, 587–9 (1977).
27. F. W. Went. *Cold Spring Harbor Symp. Quant. Biol.* **XXV**, 221–30 (1960).
28. C. S. Pittendrigh and D. H. Minis. *Proc. Natl Acad. Sci. USA* **69**, 1537–9 (1972).
29. U. St Paul and J. Aschoff. *J. Comp. Physiol.* **127**, 191–5 (1978).
30. J. Aschoff, U. St Paul and R. Wever. *Naturwissenschft* **58**, 574 (1971).
31. K. C. Hamner. *Cold Spring Harbor Symp. Quant. Biol.* **XXV**, 629–78 (1960).
32. C. S. Pittendrigh. *Proc. Natl Acad. Sci. USA* **69**, 2743–7 (1972).
33. C. S. Pittendrigh. *Z. Pflanzen.* **54**, 275–307 (1966).
34. S. D. Beck. *Biol. Bull.* **122**, 1–12 (1962).
35. D. S. Saunders. *J. Insect Physiol.* **20**, 77–88 (1974).
36. D. S. Saunders. *J. Insect Physiol.* **19**, 1941–54 (1973).
37. J. A. Watrus. *Models for Photoperiodism from Circadian Rhythms*. PhD Thesis, University of California (Riverside) (1977).
38. A. S. Danilevskii. *Photoperiodism and Seasonal Development of Insects*. Edinburgh, Oliver & Boyd, p. 180 (1965).
39. D. S. Saunders. *Insect Clocks*. London, Pergamon Press (1976).
40. T. Page, D. Page and C. S. Pittendrigh. Unpublished observations (1980).
41. W. M. Hamner. *Science* **142**, 1294–5 (1963).
42. J. A. Elliott, M. H. Stetson and M. Menaker. *Science* **178**, 771–3 (1972).

PHOTORECEPTION AND THE CLOCK

Wolfgang Engelmann

Institut für Biologie I, Universität Tubingen, Federal Republic of Germany

Summary

Although light is the main synchronizer of circadian rhythms, photoreceptors are probably not an essential part of the underlying clock itself. Nevertheless, light is still useful in attempting to understand the clock mechanism, and in formal descriptions, as well as in more physiological models, the action of light on circadian rhythms is usually incorporated. Light and temperature appear to interact before they reach the clock mechanism. In this paper, the role of calcium ions in photoreception as well as their influence on circadian rhythms is stressed. In higher organisms light influences a multi-oscillator system with occasional feedbacks also existing from the oscillators to the photoreceptors. Finally, the control of photoperiodic events by light and the circadian system is briefly discussed.

Was ist Biorhythmik? Auskunft: Tel. 0711/871433.

(Advertisement from a newspaper in Southern Germany.)

The widespread occurrence of circadian rhythms in the living world is well documented, and several recent books[1-5] and reviews[6-8] are available. In most cases (from unicellular organisms to fungi, plants and animals) light plays the dominant role in synchronizing these rhythms to the natural 24-hour day.[9,10] Different photoreceptors are often involved in perceiving the light controlling circadian rhythms. Furthermore, the photoreceptors do not seem to be an essential part of the clock mechanism. This is suggested by the finding that, for a number of circadian rhythms, light is not a zeitgeber. For example, this applies to the zonation rhythm in *Podospora*,[11] to various drinking, urinary, peristaltic and adrenal rhythms[12] all of which are synchronized by feeding cycles operating through chemoreceptors, to the oxygen consumption rhythm of *Megachile*, which is synchronized by temperature cycles,[13] and to tidal rhythms that are claimed to be circadian and yet controlled by tidal cues.[14] Man's own circadian rhythms also seem to respond poorly to light/dark cycles.[2]

It also seems that the signal induced by the light can be influenced by temperature en route to the clock. Hamm et al.[15] first found this in the circadian rhythm regulating eclosion in *Drosophila*. By comparing the phase response curves to light pulses with and without lowering the ambient temperature from 20 to 6 °C for 2 h, they showed that the effect of the light pulse could be retarded by about 1 h. This was confirmed by using a rhythm-annihilating light pulse. The conclusions

Biological Clocks in Seasonal Reproductive Cycles 37–43 (1981) (ed. B. K. and D. E. Follett: Bristol, Wright).

of Hamm et al.[15] have been tested further by using light-pulse phase response curves during and after low temperature pulses both on the *Drosophila* eclosion and rhythm (A. Schmiedt, A. Loskill, G. Götzl and W. Engelmann, unpublished observations), and on the petal movement rhythm of *Kalanchoe*[16] (*also* G. Wäscher, M. Heilemann and W. Engelmann, unpublished observations). Light can also be replaced by 'photomimetics' such as carbachol (a cholinergic agonist). This can affect (figure 1) the N-acetyl transferase rhythm in the rat's pineal and produces

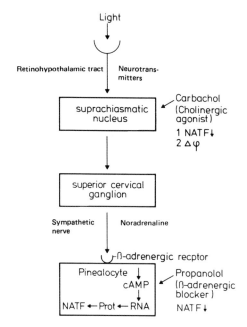

Figure 1. Intraventricular injections of carbachol in rats mimic both the phase-shifting and the activity-reducing effects of light on pineal N-acetyl transferase. (After Zatz and Brownstein.[17]).

phase shifts similar to those found with light pulses.[17] In plants, the photoperiodic effect of skeleton photoperiods can be substituted by salicylic acid, EDTA or propanolol, all of which influence membrane permeability, as well as by cyclic AMP and isoproterenol (figure 2). These compounds have to be applied 6 h after the light pulse suggesting that processes take a finite time to occur between light perception and the effect on the clock. Some fungi can be 'photosensibilized'. Thus, the zonation rhythm of *Podospora anserina* becomes dependent on the light/dark cycle if ionophores (monensin, nigericin, nonactin) are added.[11] A mutant of *Penicillium claviforme*, which only shows zonation in light/dark cycles, can be induced to exhibit rhythmic zonation in continuous light or darkness by adding hyamin 1622, a compound that influences membrane permeability.[19]

All these findings suggest that for an understanding of the mechanisms underlying circadian rhythms it may not be absolutely necessary to study the properties of the photoreceptor itself, although, as will be pointed out below, it may be helpful to do so.

Various strategies have been used to study the functioning of circadian rhythms. The 'black-box' approach tries to draw conclusions concerning the clock mechanism by observing the reactions of the system after specific treatments. This

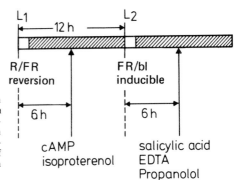

Figure 2. Photomimetica for photoperiodic timing in *Lemna gibba* (strain G3). The photoperiodic effect of an LD 12 : 12 cycle can be simulated by a skeleton photoperiod of 1 h light, 11 h darkness, 1 h light and 11 h darkness. The substances shown can replace the corresponding light pulses. Note, however, that the time of application must be 6 h after the pulse. (Data from Oota and Nakashima.[18])

approach has been used widely and the resulting models specify the ways in which light influences them.[2,3,20-22] Experiments to test these models include phase response curves to light pulses, the reactions to continuous light of different intensities, the effects of light steps and of sinusoidal light treatments etc. Well designed models have led to non-trivial predictions that are testable.

Another strategy is to try to follow up the physiological causes for observed circadian oscillations, e.g. leaf movement rhythms in plants, hoping to arrive at the clock mechanism itself (figure 2). It is, however, difficult to distinguish between the controlled process and the controlling clock mechanism.

A corresponding approach is to follow the path from the absorption of light by a pigment, via its transformation into chemical signals, down to the clock mechanism itself. If this path happens to contain fewer or less complicated steps as compared with the approach mentioned before, it may be a more advantageous strategy. However, complications can arise if this path is itself influenced by the circadian system. Thus, it is known that visual cells undergo circadian variations in their sensitivity to light,[23] that migration of pigments in the lateral eye of scorpions is controlled in a circadian fashion via neurosecretory axons of the optic nerve[24] and that the turnover of photoreceptor membranes in crabs and other invertebrates,[25] as well as in vertebrates from fish to man,[26,27] is under the influence of the circadian system (figure 3).

Besides its synchronizing and resynchronizing action light is known to influence circadian rhythms in a number of other ways. Thus, continuous light can influence the amplitude of the oscillation, leading in some cases to complete damping (e.g. plants,[10] insects,[29] *Aplysia*,[30] birds[28] and rodents[9]). Arrhythmicity can also be induced by a very special arrangement of light pulses that sends the oscillator into singularity or 'primary arrhythmicity',[4,31] a situation different from the 'conditional arrhythmicity' induced by continuous light.[32] Rules were found to describe the influence of the intensity of continuous light on period length, activity time and rest time of both diurnal and nocturnal animals,[33] and several theoretical studies have dwelt on this subject.[2,3,20,34] The phasic effects of light that are seen in phase response curves can be used to describe many findings[35] but the underlying mechanisms are still poorly understood. Thus, it is still not known whether light pulses leading to delays or advances in the rhythm are absorbed by different photoreceptors,[36-38] or whether phasic and tonic effects are sensed by different pigments.[39]

The lights-on and lights-off portions of the daily cycle have been suggested to act

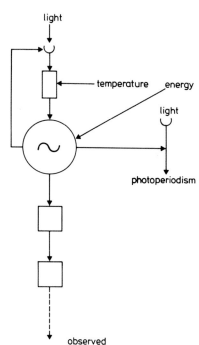

Figure 3. Schematic representation of the various inputs and outputs of a circadian oscillator. Light is received by a pigment in the photoreceptor and transduced into a chemical signal. The speed of its transfer to the oscillator can be influenced by temperature. The oscillator itself has several outputs, one of which drives processes leading to observed rhythms such as leaf movement. Another controls photoperiodic events. Finally, the oscillator can also influence the sensitivity of the photoreceptor to light. The oscillator has, of course, a requirement for energy.

upon different rhythms which, by superposition, would lead to photoperiodic induction. The latter would occur only if the phase relationship between the lights-on and lights-off rhythms was correct.[40–42] Again, it is not known whether two different photoreceptors are involved in the reception of the lights-on and lights-off signals in these internal coincidence models. The period lengths of both rhythms seem to depend differently on the light intensity, and the different phase responses might result from this.[20]

A number of models have tried to bridge the gap between the more formal descriptions of oscillations and the kind of model the physiologist seeks. Thus, Benson,[43] Gander,[34] Njus et al.,[44] Sweeney[45] and Robinson[46] have all used feedback model ideas[21] to meet experimental findings. Enright has developed a special model which takes care of the experimental findings on sleep/wakefulness patterns of birds and mammals.[3] All discuss the action of light on the circadian system.

Energetics have been stressed in the models of Bünning,[1] Satter,[47] Jones,[48] Wagner,[49] Brinkmann[50] and Sel'kov.[51] That of Sel'kov is particularly interesting since it strips away another 'essential' feature of circadian rhythms, namely their circadian nature (*see* figure 4). An illustration of this idea might be the leaf movement rhythm of *Desmodium gyrans*, which exhibits a circadian period in the terminal leaflet and a high frequency rhythm in the lateral leaflets. We have observed that the pulvini responsible for the circadian movement contain considerable amounts of starch, whereas the pulvini of the lateral leaflets contain only small amounts. This would be expected if Sel'kov's explanation for the long circadian periods holds true (compare figure 4). Since the high frequency oscillation in

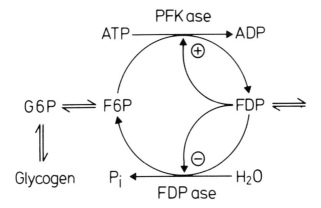

Figure 4. Model of a biochemical oscillation. A high frequency oscillation in the minute range is produced by the forward- (upper part) and backward- (lower part) reaction of fructose-6-phosphate (F6P) and fructosediphosphate (FDP). Both reactions are mutually exclusive but are influenced by the concentration of FDP. At high concentrations, phosphofructokinase (PFKase) is activated, and fructose-diphosphatase (FDPase) is inhibited. At low concentrations the reverse occurs. The coupling of this oscillator to a glycogen pool via glucose-6-phosphate (G6P) increases the period length according to the ratio of glycogen : FDP-concentrations. This can lead to circadian frequencies. (Adapted from Sel'kov.[51])

Desmodium is not affected by light, whereas the circadian oscillation is, we would expect light to enter the model system at the 'frequency slowing part' (*see* the left hand portion of figure 4).

Calcium ions may form a link between light and its action on the circadian system. The role of calcium in light perception is well known,[52] as is its role in energy metabolism, especially within the mitochondria.[53] Calcium ions also influence circadian rhythms,[54] whilst it has been suggested that some other chemicals influencing circadian rhythms may well exert their action via Ca^{++}.[55] The body temperature rhythm in mammals, which arises because of the influence of the circadian system on the temperature-regulating mechanism, might also operate via Ca^{++}.[56] Recent results with ATPase inhibitors such as azide and vanadate ions[57] as well as other findings,[49,50,58] point towards mitochondria being important in circadian mechanisms.

As pointed out by Pittendrigh, we must accept the possibility that the mechanism of circadian rhythms is not a uniform one, but that nature has used various concepts.[59] If so, this will increase considerably the difficulties in trying to unravel the timing devices. In high organisms the multi-oscillator system introduces further complications.[60] Different oscillators might use different mechanisms as well as different input devices, and light may influence only certain oscillators and even these via different pigment/receptor routes (*see* the literature on extra-retinal light perception[7]). The kind of hierarchy used by higher organisms is still open to debate.[61] Coupling between different oscillators may also be important for the production of circadian rhythms.[51,62]

In discussing the influence of light on circadian rhythms we must also mention photoperiodism. Long ago, Bünning suggested the circadian system to be the time-measuring device for photoperiodic reactions.[63] To predict photoperiodic induction we have used a feedback model[21] which specifies the entrance of light into a feedback loop.[64,65] This was done on the basis of the phase relationship that exists between the oscillation and the dark/light transition. Photoperiodic induction could be quite successfully predicted for a broad range of LD cycles, but more critical tests using low light intensities instead of darkness during the dark period (G. Schmiedt, unpublished observations), or skeleton photoperiods (G. Wiest, unpublished observations), have yielded results that do not meet the predictions of this model. In addition, Bollig[66] has provided arguments that are in favour of separate oscillators governing the photoperiodic mechanism and leaf movement rhythms in *Pharbitis nil*.

REFERENCES

1. E. Bünning. *Die physiologische Uhr*, 3rd ed. Berlin, Springer-Verlag (1977).
2. R. Wever, *The Circadian System of Man*. New York, Springer-Verlag, pp. 184–91 (1979).
3. J. T. Enright. *The Timing of Sleep and Wakefulness*. Berlin, Springer-Verlag (1979).
4. A. T. Winfree. *The Geometry of Biological Time*. Berlin, Springer-Verlag (1979).
5. J. W. Hastings and H.-G. Schweiger. *The Molecular Basis of Circadian Rhythms*. Berlin, Abakon Verlag (1976).
6. M. Menaker. *Ann. Rev. Physiol.* **40**, 501–26 (1978).
7. B. Rusak and I. Zucker. *Physiol. Rev.* **59**, 449–526 (1979).
8. W. Engelmann and M. Schrempf. *Photochem. Photobiol. Rev.* **5**, 49–86 (1979).
9. A. A. Borbely. *Progr. Neurobiol.* **10**, 1–31 (1978).
10. N. Ninnemann. *Photochem. Photobiol. Rev.* **4**, 207–66 (1979).
11. G. Lysek and H. von Witsch. *Arch. Mikrobiol.* **97**, 227–37 (1974).
12. E. Bünning. *Physiol. Veg.* **16**. 799–804 (1978).
13. D. G. Tweedy and W. P. Stephen. *Comp. Biochem. Physiol.* **38A**, 213–31 (1971).
14. H.-W. Honegger. *Marine Biol.* **18**, 19–31 (1973).
15. U. Hamm. M. K. Chandrashekaran and W. Engelmann. *Z. Naturf.* **30c**, 240–4 (1975).
16. W. Engelmann and M. Heilemann. *Chronobiologia* **4**, 109–10 (1977).
17. M. Zatz and M. J. Brownstein. *Science* **203**, 358–61 (1979).
18. Y. Oota and H. Nakashima. *Bot. Mag. Tokyo*, Special Issue 1, pp. 177–98 (1978).
19. A.-G. F. Salman. *Biochem. Physiol. Pflanzen*. **162**, 470–3 (1971).
20. C. S. Pittendrigh and S. Daan. *J. Comp. Physiol.* **106**, 333–55 (1976).
21. A. Hohnsson and H. G. Karlsson. *J. Theor. Biol.* **36**, 152–74 (1972).
22. L. N. Edmunds and K. J. Adams. *Science* (in the press) (1980).
23. W. Kaiser.*Verh. Dtsch. Zool. Ges.* **72**, 211 (1979).
24. G. Fleissner and G. Fleissner. *Comp. Biochem. Physiol.* **61A**, 69–71 (1978).
25. D. R. Nässel and T. H. Waterman. *J. Comp. Physiol.* **131**, 205–16 (1979).
26. R. W. Young. *Vision Res.* **18**, 573–8 (1978).
27. J. C. Besharse, J. G. Hollyfield and M. E. Rayborn. *J. Cell Biol.* **75**, 507–27 (1977).
28. S. Binkley. *Physiol. Zool.* **50**, 170–81 (1978).
29. A. T. Winfree. *Science* **183**, 970–2 (1974).
30. J. A. Benson and J. W. Jacklet. *J. Exp. Biol.* **70**, 183–94 (1977).
31. E. L. Peterson and M. D. R. Jones. *Nature* **280**, 677–9 (1979).
32. D. Njus, L. Mcmurry and J. W. Hastings. *J. Comp. Physiol.* **117B**, 335–44 (1977).
33. J. Aschoff, U. Gerecke, A. Kureck, H. Pohl, P. Rieger, U. St Paul and R. Wever. In: M. Menaker (ed.) *Biochronometry*. Washington, DC, National Academy of Sciences, pp. 3–29 (1971).
34. P. H. Gander. Thesis, University of Auckland, New Zealand (1976).
35. S. Daan and C. S. Pittendrigh. *J. Comp. Physiol.* **106**, 253–66 (1976).
36. E. Bünning and G. Jörrens. *Z. Naturf.* **15B**, 205–13 (1960).
37. R. Halaban. *Pflanzenphysiol.* **44**, 973–7 (1969).
38. M. K. Chandrashekaran. *Z. Vergl. Physiol.* **56**, 163–70 (1967).
39. H. Underwood and M. Menaker. *Photochem. Photobiol.* **23**, 227–43 (1976).
40. W. Engelmann. *Experientia* **22**, 606–8 (1966).
41. V. P. Tyshchenko. *Zh. Obshch. Biol.* **27**, 209–22 (1966).
42. K. C. Hamner and T. Hoshizaki. *Bioscience* **24**, 407–14 (1974).
43. J. A. Benson and J. W. Jacklet. *J. Exp. Biol.* **70**, 195–211 (1978).
44. D. Njus, F. M. Sulzman and J. W. Hastings. *Nature* **248**, 116–20 (1974).
45. B. M. Sweeney. *Int. J. Chronobiol.* **2**, 25–33 (1974).
46. G. R. Robinson. Thesis, University of Tasmania, Hobart (1979).

47. R. L. Satter and A. W. Galston. *Bioscience* **23**, 407–16 (1973).
48. P. C. T. Jones. *J. Theor. Biol.* **34**, 1–13 (1972).
49. E. Wagner. *Soc. Exp. Biol.* **31**, 33–72 (1977).
50. K. Brinkmann. In: Chance et al. (ed.) *Biological and Biochemical Oscillators.* New York, Academic Press, pp. 513–521 (1973).
51. E. E. Sel'kov. In: H. Haken (ed.) *Pattern Formation by Dynamic Systems and Pattern Recognition.* Berlin, Springer-Verlag (1979).
52. W. A. Hagins and S. Yoshikami. In: H. B. Barlow and P. Fatt (ed.) *Vertebrate Photoreception.* (Rank Prize Funds Optoelectronics Biennial Symposia.) London, Academic Press, pp. 97–139 (1977).
53. L. Lehninger and E. Carafoll. *J. Biol. Chem.* **243**, 506 (1971).
54. A. Eskin and G. Corrent. *J. Comp. Physiol.* **117**, 1–21 (1977).
55. I. C. Bollig, K. Mayer, W.-E. Mayer and W. Engelmann. *Planta* **141**, 225–30 (1978).
56. J. F. Manery. In: P. Lomax and E. Schönbaum (ed.) *Modern Pharmacology and Toxicology.* New York, Dekker, pp. 119–144 (1979).
57. H. Saxe and R. Satter. *Plant Physiol.* **64**, 905–7 (1979).
58. C. Dieckman and S. Brody. *Science* **207**, 896–8 (1980).
59. C. S. Pittendrigh. *Harvey Lectures* **56**, 93–125 (1961).
60. J. Aschoff and R. Wever. *Fed. Proc.* **35**, 2326 (1976)
61. D. C. Klein. In: D. T. Krieger (ed.) *Endocrine Rhythms: Comprehensive Endocrinology.* New York, Raven Press, pp. 203–23 (1979).
62. T. Pavlidis. *Bull. Math. Biol.* **40**, 675–92 (1978).
63. E. Bünning. *Berl. Dtsch. Bot. Ges.* **54**, 590–607 (1936).
64. W. Engelmann, H. G. Karlsson and A. Johnsson. *Int. J. Chronobiol.* **1**, 147–56 (1973).
65. I. C. Bollig, M. K. Chandrashekaran, W. Engelmann and A. Johnsson. *Int. J. Chronobiol.* **4**, 83–96 (1976).
66. I. C. Bollig. *Z. Pflanzen.* **77**, 54–69 (1975).

DISCUSSION

Turek: I am concerned with your assumption as to the site of action of various agents on the circadian system. For example, your slide showing the effects of carbachol on pineal NAT activity indicated that the effect was directly on the presumed biological clock, the SCN. However, the effects of carbachol are just as readily explained on the basis that it alters cholinergic neurones which innervate the SCN. On another issue, what is the experimental evidence to indicate that temperature alters the speed at which light information reaches the clock, as opposed to the temperature altering the response of the clock to light information?

Engelmann: Yes, I agree. However carbachol still acts as a photomimetic. The term has been used by Zatz and Brownstein in this sense. As to your second question, it is possible to differentiate between actions of temperature on the clock itself, on the processes between the clock and the hands and on the processes between photoreceptor and clock.

FLOWERING HORMONES AND INHIBITORS IN PHOTOPERIODISM

W. W. Schwabe

Department of Horticulture, Wye College, University of London, Ashford, Kent

Summary

Perhaps in contrast to other papers in this session, this paper will be less concerned with a direct assessment of the working of the biological clock mechanism than with the reaction mechanism through which the timing effects may be translated into biochemical, and ultimately morphological, changes in the leaf leading to flowering. The basic model has been derived from work with both short-day- and long-day-requiring species and emphasis will be laid on the progress made so far in testing experimentally various steps in the hypothetical reaction sequences.

While the flowering plant living in a rhythmically changing environment of light and dark may be steered in its circadian rhythms by these environmental changes, it seems highly probable that light and dark also have specific effects on the flower-inducing mechanism. The ultimate result of these reactions is believed to be the production of a flower-promoting substance which is transmissible from the leaf to the shoot apex where the transformation into floral organs takes place. Since all kinds of obligate, qualitative, quantitative and neutral response types seem to occur in plants, without recognizable relation to their taxonomic position, it seems probable that the various response types do not differ fundamentally in their metabolic mechanisms for flowering and that such differences may be due to quantitative differences in the reaction mechanisms. This is also confirmed by the wide range of types between which graft transmission of the flowering stimulus can occur.

A theoretical scheme elaborated a few years ago allows the various types of flowering response to be fitted into a simple model which can account for (a) the factors leading to flower promotion, (b) inhibition and (c) the build-up with increasing inductive treatment. The possible role of biological rhythms in this scheme seems most probably to be associated with the state and formation of phytochrome. The scheme will be discussed briefly and the research described which has aimed at the further elucidation of some possible steps in the proposed reaction sequence. These include factors involved in the reactions believed to occur in the leaf, such as the importance of the control of gas exchange by the epidermis and the possible role of carbon dioxide supply during the light period of inductive treatment. Specific evidence is offered for the formation of inhibiting substances, confirming their existence although they have not yet been chemically characterized.

Biological Clocks in Seasonal Reproductive Cycles 45–55 (1981) (ed. B. K. and D. E. Follett: Bristol, Wright).

INTRODUCTION

Under natural conditions the flowering plant lives in a rhythmically changing environment which causes profound diurnal changes in metabolism between the light and dark parts of each day, affecting carbohydrate turnover, water and gas exchange etc. These diurnal environmental changes, which under natural conditions also comprise temperature modulation, not only control the gross overall net carbon fixation and loss, but almost certainly steer the internal circadian rhythms even when the endogenous oscillator is operating freely. However, the metabolic switching during each cycle of light and dark is also likely to affect the photoperiodic flowering response of plants *directly*, and not merely through control of circadian rhythms.

A theoretical schema (figure 1), which takes account of this and which may be regarded as a series of biochemical steps leading to the production of a final

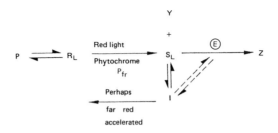

Conditions for each reaction step, rates of each reaction and limitations on product accumulation

Reaction	Condition for reaction	Rate of reaction	Limitation on product accumulation
$P \rightleftharpoons R_L$	none	Sigmoid	Up to maximum R_L
$R \rightarrow S$	Red light, phytochrome P_{fr}	Generally very rapid	S/I equilibrium with I predominating heavily
$S \rightleftharpoons I$	none	Very rapid	S_L and I_L
Loss of I	Perhaps far red light accelerated	Roughly constant	Until I exhausted
$Y + S \rightarrow Z$	Presence of free enzyme E	Dependent on amount of E	As E increases Z will exceed threshold level and accumulate until full flowering
$E + I \rightleftharpoons EI$	none	Rapid, constant	Competitive combination and reversal as level of I declines in dark
E increase	Free E functioning	Dependent on activity	Increasing with activity until full induction
Y	Perhaps dependent on photosynthesis	Unspecified	May become limiting in very long dark periods

Figure 1. Hypothetical reaction schema for the production of a flowering substance in long- and short-day plants.

flowering hormone, was proposed as a model for the processes involved in flower induction.[1] This substance might represent florigen, the name for such a substance suggested almost half a century ago by Chaijlakjan,[2] or anthesin (Moshkov's name).[3] I have designated it as Z, i.e. the final product. There can be little doubt that some such substance exists in view of the many successful grafting experiments transmitting a positive stimulus from an induced donor to a non-induced receptor (for example, the work of Zeevaart[4]), yet positive proof of an actual flower-promoting substance still eludes us, although tantalizing pieces of evidence are to hand. Quite recently Chaijlakjan and co-workers have found biological activity in extracts made from Maryland Mammoth tobacco, which is capable of inducing flowering in seedlings of young plants of *Chenopodium rubrum*.[5] Before briefly describing the schema, which fits both long- and short-day plants, I would like to draw attention to a few other significant facts.

1. While the photoperiodic flowering responses range from an obligate short-day requirement to an obligate long-day requirement, there are many intermediate types with requirements for either daylength. These grade into the so-called day-neutral plants which occupy the middle ground, but *taxonomically* are unconnected through any evolutionary relationships. This suggests that even the extremes do not differ in fundamental characteristics. This fact that extreme long- and short-day-requiring species or cultivars may be closely related suggests that all flowering may be controlled in a very similar manner. Moreover, graft transmission of the stimulus between extreme opposites points in the same direction and one may conclude that quantitative differences alone could account for the observed differences in behaviour.

2. While long- and short-day requirements appear in many ways to be direct mirror-images of one another, there are some significant differences. Thus, short-day plants which have a requirement for a series of short-day cycles, may be seriously disrupted in flower induction by intercalating long days or light-breaks in such short-day cycles. Long-day plants are very much less affected, if at all, by intercalating unfavourable short-day cycles.

3. Similarly, long-day plants and short-day plants may respond to interruptions of a long dark period with a light-break by becoming induced, or prevented from flowering, respectively. The periods needed are usually very different, however, short-day plants responding to perhaps a fraction of one second of light provided that the intensity is high enough, while long-day plants often need light-breaks of several hours' duration before becoming induced.

The model

While this schema applies to long-day and short-day plants as well as to the other categories, two points are envisaged at which the progress of the reaction to flowering may be blocked, one being generally operative for short-day plants and the other in long-day plants. The basic evidence and justification for the underlying assumptions was derived from work with *Kalanchoe*,[6] which showed that in this short-day plant, flowering can be partially or entirely suppressed by long-day treatment, by light-breaks or by intercalating long days among short days. The evidence strongly suggested the involvement of a flowering inhibitor which, however, does not accumulate above a maximum level that is reached after 2–3 long days, i.e. no counter-induction is possible. Data for long-day plants were derived from numerous experiments in the literature as well as unpublished work with *Epilobium adenocaulon* and other long-day species, all of which indicated that even quite short dark periods in a cycle can diminish the effect of a long day. Thus, a period of 20–30 min of dark per 24 hours significantly increases the number of cycles required to give floral induction, roughly parallelling the light-break effect in short-day plants.

Very briefly the model is as follows. The hypothetical reaction scheme (figure 1) postulates an adaptive enzyme (E) catalysing a reaction between a photosynthesis-related substance (Y) and a product of another light reaction (S) to yield the flowering promoter (Z); the substrate (S) spontaneously reverts to a competitive inhibitor molecule (I) capable of blocking E. This inhibitor, it is postulated, disappears in the dark and is in equilibrium with S. The substrate S itself is formed from another substance R by way of a phytochrome-mediated reaction, and it is believed that this is subject to modification by rhythmic phenomena. R itself is limited in the amount to which it can accumulate and is formed from a precursor (P).

In short-day plants it is postulated that an initially very low level of E controls the production of Z, but as E increases in amount and activity, the progressively greater effect of induction (e.g. a doubling of flower numbers for each additional inductive cycle in *Kalanchoe*), is explained. In long-day plants it is suggested that the production of R from P is an accelerating process with a lag period, but that the rate is depressed again to very low levels if R is accumulated (as would be the case in darkness), instead of being converted to S immediately.

The scheme has been published in full[1] and has also been tested as a computer model for internal consistency, yielding results fully compatible with quantitative data available. It seems unnecessary to give further details of the computer tests here.

Thus, the crucial requirement for a short-day plant is the development of enough enzyme to catalyse the final stages in the production of Z, although a lack of Y could also be limiting in special circumstances. In the case of a long-day plant, it is envisaged that the bottleneck normally occurs at the earlier stage of precursor (P) production.

The role of circadian rhythms

The involvement of rhythmic activity in the plant is believed to centre on phytochrome production/function, perhaps measured in terms of P_{fr}, which is believed to operate on the conversion of P to S, the substrate for the enzyme. The inhibitor produced in the absence of adequate amounts of enzyme E meets all the experimental evidence from short-day plants. There is also cogent evidence from long-day plants that an inhibitor may be operative.[7] Its role is not yet clearly defined, although it would seem to be transmissible.

The role of the epidermis

Since it is assumed that the schema outlined operates in the leaf, it seemed of interest to look at several aspects of this. First of all, where in the leaf do the photoperiodic reactions occur? The epidermis with its stomatal apparatus would seem to be a plausible site for some or all of the reactions, and some years ago, an attempt was made to identify its role.[8] In *Kalanchoe* it is possible to strip off the epidermis from the upper or lower leaf surfaces without causing so much damage that the leaf is abscinded (debladed petioles or severely damaged leaves are dropped rapidly). The quantitative effect on flowering of leaves without one or other of the epidermal surfaces was clear-cut: virtually no flower induction took place—even partial removal of the epidermis proved detrimental (figure 2).

The function of the epidermis could be restricted to the control of the internal gaseous atmosphere of the leaf, via the stomata, but it could also be involved

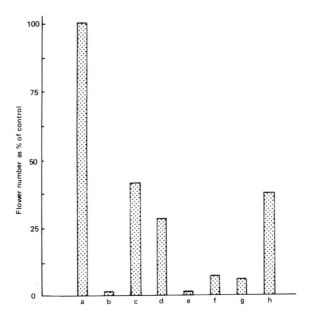

Figure 2. Effects of interference with epidermal function on ability of treated leaves to induce flowering in *Kalanchoe*. The number of flowers produced following 2 weeks of short-day exposure (LD 8 : 16), expressed as percentages of controls. **a**, intact control; **b**, debladed petiole; **c**, stomata on upper epidermis blocked with petroleum jelly; **d**, stomata on lower epidermis blocked with petroleum jelly; **e**, stomata on both surfaces blocked with petroleum jelly; **f**, upper epidermis stripped from leaf; **g**, lower epidermis stripped from leaf; **h**, lower epidermis stripped from proximal half of leaf.

directly in the chain reaction leading to floral promoter products *and*, of course, both functions could be linked. Eliminating the stomatal function either by blocking some or all of the stomata or by removing epidermis so as to allow free access of the outside air to the internal spaces of the leaf leads in both cases to leaves incapable of promoting floral induction. These techniques, however, have not yet permitted a distinction to be made between the two possible modes of action.

Another possible way of identifying the involvement of the leaf epidermis in the flower-inducing process would be the demonstration that phytochrome is present in epidermal cells. The determination of phytochrome in the epidermis should be easier since chloroplasts are normally absent from the epidermis except in the stomatal guard cells. Attempts to discover phytochrome in detached epidermal strips were kindly carried out for me by Dr Frankland at Queen Mary College, but with the techniques available to him at the time, he was unable to detect any phytochrome at all in the epidermal strips. While this may not perhaps be complete proof of the absence of phytochrome in the epidermis of *Kalanchoe*, it would seem to rule out the presence of relatively large amounts.

Involvement of carbon dioxide supply

The marked effect of preventing floral induction by stomatal blockage relates also to the CO_2 requirement during the light period of photoinductive cycles that has been recorded for numerous species in the past. In general, this has been interpreted as a requirement for photosynthate so as to make possible the morphogenetic responses. However, there is also evidence from Harder's work,[9] repeated by myself,[6] that in the succulent *Kalanchoe blossfeldiana* the daily light period may be as short as one second a day, and yet this would be adequate for floral induction, whereas total darkness fails to cause flower promotion. It would seem highly improbable that this response could be due to gross carbohydrate synthesis. Attempts have therefore been made by Dr Ireland and myself[10] to discover whether the CO_2 requirement in the leaf could be replaced by supplying carbohydrate. The plant material used for this purpose was again *K. blossfeldiana*, and enclosing a single leaf near the top of the plant in a CO_2-free atmosphere made such leaves ineffective as photoperiodic inducers, whereas control leaves in normal air were fully inductive. In both cases the other leaves on the plant below the induced leaf were left intact and in normal air throughout, so that there was no overall carbohydrate shortage in these plants. The application of CO_2 substitutes was made by injecting material in solution into the intercellular spaces of the leaf. Use of labelled material (sucrose and amino acids) showed that not only was the material taken up by the leaf cells, but that it was also satisfactorily translocated to the growing point (table 1). Thus, although labelled material could

Table 1. Proportion of total detected ^{14}C label found outside injected leaf 72 h after application of ^{14}C-labelled compounds in *Kalanchoe blossfeldiana*

^{14}C glucose	20·0%
^{14}C citric acid	20·9%
^{14}C leucine	12·0%

not be tested for all the substances applied, there is little doubt that it must have entered the mesophyll cells bathed in the material, restoring the gaseous atmosphere in the leaf. A considerable number of substances representing either final products of the photosynthetic processes or possible intermediates were tried with *Kalanchoe*. The results obtained with some of these are shown in table 2, but in no case could substitution for CO_2 be effected.

Table 2. Effect on flowering of some metabolites injected into single *Kalanchoe blossfeldiana* leaves during short-day induction in normal and CO_2-free air (6 replicates)

	Mean number of flowers	
Substance	Normal air	CO_2-free air
Sucrose 0·02%	14·0±4·5	0·0
Malate 0·02%	11·7±4·4	0·83±0·54 n.s.
Citrate 0·02%	14·2±3·3	0·0
Oxaloacetate 0·05%	15·2±2·6	0·92±0·82 n.s.
ATP 40 parts/10^6	14·7±3·0	0·10±0·10
H_2O control	16·7±3·2	0·0
Glutamine (10^{-4}M)	62·7±10·5	0·67±0·48
Serine (10^{-4}M)	86·0±26·6	0·42±0·17
Tryptophan (10^{-4}M)	85·7±24·0	0·08±0·08
H_2O control	90·3±9·2	0·25±0·1

As a second line of attack, leaves given inductive short days were held in ordinary air but injected with photosynthetic inhibitors such as DCMU etc., and in this case, in spite of the presence of CO_2, the leaves failed to induce flowering (table 3). Equally, if chlorophyll synthesis was prevented by application of

Table 3. *Kalanchoe blossfeldiana*. Photosynthetic light re-action inhibitors. Plants (6 replicates) were treated with 15 short days. The flower number is shown after 60 days

Treatment	Mean no. of flowers
H_2O (control)	$20 \cdot 87 \pm 6.61$
Aq. ethanol $0 \cdot 5\%$	$22 \cdot 17 \pm 4.71$
DCMU 10^{-5}M	$1.58 \pm 0 \cdot 45$ $P<0 \cdot 01$
DCMU 10^{-6}M	$20 \cdot 20 \pm 3.23$
Antimycin A* 10^{-6}M	$9 \cdot 40 \pm 2 \cdot 17$ $P<0 \cdot 05$
Antimycin A* 10^{-7}M	$16 \cdot 33 \pm 3 \cdot 59$

* Prepared in $0 \cdot 5\%$ aq. ethanol.

streptomycin, the leaf was photoperiodically ineffective. This was done with *Xanthium strumarium* (table 4). Thus, photosynthesis still seems to be the process affected by the CO_2 supply, and yet at the same time none of the major

Table 4. Streptomycin-induced leaf bleaching. Plants (6 repli-cates) were exposed to 2 short days. The floral index is shown after 15 days. Streptomycin was applied pre-inductively

Treatment	Floral index
Streptomycin	
$0 \cdot 2\%$	$0 \cdot 0 \pm 0 \cdot 0$ $P = 0 \cdot 1$
$0 \cdot 1\%$	$0 \cdot 0 \pm 0 \cdot 0$ $P = 0 \cdot 1$
$0 \cdot 05\%$	$4 \cdot 8 \pm 0 \cdot 80$ n.s.
$0 \cdot 01\%$	$5 \cdot 8 \pm 0 \cdot 63$ n.s.
$0 \cdot 001\%$	$6 \cdot 4 \pm 0 \cdot 75$ n.s.
H_2O (control)	$6 \cdot 8 \pm 0 \cdot 40$

photosynthetic products could substitute for CO_2 as such (table 2). While it is impossible here to go into these investigations, the conclusion arrived at was that in all probability some minor or side-product of the photosynthetic mechanism led to a product essential for the flower-inducing process. In the theoretical scheme devised before these studies this input had been represented by Y and it is believed to be one of the factors acted upon by the adaptive enzyme. Thus, while we are none the wiser about the nature of this product, the necessity for it seems to be established.

Test for the presence of the adaptive enzyme

One of the corollaries of the scheme is that with partial induction in short-day plants there should be some lengthening of the critical dark period. Tests were carried out to see whether this could, in fact, be detected, and these gave positive results. Thus, while control plants failed to become induced by 13 h days—the critical daylength for *Kalanchoe* is approximately 12 h—it is possible to increase (almost to double) flower production in partially induced plants with 13 h days, and this would not contradict, therefore, the suggestion of a multiplication of the hypothetical enzyme. In any case, it is known that many short-day plants, after receiving more or less complete inductive treatment, continue to flower readily under conditions of continuous illumination.

The flowering inhibitor

Tests have also been carried out with the objective of detecting the hypothetical inhibitor. Initially these were carried out with quite crude extracts expressed from the sap of leaves. The extracts were re-injected, without being concentrated, into leaves of the assay plants, which were simultaneously exposed to short days. Two types of control were run; first, similar extracts from leaves that had been exposed to short days for a long period, and secondly, water controls. These experiments revealed quite conclusively that leaves held in non-inductive long-day conditions contained substances which, when re-injected into assay plants, caused flower inhibition, whereas short-day-treated (induced) leaves had no such effect, or only a slight inhibition comparable with that of water injection. Two injections in the course of some 12–14 short-day cycles caused an average reduction in flower production of just under 60 per cent (table 5). Thus, there is cogent evidence for

Table 5. Effect of injection of single induced leaves with a crude or centrifuged sap extract on the fifth and tenth days of treatment with 24 short days. Flower numbers are expressed as a percentage of controls injected with non-inhibitory extracts from short-day leaves

Crude extract (sap)		Partly purified extract	
Expt no.	Flowering	Expt no.	Flowering
1	63	11	54
2	58	12	10
3	54	13	23
4	29	14	6
5	26	15	66
6	13	16	107
7	43	17	52
8	73	18	0
9	23	19	41
10	38	20	65
Mean	42·0	Mean	42·4

the presence of an inhibitory substance. We are now in the process of trying to elucidate the chemical nature of this substance, but regrettably not in a position as yet to name a specific inhibitory substance. Partial purification by centrifugation etc. has given closely similar results (table 6). A number of known plant hormones have been shown to be inhibitory to flowering in *Kalanchoe* and these include indole-3yl-acetic acid, gibberellin, abscisin and xanthoxin. The claim made by Pryce[11] that gallic acid is inhibitory has, however, been proved false and gallic acid, even at a concentration of 1000 parts/10⁶ does not inhibit flowering. When injected

Table 6. Effect of injection of single induced leaves with a crude sap extract (testing dialysate and non-diffusible residue after removal of particulate matter by centrifugation) on the fifth and tenth days of an induction period lasting 14 days. Flower numbers are expressed as a percentage of controls, which were not injected

Expt no.	SD diff.	LD diff.	SD non-diff.	LD non-diff.
1	114·0	45·8	37·2	35·1
2	96·0	52·5	31·8	37·3
3	59·2	22·6	–	–
Mean	89·73	40·3	34·5	36·2

SD, Short day; LD, Long day; diff., diffusible.

together with an unconcentrated crude extract of long-day leaves, it is only as inhibitory as the long-day extract. We have some evidence that the molecule involved may be small. It is not as yet clear whether a new specific flowering inhibitor is involved; it could still be one of the known natural growth regulators,[12] though this possibility is becoming less likely in the light of recent work.

Translocation

It has been claimed that the inhibitory effects recorded in the past could be explained merely in terms of alterations of the translocation pattern brought about in the plant. Although the injection experiments are definite enough, we have tested this transport hypothesis further by applying labelled sucrose to leaves in different positions as regards the translocation pathways between them and apices, again using K. blossfeldiana. These results suggest that, while translocation may be modified quantitatively by different treatments, it seems unlikely that the co-transport of flower-promoting substances and carbohydrates which is likely to be involved is, in fact, interdicted by particular treatments, and therefore effects acting purely on flower-promoter transport from the leaves are not very likely.

It must be realized that the above scheme applies largely to what goes on in the leaf and there has been no need to postulate that the flowering inhibitor itself is transmissible. The more recent work with long-day plants by Lang et al.[7] suggests that here transmissibility is operative.

In summary, therefore, it seems that, regardless of whether the scheme and its consequences presented here should ultimately prove to be right or wrong, it is likely that a common reaction pattern must obtain for the great majority of flowering plants, and that both promotion and inhibition are involved in each.

REFERENCES

1. W. W. Schwabe and R. H. Wimble. In: N. Sunderland (ed.) Perspectives in Experimental Biology, vol. 2. Oxford, Pergamon, pp. 41–57 (1976).
2. M. C. Chaijlakjan. CR (Dokl.) Acad. Sci. URSS 3, 443 (1936).
3. B. S. Moshkov. Bull. Appl. Bot. Gen. Plant Breed. Ser. A. 17, 25–30 (1936).
4. J. A. D. Zeevaart. Planta 140, 289–91 (1978).
5. M. C. Chaijlakjan, N. Y. Grigoryeva and V. N. Lozhnikova. Dokl. Akad. Nauk. SSSR 236, 773–6 (1977).

6. W. W. Schwabe. *J. Exp. Bot.* **10**, 317–29 (1959).
7. A. Lang, M. C. Chailakhyan and I. A. Frolova. *Proc. Natl Acad. Sci. USA* **74**, 2412–16 (1977).
8. W. W. Schwabe. *J. Exp. Bot.* **19**, 108–13 (1968).
9. R. Harder and G. Gümmer. *Planta* **35**, 88–99 (1947).
10. C. Ireland. PhD thesis, University of London (1980).
11. R. J. Pryce. *Phytochemistry* **11**, 1911–18 (1972).
12. W. W. Schwabe. *Planta* **103**, 18–23 (1972).

DISCUSSION

Brinkmann: During its individual life span a plant may pass through a sequence of different photoperiodic decisions, all of which may be mediated via phytochrome (spring long-day induction of vegetative growth, summer long-day induction of flowering, autumnal short-day induction of winter buds). That means a particular plant cannot be a 'short-day type' or a 'long-day type' by its constitution. Would you consider that your model of hormone production is still valid for all the seasonal sequences of photoperiodic events, simply by changing some properties with season?

Schwabe: As far as I am aware, some recognizable morphogenetic response is elicited by photoperiod—be it short or long—in all species, but this may be quite unrelated to flowering. Thus, it would seem that short-day treatment may cause the onset of dormancy as well as the induction of flowering in short-day plants, while in long-day-requiring species it may still induce dormancy but in this instance inhibit flowering. Long-day conditions may equally have a variety of morphogenetic effects. My model is purely concerned with the promotion or inhibition of flowering, i.e. the conditions leading to the production of a transmissible morphogen Z, triggering the transition to floral organ production. I would not think that a plant requiring short days for flowering must necessarily have the same daylength response for other morphogenetic reactions. It may well be a long-day type for, say, vegetative growth and a short-day type for tuber production at the same time, as is the case, for instance, in the potato.

Follett: Many plants do, of course, respond to a single short (or long) day and flower some time later. Is this due to a long-term effect on the photoperiodic clock, or to a long-term effect on the physiological system triggering florigen formation?

Schwabe: Where a single inductive cycle is effective in causing flowering, it is generally held that the triggering occurs almost within hours of the cycle end, and the morphogenetic effects are merely the delayed expression of the effect, i.e. if the clock operated like a time-switch, it would set in motion the morphogenetic response as soon as the circuit had been closed.

Nicholls: A single inductive night-period in *Kalanchoe* changes the critical duration of darkness for induction in subsequent cycles. Why? Are comparable changes in critical daylength found in long-day species or in animals?

Schwabe: While the shortening of the critical dark period in *Kalanchoe* found with partial induction usually requires more than a single cycle in order to be detected quantitatively, it is assumed that there is summation of this effect and in the reaction scheme this would be explained by an increment in the amount of enzyme E. I do not know of comparable experiments with long-day plants or animal species.

Aschoff (comment): With reference to the fact that *one* exposure to a short photoperiod is effective in flower induction, Follett asked whether this is due to a permanent change in the clock or in the production of a substance, the clock remaining unchanged. Answering a question by Nicholls, Schwabe mentioned that *one* exposure to a short photoperiod can change the *critical* photoperiod measured afterwards. From this observation I conclude that *one* exposure to a short photoperiod in fact changes the clock. If this interpretation is correct, it has a bearing on the following statement by Schwabe: '12 consecutive short photoperiods (SD) produce 100 per cent flowering but the same number of short days interrupted by one long photoperiod (6SD, 1LD, 6SD) results in only 50 per cent flowering'. Schwabe inferred from this the production of an inhibitory substance by the long photoperiod. This conclusion does not seem stringent because if the one long photoperiod has changed the clock, the second series of six short days is 'seen' by the plant in a different way from the first six short days.

Schwabe: It is not strictly correct to say that the change of critical photoperiod has been detected after a single short-day exposure, a quantitative effect required more than one inductive cycle. It should also be borne in mind that the first set of 6 short days was preceded by long days in the same manner as the second set was, following the interrupting long day.

Turek: What is known about the flowering hormone?

Bünning: We know that after photoperiod induction of a leaf, information leading to flower initiation goes to the bud. The assumed substance responsible for this information is called 'florigen'. However, all efforts to isolate this substance have been without success so far. Contrary to other hormones (for example auxin), the 'florigen' does not pass if living contact is interrupted. After such an interruption (cutting) contact between the living protoplasts must be re-established before the information can be transmitted. Thus, it is doubtful whether we are dealing with a single substance.

PHOTOPERIODIC CONTROL OF TOCOPHEROL OXIDASE IN *XANTHIUM*: AN *IN VITRO* MODEL

J. K. Gaunt and E. Susan Plumpton

Department of Biochemistry and Soil Science,
University College of North Wales, Bangor, Gwynedd

Summary

Floral induction in *Xanthium* is under very precise photoperiodic control. *Xanthium* is a short-day plant in which the critical dark period is about 8·5 h. Studies of the metabolism of vitamin E in the leaves of this species have revealed photoperiodic control of an enzyme—tocopherol oxidase. This enzyme has several regulatory properties. It is reversibly inhibited by brief exposure to darkness and by certain of the plant hormones. During a prolonged exposure to darkness a series of changes occurs in the properties of the system. These include a major alteration at between 8 and 9 h which correlates with the critical dark period for floral induction. The sequence of changes has been demonstrated both *in vivo* and *in vitro*. Both crude cell-free extracts from *Xanthium* leaves and partially purified preparations show the same sequence of changes over a 12 h dark period. This system clearly affords a model in which to study the mechanism of photoperiodism. It can be manipulated *in vitro* and appears to be sufficiently robust for the application of biochemical techniques to its analysis.

INTRODUCTION

A major obstacle to resolving the mechanism of the biological clock has been the absence of a cell-free system in which the timing process could be studied. We have recently found such a system, the properties of which are described in this paper.

The experimental plant is *Xanthium strumarium* L. This is a short-day plant in which the critical dark period for floral induction is around 8·5 h. The induction of flowering in this species has been the subject of extensive investigation.[1]

The enzyme tocopherol oxidase is widely distributed in plants.[2] It catalyses the oxidation of members of the vitamin E family of compounds. The basic characteristics of this enzyme were first elaborated by Barlow and Gaunt.[3] Recently the enzyme has been found to be subject to a variety of very striking control functions which may all be demonstrated *in vitro*. 1. It is inactivated by far-red light ($\lambda \geqslant 730$ nm). This is freely reversed by red light and phytochrome is believed to be the photoreceptor involved.[4] 2. It is rapidly inactivated by darkness (3–4 min) and again this can be reversed by red light.[4] As yet it is not known which photoreceptor is involved here. The reversible inhibition by far-red light and darkness is termed photocontrol. 3. It is regulated by plant hormones.[5] 4. The enzyme isolated from

Biological Clocks in Seasonal Reproductive Cycles 57–66 (1981) (ed. B. K. and D. E. Follett: Bristol, Wright).

leaves of *Xanthium* has now been found to show photoperiodic control, again *in vitro*.[6]

MATERIALS AND METHODS

Plant material

Plants of *Xanthium strumarium* L. were grown from seed in John Innes compost in a controlled-environment room at a constant day and night temperature of 20–22 °C. Vegetative plants were maintained on an 18 h photoperiod. Plants were generally used after about 8 weeks for photoperiodic experiments.

Enzyme preparation and assay

Leaves were homogenized in a mortar and pestle with buffer using 1–5 ml/g fresh weight of tissue. Sodium phosphate-citrate buffer, pH 5·5 (0·1 M-phosphate, 0·02 M-citrate) was used routinely. The plant homogenate was filtered through muslin and then centrifuged at about 500 g for 5 min. The supernatant fraction is referred to as the crude cell-free extract.

A particulate fraction was prepared as follows. Crude cell-free extract (10–15 ml) was layered on to 20 ml 15 per cent sucrose in buffer and centrifuged at 38 000 g for 15 min. The pellet was suspended gently in buffer to give a concentration equivalent to 1 g fresh weight of original tissue per millilitre.

A soluble protein preparation was obtained by centrifugation of the crude cell-free extract at 38 000 g for 15 min, followed by a further centrifugation of the supernatant fraction at 100 000 g for 1 h. Ammonium sulphate (heavy metal free) was then added to the supernatant fraction to 70 per cent saturation and the precipitated proteins redissolved in buffer at a concentration equivalent to 1 g fresh weight of original tissue per ml.

Two methods of assaying tocopherol oxidase were used, each a modification of the basic procedure developed by Barlow and Gaunt.[3] *Method 1* (for use with crude cell-free extracts): To 3·2 ml of a suitable dilution of cell-free extract was added 0·8 ml of α-tocopherol in ethanol (2 mg/ml). After gentle mixing the mixture was incubated at 20 °C. Subsamples (0·5 ml) were taken at 15 s, 1 min, 2 min and at either 3 min or 5 min. Each sample was mixed with isopropanol (2 ml) to stop the reaction. To each was then added 10 ml light petroleum (boiling point 40–60 °C) followed by about 20 ml water and 0·2 ml ethanol. Exactly 5·0 ml of the upper phase containing the tocopherol was evaporated to dryness under reduced pressure. Estimation of tocopherol was by the Emmerie–Engel procedure.[7] A solution of α,α′-bipyridyl (3·5 ml, 0·07 per cent in ethanol) was then added followed by ferric chloride (0·5 ml, 0·2 per cent in ethanol). Absorbance was read at 520 nm after exactly 2 min.

Method 2 (suitable for partially purified enzyme): To 0·9 ml enzyme preparation was added 0·1 ml α-tocopherol (20 mg/ml in ethanol) at 20 °C. Subsamples (0·1 ml) were taken at 15 s, 1 min, 2 min and 3 min and each was mixed with 0·5 ml isopropanol. Exactly 1 ml light petroleum (boiling point 60–80 °C) was added followed by water (4·5 ml) and ethanol (0·1 ml). Exactly 0·5 ml of the upper phase was taken and mixed with the colour reagent as described in Method 1.

Data from each assay was plotted in the form of a progress curve. Since the majority of assays showed the reaction to slow rapidly during the course of an incubation, initial reaction rates could not be ascertained with accuracy. Quantitative measurements of activity were taken as the total fall in absorbance between initial and final assay readings.

It must be recorded that considerable care and practice is needed in order to obtain good data from assays since the activity is often low and may involve measurement of a drop in absorbance of only 10 per cent of the original value. However, in our hands the procedure has proved highly reproducible.

The enzyme shows unusual behaviour in several respects that will be reported elsewhere, but it is important to recognize one of these in any attempt to obtain reliable assay data. This is that the enzyme may only show activity over a narrow range of enzyme concentration. For simplicity, enzyme concentration is referred back to the original fresh weight of leaf tissue used in enzyme preparation. In our hands a crude cell-free preparation from 0·1 to 0·2 g fresh weight per ml gives satisfactory results as do mixtures containing membranes from 0·2 to 0·4 g fresh weight and soluble proteins from 0·1 to 0·2 g fresh weight per millilitre.

Photoperiodic experiments

These were conducted as described in the text. All 'dark' manipulations were performed with the aid of a low intensity green safelight. Various light sources were used: daylight; white light obtained from warm white fluorescent tubes at a quantum flux density of about 150 $\mu E \ s^{-1} \ m^{-2}$; green, red, blue and far-red sources were prepared with the use of Cinemoid filters.[8] Red and blue light was used at a quantum flux density of 2 $\mu E \ s^{-1} \ m^{-2}$.

All experiments were conducted at 20 ± 2 °C unless otherwise specified.

RESULTS AND DISCUSSION

Photoperiodic behaviour of tocopherol oxidase *in vivo*

Three types of experiment were performed.

1. Plants were placed in continuous darkness for up to 16 h. At intervals leaves were harvested, a crude cell-free extract prepared and tocopherol oxidase activity measured—all in the dark. In samples taken during the first 12 h of darkness no activity was detected. This was expected since the enzyme is known to be rapidly inactivated in darkness. What was not expected was the sudden appearance of activity at between 12 and 13 h (figure 1).

2. In a parallel experiment leaves were harvested from plants which had been exposed to darkness for different periods and were given white light (5–15 min) prior to enzyme preparation and assay in daylight. In such samples the pattern of activity is as shown in figure 1. For the first 4·5 h of darkness the enzyme was found to be active, as expected. However, samples taken at 5 and 5·5 h were without activity. There followed a period when activity was detectable until at between 8·5 and 9 h activity was once again missing. This time the enzyme remained inactive for about 3 h and in all samples taken between 9 and 12 h the enzyme was inactive. After 12 h activity was again found. The level of activity over the periods 0–4 h and 6–8 h of darkness was approximately the same and there was no evidence for a slow and steady decrease. Activity declined rather rapidly to zero between 4 and 5 h and between 8 and 9 h. When activity reappeared after 12 h the level was more variable but without any discernible pattern.

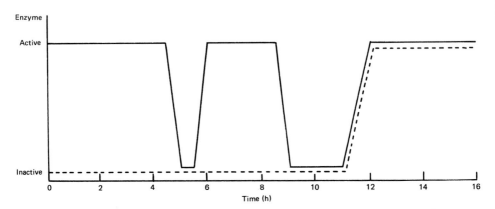

Figure 1. Tocopherol oxidase activity in *Xanthium* leaf extracts during 16 h exposure to darkness. Leaf samples taken at intervals during a continuous dark treatment were either assayed in darkness (broken lines) or after a brief exposure to white light (solid lines). Tocopherol oxidase activity is recorded as either present or absent.

Thus, tocopherol oxidase apparently exists in several states during 16 h of continuous darkness. These states are identified by the activity of the enzyme in light and in darkness. Three basic conditions have been found.

(i) Enzyme showing normal photocontrol, that is to say it is active in light and inactive in darkness. This state occurs between 0 and 4·5 h and between 6 and 8·5 h of darkness.

(ii) Enzyme showing no activity, irrespective of lighting conditions. This was observed between 5 and 5·5 h and between 9 and 12 h of darkness.

(iii) Enzyme active but without photocontrol, that is to say active in the light and in the dark. This was the case between 13 and 16 h of darkness.

3. Plants that had been maintained in continuous darkness for between 9 and 16 h were placed in white light for 1 h prior to enzyme preparation and assay. In all cases the enzyme showed activity and normal photocontrol—the system appeared to have returned to the original state that pertained before placing the plants into darkness.

These results suggest that tocopherol oxidase is regulated by signals coming from a timing mechanism that is triggered by darkness. A long exposure to light may be needed before the clock can be reset.

Photoperiodic behaviour of tocopherol oxidase in crude cell-free extracts

A crude cell-free extract was placed in continuous darkness for 16 h. During this period samples were tested for enzyme activity either in the dark or after a brief exposure to daylight (2–5 min). Levels of activity in dark assays were zero within 3 min. After this the pattern of dark and light activity was exactly as is depicted in figure 1. The *in vitro* behaviour of the system appeared to be identical to that found *in vivo*. Subsequent studies have revealed one minor difference. Crude cell-free extracts placed in daylight after 9 h of darkness require much more than 1 h of light to restore them to the starting condition.

This result shows that the timing mechanism that is able to regulate tocopherol oxidase retains its normal function *in vitro*.

Photoperiodic behaviour in partially purified cell-free preparations

Previous studies have shown that the tocopherol oxidase system can be partially purified with retention of its photocontrol properties.[4] The system requires two components—a washed particulate fraction and a soluble protein preparation. The membranes are an essential cofactor for the enzyme[3] and also contain photoreceptor molecules that regulate its activity. The soluble proteins contain the enzyme and a further supply of photoreceptor. The membranes are very sensitive to mechanical damage and it is essential to treat them gently in order to retain normal control properties.

A membrane fraction (sedimenting at between 500 and 38 000 g) and a soluble protein preparation were mixed in equal proportions and examined for photoperiodic behaviour. It is important to state that 'equal proportions' refers to the fresh weight of leaf tissue from which each component was isolated. That is, membranes from 1 g of leaf tissue were mixed with soluble proteins isolated from 1 g of leaf tissue. The mixture was placed in continuous darkness and tocopherol oxidase activity determined at intervals both in darkness and after a brief exposure to daylight (2–5 min). Behaviour of the system over the first 12 h was exactly as shown in figure 1. However, activity did not return after this time. The clock is apparently still working normally over a 12 h period. This suggests that the timing mechanism can have only two types of component—membrane-associated and soluble protein. It also demonstrates that the clock is sufficiently robust to permit biochemical manipulation.

The effect of temperature on photoperiodic behaviour *in vitro*

A 1 : 1 mixture of membranes and soluble proteins was maintained in the dark at 10 and at 20 °C. Assays of tocopherol oxidase activity were made at between 8 and 9 h of darkness. They showed almost identical behaviour at each temperature. At between 8·5 and 9 h the enzyme lost its ability to be activated by light. The overall timing mechanism shows the relative insensitivity to temperature that is so characteristic of photoperiodic clocks.

Behaviour of the membrane fraction in darkness

A membrane fraction was placed in continuous darkness. From this, samples were taken at intervals and mixed with soluble proteins that had been maintained under daylight up to this point. The mixture was immediately exposed to daylight (5 min) before assay. Assays involving membranes that had been exposed to 10 and 20 min of darkness were always active. Those from membranes that had received 30 min of darkness were always inactive. This experiment shows that some sort of timing device is associated with the membrane fraction and that it appears to function in the absence of other cellular components.

Behaviour of soluble proteins in darkness

A soluble protein preparation was exposed to continuous darkness for 3 min, 15 min or 1 h. At the end of each period a sample was treated with red light or daylight before mixing with membranes that had been maintained in the light. After 3 min of darkness the soluble protein preparation required 1·5 min of red light to restore activity fully. After 15 min of darkness activity was not restored by either 30 min of red light or by 5 min of daylight. After 60 min of darkness activity was restored by less than 5 min of red light. Clearly a sequence of changes occurs in the soluble protein fraction in the absence of other cellular components. The full extent of these changes has yet to be elucidated.

The effect of varying the ratio of soluble proteins and membranes

Since membranes and soluble proteins behave differently in the dark compared with a combined system, it seems that both components are required for normal functioning of the clock. To test the hypothesis that an interaction occurs between the two, mixtures were prepared in which the ratio of the two components was varied. A series was made up with the soluble proteins representing from 25 to 100 per cent of the membrane fraction, again referred back to the fresh weight of the tissue. These mixtures were placed in darkness and at intervals subsamples were removed and assayed after 5 min of daylight. Results are shown in figure 2. It is clear that the sequence of changes in each case is identical to those previously described for 1 : 1 mixtures but the timing of these changes depends upon the proportion of soluble protein used.

The amounts of red and blue light needed to activate tocopherol oxidase after different periods of darkness

After 4 min of darkness a crude cell-free preparation required 15 min of red or 25 min of blue light to restore enzyme activity completely. After 15 min of darkness, only 5 min of red or 10 min of blue light was required. This shows that there is a major difference in the total light energy needed for activation and a minor alteration in the relative effectiveness of red and blue light. This clearly shows that the state of the photocontrol system at 15 min of darkness is different from that at 4 min. It supports the idea that rapid changes occur within the system once it is exposed to darkness. Exactly how many different states are involved has not yet been investigated.

GENERAL DISCUSSION

Certain properties of tocopherol oxidase in the leaves of *Xanthium* clearly alter according to the length of exposure to darkness. The system appears to pass through a sequence of discrete stages. At present it looks as though the same sequence occurs in crude cell-free extracts and in the intact plant, at least over the first 16 h of darkness. As the system is purified alterations are introduced and the membrane/soluble protein system shows different properties after the first 12 h of darkness. Nevertheless, up to this point timing is the same.

The clock has two major types of component, one is membrane-associated and the other a soluble protein. Changes occur within each of these fractions when placed alone in darkness. The nature of the changes and of the components involved is not known. In mixtures of membranes and soluble proteins the timing properties depend on their ratio. This indicates an interaction between them, but again of unknown nature. A model that could explain such an interaction is presented later.

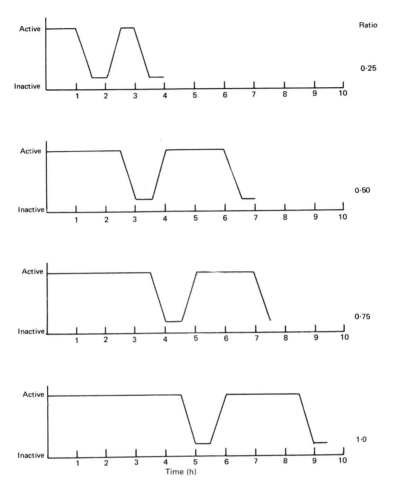

Figure 2. The effect of the ratio of soluble protein to membranes on timing properties of tocopherol oxidase. Various mixtures of membranes and soluble proteins were placed in continuous darkness. At intervals samples were given daylight and assayed for tocopherol oxidase activity.

Two types of reaction could be envisaged to take place in darkness.

1. One or more photoreceptors exists in an unstable form in the light and slowly reverts to a more stable form in darkness. Such a reaction is believed to exist for phytochrome.

2. Photosynthetic electron transport stops in darkness with consequential effects upon the redox state of the cell-free system. Enzymic or chemical oxidation of cellular components could alter their activity. This is possible in the tocopherol oxidase system from *Xanthium* leaves since the membranes used contain photosynthetic pigments.

It is known that tocopherol oxidase shows dark inhibition in systems that contain no photosynthetic pigments. Thus, some dark changes appear to be photoreceptor-mediated rather than linked to photosynthesis. However, some involvement of photosynthesis cannot yet be eliminated.

Previous work on the far-red and dark inhibition of tocopherol oxidase has shown that membrane-associated photoreceptors are important in regulation.[4] The red/far-red control of the enzyme almost certainly involves phytochrome. However, it is by no means clear whether this or some other photoreceptor is involved in the dark responses of the system. Phytochrome does show change in darkness, but this is considered to be relatively slow.[8] Either phytochrome shows some previously undetected changes in darkness or some other photoreceptor is present that is responsible for rapid dark inhibition. At present it is not possible to say how many photoreceptors are able to control tocopherol oxidase. Nor is it possible to say what is responsible for the progression of changes occurring during dark exposure. The total number of identifiable stages seems likely to be large. During the first 15 min of darkness two have been detected on the basis of differing sensitivity to light activation. Between 5 and 5·5 h of darkness at least one further stage is evident. At least one more exists between 6 and 8 h, another between 9 and 12 h and yet another after this. It is certain that others will be found when a more detailed survey is made.

These stages presumably represent changes in the clock components, that is, they represent the timing process. When we understand their nature we shall understand the clock mechanism.

There are two ways of looking at the sequence of changes.

1. It solely represents the machinery of the timing mechanism and somehow generates a signal, recognized within the plant, at only one point, probably between 8 and 9 h of darkness.

2. It provides a continuing source of reference for the plant. Each stage may be recognized and interpreted in some way and thus provide information to indicate the time of dark exposure. Exactly how this could be achieved is impossible to predict.

Whichever of these possibilities is correct, there should be at least one signal emanating from the system. What is the nature of this? Tocopherol oxidase can exist in two states—active and inactive. In continual darkness it becomes inactive after 3 min and remains so for the next 12 h. Then activity reappears and may thus directly generate a signal by the sudden catalysis of a reaction. But could the system generate a signal at between 8 and 9 h? This possibility will not be discussed here.

Our present view of the system is that the photoreceptor(s) and other components that comprise the clock pass through a series of different interactions driven by altered conformations of the photoreceptor(s). As far as tocopherol oxidase is concerned the important changes are those of membrane-associated components as it is the state of these that determines the properties of the enzyme. The role of the soluble proteins is to interact with membrane-components and so extend the timing process. A model that would account for this is shown in figure 3. It requires a photoreceptor that is both associated with membranes and a soluble protein. Imagine that a dark-induced conformational change occurs in the membrane-associated photoreceptor that gives it a reduced affinity for the membrane. An exchange could then occur between it and soluble photoreceptor. This would 'recharge' the membrane with photoreceptor in its original form and so the cycle could start again. This would continue until all of the soluble photoreceptor is in the changed state. Thus a fairly rapid change affecting only membrane photoreceptor could take a long time to alter the entire photoreceptor population and this would depend upon the relative proportions of membrane and soluble proteins present. The principle of a photoreceptor existing in both soluble and membrane-associated

Time in darkness →

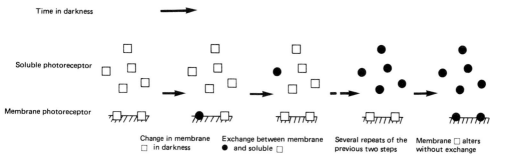

Figure 3. A model showing a possible interaction between membrane and soluble photoreceptor molecules. The model makes four assumptions. 1. The photoreceptor exists in two states (□ and ●). 2. Membrane-bound □ can be converted to ● in the dark, while soluble □ cannot. 3. The affinity of ● for the membrane is less than that of □, permitting an exchange. 4. The properties of the system depend upon the state of the membrane-bound photoreceptor.

forms in a plant cell is long established. Phytochrome is distributed in this way and there has been much speculation about its significance. The above model offers an answer.

If this model correctly describes the situation, it will have to be applied more than once during the timing process. Nevertheless, despite the complexity of changes occurring, it does offer an explanation for the interaction between the two types of clock component.

The physiological significance of photoperiodic control of tocopherol oxidase is quite unknown. A major obstacle here is lack of understanding of the role of tocopherol oxidase in plants. Does it have a role in the regulation of tocopherol levels in cell membranes? If so, what is the biochemical function of tocopherols in membranes? It seems likely to be important and specific, rather than the general antioxidant role so commonly ascribed to this molecule. Unfortunately there is no information on this. Despite this, the system still serves as an *in vitro* photoperiodic model which is sufficiently robust to permit the biochemical manipulation necessary to investigate the components and interactions that constitute a biological clock.

REFERENCES

1. F. B. Salisbury. In: L. T. Evans (ed.) *The Induction of Flowering*. Sydney, Macmillan, pp. 14–61 (1969).
2. E. Murillo, E. S. Plumpton and J. K. Gaunt. *Biochem. Soc. Trans.* **4**, 486–7 (1976).
3. S. M. Barlow and J. K. Gaunt. *Phytochemistry* **11**, 2161–70 (1972).
4. J. K. Gaunt and E. S. Plumpton. *Biochem. Soc. Trans.* **6**, 143–5 (1978).
5. J. K. Gaunt, G. M. Matthews and E. S. Plumpton. *Biochem. Soc. Trans.* **8**, 186–7 (1980).
6. J. K. Gaunt and E. S. Plumpton. *Biochem. Soc. Trans.* **8**, 187–8 (1980).
7. Analytical Methods Committee: Vitamin E Panel. *Analyst* **84**, 356–72 (1959).
8. H. Smith. *Phytochrome and Photomorphogenesis*. London, McGraw Hill, pp. 83–84, 97–107, 214–222 (1975).

DISCUSSION

Brady: What was the resolution of your sampling schedule?

Gaunt: Most samples were taken at hourly intervals, but at critical periods, when changes were apparent, we sampled each 30 min.

Engelmann: The change in the timing frequency after the change in the ratio of the two constituents in your system reminds me of the change of frequency in Sel'kov's proposal. This would imply that the ratio is important for the circadian component.

Gaunt: As I understand Sel'kov's model, it depends upon the presence of low molecular weight substrates and may explain rather high frequency rhythms. The tocopherol oxidase system in *Xanthium* does not show rhythmic changes during 16 h of continuous darkness. We have not yet looked at the system during longer periods to see whether a circadian rhythm exists. In partially purified enzyme preparations we have eliminated low molecular weight, water-soluble substances yet still find normal timing properties. Thus I do not believe that a model such as that of Sel'kov will explain timing in the tocopherol oxidase system.

Brinkmann: Have the nuclear envelopes been excluded by the 500 g centrifugation step? I ask because it has been suggested that the lack of circadian rhythmicity in prokaryotes may be due to the lack of nuclear envelopes.

Gaunt: I would expect all nuclei to be removed from the system by centrifugation at 500 g. It is possible, however, that fragments of nuclear membrane do remain in the crude cell-free preparations used.

Turek: It appears that timing of the enzyme activity is due to transition from light to darkness, i.e. the onset of darkness appears to initiate the timing process. Is there any evidence that this enzymatic process *in vitro* shows a circadian variation in DD, and is a circadian rhythm involved in the photoperiodic response of flowering in *Xanthium*?

Gaunt: Timing certainly appears to be initiated by the start of the dark period. Irrespective of the length of the preceding light period the changes of the system in darkness are the same. In resonance experiments with *Xanthium* no circadian rhythm has been found for floral induction. We have not investigated the tocopherol oxidase system over periods of more than 16 h of continuous darkness and thus cannot say whether there are circadian rhythms in the control of its properties.

INSECT PHOTOPERIODISM: ENTRAINMENT WITHIN THE CIRCADIAN SYSTEM AS A BASIS FOR TIME MEASUREMENT

D. S. Saunders

Department of Zoology, University of Edinburgh

Summary

The induction of pupal diapause in the long-day insect *Sarcophaga argyrostoma* is analysed in terms of entrainment within the insect's circadian system, and within the framework of the 'external coincidence' model. Using phase response curve data derived from single-pulse resetting experiments on the pupal eclosion rhythm, the phase relationships of the putative photo-inducible phase (ϕ_i) to the entraining light cycle were computed for a variety of 'complete' and 'skeleton' photoperiods, the latter including symmetrical and asymmetrical skeletons, symmetrical skeletons in the 'zone of bistability' and one- and two-pulse régimes in cycles longer or shorter than 24 hours. This cybernetic approach was then compared with diapause-induction experiments in identical lighting cycles. In all cases, ϕ_i falling in the dark led to induction of the diapause state, whereas illumination of ϕ_i led to non-diapause or continuous development, suggesting that the external coincidence model is appropriate for this species. Although the photoperiodic clock in *S. argyrostoma* is oscillatory (circadian) its 'hour-glass' properties in diel light cycles are stressed.

INTRODUCTION

Approaches to the problem of insect photoperiodism are still largely restricted to 'black-box' experiments, and to the interpretation of these experiments by formal or even abstract models. A considerable amount is thus known about the formal properties of these clocks but next to nothing about their concrete physiology. This has inevitably given rise to a large number of clock models, the best of which being those based on the largest accumulations of experimental data, or on the soundest theoretical background. Some insects, for example, appear to measure night-length with a non-oscillatory 'hour-glass' or 'interval timer',[1] whilst others—like flowering plants, birds and mammals—may use a clock based on the circadian system.[2] Among the circadian propositions are those that include a *single* oscillation interacting with the environmental light cycle (='external coincidence')[3,4] or those that include *two*—'dawn' and 'dusk'—oscillations whose mutual phase relationship changes with photoperiod and hence with season (='internal coincidence', and

Biological Clocks in Seasonal Reproductive Cycles 67–81 (1981) (ed. B. K. and D. E. Follett: Bristol, Wright).

others).[5-7] These models and their various merits have been reviewed elsewhere:[8,9] here it is sufficient to note that they probably reflect genuine diversity of response within the insects.

The flesh fly *Sarcophaga argyrostoma* is a 'long-day' insect with a pupal diapause induced by short days (or long nights) accumulated during the larval sensitive period.[10] The investigation of this clock has passed through three stages: (*a*) the demonstration that the clock is a function of the insect's circadian system;[11-13] (*b*) the demonstration that the photoperiodic oscillation and the circadian rhythm of pupal eclosion are sufficiently similar to allow the latter to be used as overt 'hands of the clock' and (*c*) the analysis of photoperiodic induction in terms of entrainment to light pulses and cycles, using the overt indicator rhythm (eclosion) as a measure of phase.[8,14,15] It is the last of these three stages which will be reviewed here.

THE EXTERNAL COINCIDENCE MODEL

The properties of the photoperiodic clock in *S. argyrostoma* are best described by the 'external coincidence' model[4,16] and this model has been adopted as the best working hypothesis, although it is recognized that the circadian system and the photoperiodic clock in *S. argyrostoma* must be of a more complex or 'multi-oscillatory' nature.[8,14] In its simplest form (figure 1), the model consists of a circadian oscillation phase-set by the light cycle in a manner exemplified by the eclosion rhythm of *Drosophila pseudoobscura*.[4] The crux of the scheme is the observation that the oscillation is reset to a particular phase (called circadian time—CT, 12) at the end of the light component once the latter exceeds 12 h; the oscillation then free-runs in darkness. Consequently a putative light-sensitive or 'photo-inducible' phase (ϕ_i), thought to occur about $9\frac{1}{2}$ h later (i.e. at CT 21·5), falls in the *dark* of a long night (e.g. LD 12 : 12) but *after dawn* in a short-night régime (e.g. LD 16 : 8). Illumination or non-illumination of ϕ_i thus provides a ready explanation for the critical night length, but the model, as we shall see later, cannot explain other features of the response without the inclusion of further aspects of the entrainment phenomenon.

THE USE OF PUPAL ECLOSION AS 'HANDS OF THE CLOCK'

The eclosion rhythm and the photoperiodic oscillation are both reset to a phase close to CT 12 once the photoperiod exceeds about 12 h (figure 2). This common property encourages the view that the external coincidence model is appropriate for the *Sarcophaga* case, and that eclosion may be used as the overt 'hands'. The analysis that follows rests heavily on the assumption that these two components of the circadian system are sufficiently similar; in my view, the results to be described support that assumption. The only other important assumption, namely that a 'light-sensitive phase' (ϕ_i) occurs at the end of the critical night (at CT 21·5), is also vindicated by the results described below, although they provide no indication as to the *nature* of that particular phase.

The eclosion rhythm in *S. argyrostoma*, like that in *D. pseudoobscura*, free-runs in DD producing peaks of emerging adults at roughly 24 h intervals. At 25 °C and after a step-down from light to dark (LL/DD) as newly deposited larvae, flies emerge in five or six consecutive peaks 16–21 days after larviposition (figure 3).

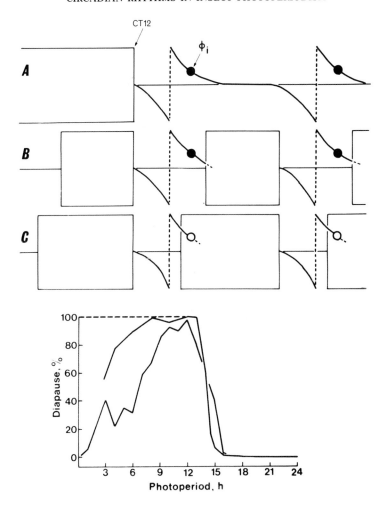

Figure 1. The 'external coincidence' model for the photoperiodic clock (after Pittendrigh[4]). The photoperiodic oscillation, shown here as the phase response curve (15 min light pulses, *Drosophila pseudoobscura*: *see* figure 4 for examples), commences its dark-period kinetics at CT 12 following exposure to continuous light (in **A**), LD 12 : 12 (**B**) or LD 16 : 8 (**C**). In **A**(LL to DD) the oscillation free-runs in darkness. The photo-inducible phase (ϕ_i) is shown lying $9\frac{1}{2}$ h into the night, at CT 21·5. In long nights (LD 12 : 12, **B**) ϕ_i falls in the dark, whereas in short nights (LD 16 : 8, **C**) it falls after dawn. Illumination or non-illumination of ϕ_i is thought to control the critical night-length response. *Lower panel* shows two photoperiodic response curves (solid lines) for pupal diapause induction in the flesh-fly, *Sarcophaga argyrostoma*. The dotted line shows the response expected in a rigid interpretation of the model (*see* text).

Since the circadian oscillation governing eclosion is at a phase equivalent to CT 12 at the LL/DD transition, the medians of pupal eclosion may be calculated to occur at about CT 23·5, or at intervals of about 11·5 + modulo τ after transfer to darkness.[13,17] The free-running period, τ, is close to 24 h.

If this overt rhythmicity is to be used in the analysis of the photoperiodic clock, we need to know how the oscillation is reset (advanced or delayed) each time it experiences a light pulse. It is then possible to calculate the approach to steady-state entrainment when the pacemaker is exposed to trains of light pulses such as those

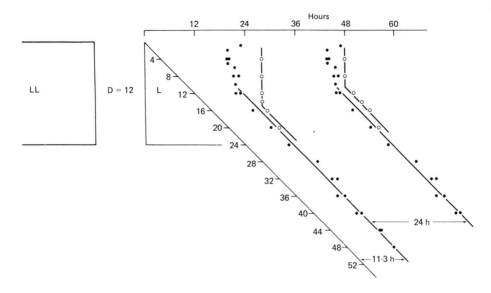

Figure 2. Sarcophaga argyrostoma. Resetting of the pupal eclosion and photoperiodic oscillations to CT 12 after protracted (>12 h) photoperiods. All cultures were taken from continuous light (LL) into a 12 h dark period, then exposed to a last photoperiod of between 2 and 50 h. ●, medians of pupal eclosion plotted modulo 24 h; ○, maxima of diapause induction in 'resonance' experiments using light periods of 4–20 h (data from Saunders[11,13]).

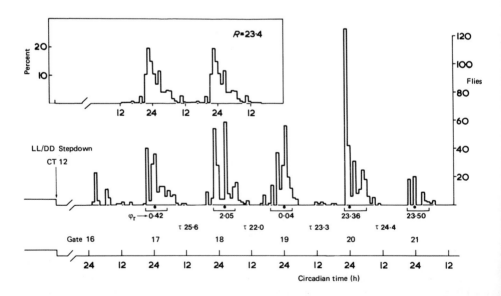

Figure 3. Sarcophaga argyrostoma. The pattern of pupal eclosion generated by a single step-wise transfer from LL to DD in the first 24 h of larval development, showing peaks of eclosion 16 to 21 days later. ●, medians of eclosion (ϕ_r); τ, estimates of free-running periods between successive peaks. The phase of the eclosion pacemaker at the LL/DD stepdown is equivalent to CT 12. *Inset:* data summed for each hour and plotted twice along the time axis, modulo 24 h. R, arrhythmicity value. (After Winfree[18].)

used in photoperiodic experiments. The necessary information was provided by phase response curves calculated for a series of light pulse durations (1 to 20 h, 240 μW/cm²), starting at all phases of the circadian oscillation (CT 01 to CT 24), using steady-state changes in the observed medians of eclosion (ϕ_r) as measures of phase delay ($-\Delta\phi$) or phase advance ($+\Delta\phi$). These curves are shown in figure 4.

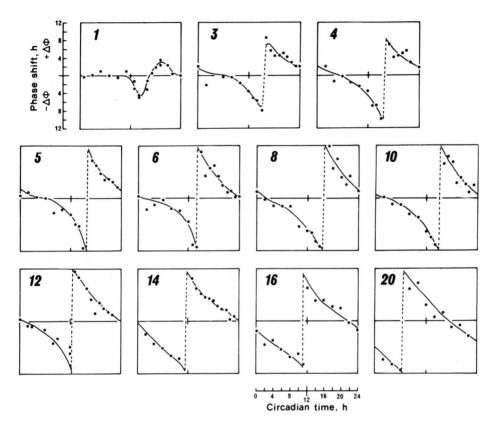

Figure 4. Sarcophaga argyrostoma. Phase response curves for 1–20 h pulses of white light (240 μWcm²) on the rhythm of pupal eclosion. Each box shows advances ($+\Delta\phi$) or delays ($-\Delta\phi$) in phase, in hours, for pulses beginning at all circadian times (CT 01 to 24).

Those for 1 and 3 h pulses were of the low-'amplitude' type 1 of Winfree,[18] while those for 5, 6, 8, 10, 12, 14, 16 and 20 h were of Winfree's high-'amplitude' type 0. Four-hour pulses, however, gave rise to an ambiguous type of resetting curve since this pulse duration was close to that of the 'singularity' which, for *S. argyrostoma* and an irradiance of 240 μW/cm², was close to T* 4 h, S* 4 h.[8]

Phase response curve data were used to calculate theoretical steady-state phase relationships (ψ-values) of the eclosion peaks (ϕ_r) to a variety of experimental light cycles, using a computer program designed to accommodate all pulse durations, up to 4 pulses per 'cycle' (of any length), with the first pulse in the train starting at all circadian times (CT 01 to CT 24), for up to 19 consecutive cycles. The calculations were based on the non-parametric entrainment model of Pittendrigh,[19] and particular attention was paid to *initial conditions* (i.e. starting phase) and the *rate* at which entrainment was achieved. In an earlier paper[8] it was shown that computed ψ-values for ϕ_r were very close to those determined experimentally. In this paper the same procedure was used to compute ψ-values for the putative photoinducible phase (ϕ_i) assuming that it occurred at CT 21·5, or about 2 h before the median of eclosion. Whether ϕ_i fell in the light or in the dark was then compared with data from parallel diapause-induction experiments, both as a test of external coincidence, and of the idea that photoperiodic induction can be analysed in terms of entrainment using only a minimum of assumptions and experimentally determined resetting data.

Computed phase relationships of ϕ_i to the light cycle and the incidence of pupal diapause

Phase relationships of ϕ_i to a variety of simple and complex light cycles were compared with results from parallel diapause-induction experiments. Examples for 'complete' and 'skeleton' photoperiods, all in cycles of $T = 24$ h, are presented here.

Figure 5 shows the results for complete photoperiods, with ψ-values calculated for trains of pulses starting at all initial phases and continued for up to 18 cycles.

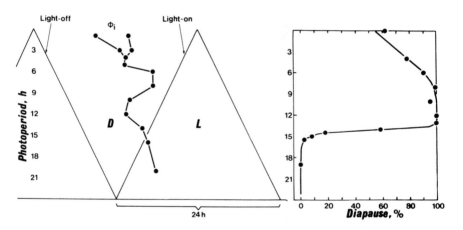

Figure 5. Sarcophaga argyrostoma. Left: Computed phase relationships of the photo-inducible phase (ϕ_i) in complete photoperiods (*T* 24), with the first pulse in the train starting at all possible circadian times (CT 01 to 24) and continued for eighteen consecutive cycles. For LD 1 : 23 and LD 3 : 21 the oscillation has still not reached steady-state by the eighteenth cycle; the two plotted points for ϕ_i show the residual range of phase relationships. *Right:* The photoperiodic response curve for pupal diapause. Note that the critical daylength coincides with the passage of ϕ_i from dark to light.

For weak light pulses (1 and 3 h) the approach to steady-state was slow and a residual scatter of ψ was still evident, even after 18 days. For the stronger pulses, however, the oscillation achieved a unique steady-state phase relationship to the zeitgeber. (Little importance is attached to the *shape* of this relationship, which merely reflects differences in the phase response curves for the various light pulses.) The important aspect is that for all photoperiods of less than 14 h ϕ_i falls in the dark, whereas with 16 and 20 h photoperiods it falls after dawn. Data from a parallel diapause induction experiment[10] show that diapause was high when the photoperiod was less than 14 h but almost negligible when greater than 16 h. The critical day length computed from the eclosion PRCs (≈ 15 h/24) was almost identical to that found by experiment ($\approx 14\frac{1}{2}$ h/24), and the remarkable similarity of these observations to the presumed events in the external coincidence model (figure 1) is obvious.

'Skeleton' photoperiods consisting of two short pulses of light per cycle are known to simulate many of the entraining effects of longer 'complete' photoperiods, and have played an important part in the development of entrainment theory.[16,19] Skeletons have also played a major role in the analysis of the photoperiodic clock, particularly as 'asymmetrical skeleton' or 'night interruption' experiments in which a longer 'main' photoperiod is coupled with a shorter supplementary pulse which scans the accompanying 'night'.

In many insect species night interruption experiments produce a characteristically bimodal response with two points of diapause-averting or short-night effects, usually referred to as points A and B[2,20] (*see* figure 6, right-hand panel). Such

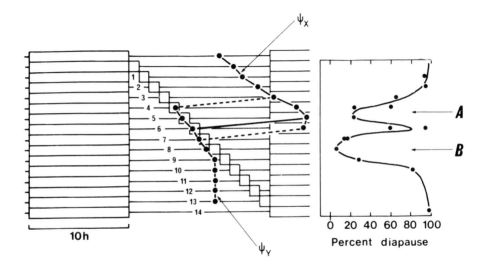

Figure 6. Sarcophaga argyrostoma. Left: Computed phase relationships of ϕ_i (●) to asymmetric skeleton photoperiods composed of a 10 h 'main' photoperiod and a 1 h supplementary or scanning pulse. ψ_X and ψ_Y are the two phase relationships, and the dotted lines the earliest and latest phase jumps between them. The solid connecting line shows the phase jump for the experimental population whose diapause responses are shown to the right. A and B are the two points of low diapause characteristics of insect night-interruption experiments. (Data from Saunders[12].)

responses provided the main impetus for the formulation of the external coincidence model by Pittendrigh and Minis.[16] Here, results for *S. argyrostoma* raised as larvae in cycles of LD 10 : 14 with the 14 h night systematically interrupted by a 1 h pulse are presented in figure 6. The left-hand panel shows the phase relationships of ϕ_i (CT 21·5) computed for each régime, and shows first phase delays ($-\Delta\phi$) and then phase advances ($+\Delta\phi$) as the supplementary pulse scans the night. Up to LD 10 : 3 : 1 : 10 and after LD 10 : 7 : 1 : 6, there are single, unique, steady-states; between LD 10 : 4 : 1 : 9 and LD 10 : 6 : 1 : 7, on the other hand, there are *two* possible steady-states, ψ_X and ψ_Y, forming a 'zone of bistability'. Which of the two modes is adopted is determined by the initial conditions, or the phase of the oscillation illuminated by the first pulse in the complex two-component train (LD

10 : x : 1 : y). For example, a 'population' starting at CT 07 to CT 10 would phase-jump to ψ_Y after LD 10 : 3 : 1 : 10, but the phase jump for a population starting between CT 16 and CT 01 would be delayed until after LD 10 : 6 : 1 : 7; these two phase jumps are shown in the figure as dotted lines. The *experimental* populations whose diapause responses are shown in the right-hand panel were transferred straight from LL into the first dark period: these show a computed phase jump between LD 10 : 5 : 1 : 8 and LD 10 : 6 : 1 : 7 (solid line).

The panel on the right of figure 6 shows that diapause incidence is high whilst ϕ_i lies in the dark, but falls to a low value in two places, once when ϕ_i is phase delayed into the 'dawn' of the main 10 h photoperiod (generating point A), and once after the phase jump when ϕ_i coincides directly with the 1 h scanning pulse (generating point B). The occurrence of a higher incidence of diapause between A and B corresponds to the position of the phase jump itself. The bimodal response is therefore explained in terms of entrainment within the circadian system, and the results are clearly consistent with the external coincidence model.

Figure 7 presents a similar analysis for populations exposed to symmetrical

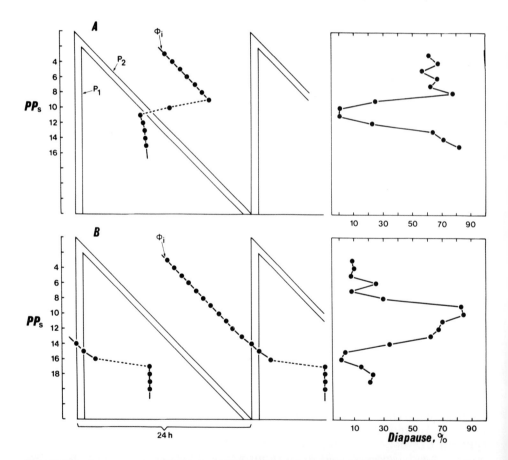

Figure 7. Sarcophaga argyrostoma. Computed phase relationships of ϕ_i to symmetrical skeleton photoperiods (PP$_s$) of two 1 h pulses, with the first pulse in the train starting (**A**) at CT 12 and (**B**) at CT 24. *Right:* The incidence of pupal diapause. Note that a low incidence of diapause is found when ϕ_i coincides either with pulse 2 *or* with pulse 1.

skeletons (of two 1 h pulses per cycle) with the first pulse in the train (P_1) starting at either CT 12 or at CT 24. Pittendrigh[4] pointed out that symmetrical skeletons were open to two 'interpretations' depending on whether the shorter or the longer interval between the pulses was taken as simulating the photoperiod. Which of the two intervals is taken depends, in turn, on (*a*) the magnitude of the interval between the pulses and (*b*) the phase of the oscillation illuminated by the first pulse in the train. For the *Sarcophaga* case all skeletons less than about 9 h (LD 1 : 7 : 1 : 15) or greater than about 16 h (LD 1 : 14 : 1 : 8) give rise to single unique steady-state phase relationships, but those in between (in the 'zone of bistability') can adopt either of the interpretations according to the initial conditions noted above. Figure 7 shows that populations in which P_1 started at CT 12 phase-jumped to the alternative mode between LD 1 : 7 : 1 : 15 and LD 1 : 9 : 1 : 13, whereas those starting at CT 24 jumped between LD 1 : 14 : 1 : 8 and LD 1 : 15 : 1 : 7. In the former case ϕ_i coincided with P_2 in the 11 h skeleton, whereas in the latter case it coincided with P_1 in the 16 h skeleton. Reference to the right-hand panels in figure 7 shows that a low incidence of pupal diapause was *exactly* correlated with these theoretical predictions.

The 'zone of bistability' is further explored in figure 8 in which the two reciprocal skeletons LD 1 : 9 : 1 : 13 and LD 1 : 13 : 1 : 9 were started at all circadian times following a transfer from LL to DD. These two régimes were selected because the 11 h skeleton, if adopted as 'day', would be diapause-inductive, whereas the 15 h skeleton, if adopted, would induce diapause-free development. The experimentally determined diapause incidences were once again in very close agreement with computations of whether ϕ_i fell in the dark or was illuminated by one of the two

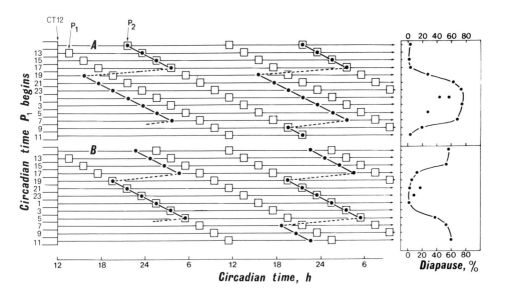

Figure 8. Sarcophaga argyrostoma. Computed phase relationships of ϕ_i to skeleton photoperiods of PP_s 11 (LD 1 : 9 : 1 : 13) and PP_s 15 (LD 1 : 13 : 1 : 9) in the 'zone of bistability', with the first pulse in the train starting at all circadian times (CT 13 to CT 11) following a transfer from LL to DD (CT 12). *Right:* the incidence of pupal diapause (data from Saunders[12]). Note that diapause is high when ϕ_i falls in the dark, but is low when ϕ_i coincides with one of the two pulses forming the skeleton.

pulses making up the skeleton. The results for the two régimes were, of course, both periodic and mirror images of each other.

THE PHOTO-INDUCIBLE PHASE: IS ϕ_i A REALITY?

Making just two assumptions—the close similarity between the eclosion and photoperiodic pacemakers, and the existence of a photo-inducible phase (ϕ_i) at CT 21·5—the foregoing section shows that photoperiodic induction may be predicted from computations based solely on phase response curve data for the system. Although it must be stressed that the results do not rule out an alternative hypothesis for the clock (i.e. internal coincidence, *see below*), or provide an insight into the physiological nature of ϕ_i, they do suggest that external coincidence and ϕ_i are relevant to the *Sarcophaga* case. The experiments described in this section further underline the probable reality of ϕ_i.

Pittendrigh and Minis[16] showed that when the circadian pacemaker of *D. pseudoobscura* (period τ) became entrained to a zeitgeber (period T) consisting of one short pulse of light per cycle, the pulse came to lie, in each cycle, on that part of the phase response curve which generated a phase shift equal to the difference between T and τ. Consequently, when T was greater than τ, the pulse fell in the early subjective night (CT 12 to CT 18) to cause phase delays in each cycle, whereas when T was *less than* τ the pulse fell in the *late* subjective night (CT 18 to CT 24) to cause phase advances. Simply by changing T, therefore, a single short pulse of light per cycle can be made to illuminate different circadian phases. This phenomenon was used by Pittendrigh and Minis[16,21] and Minis[22] to explore the subjective night of *Pectinophora gossypiella* for light-sensitive phases as a test of the external coincidence model, the strength of this so-called T experiment being the accuracy with which this probing may be done.

Females of *S. argyrostoma* were maintained at 25 °C and in LD 12 : 12 to

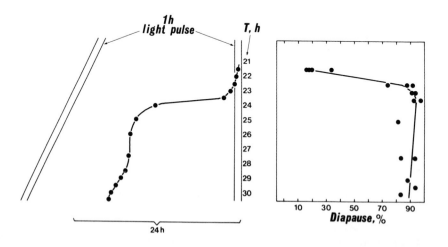

Figure 9. Sarcophaga argyrostoma. Computed phase relationships of ϕ_i to single pulses of light (1 h) in cycles from T 21·5 to T 30·5 (the range of entrainment), compared with the incidence of diapause (*right*). Diapause incidence is low only when ϕ_i is illuminated. (Data from Saunders[14].)

'programme' their larvae for subsequent pupal diapause.[23] These larvae were then set up at 18 °C in a variety of T cycles each consisting of a single 1 h pulse per cycle. Figure 9 shows the phase relationship of ϕ_i computed for T cycles between 21·5 and 30·5 h (the range of entrainment for 1 h pulses), and compares the resulting illumination or non-illumination of ϕ_i with observed diapause frequencies. Once again, diapause was high when ϕ_i fell in the dark, but was low when it coincided with the light pulse. Coincidence only occurred in the experimental T cycle of T 21·5 (LD 1 : 20·5) in which the pulse came on at CT 20 and finished at CT 23·5, in each cycle, when in steady-state. Illumination of ϕ_i (at CT 21·5) can thus lead to non-diapause development in the complete absence of any other photoperiodic influence, merely by altering the period of the zeitgeber cycle.

In early versions of the external coincidence model[16] it was thought that the photo-inducible phase, or its equivalent, could be at *either* of the two points of sensitivity in the night (i.e., at A or B in figure 6). However, Pittendrigh[4] later pointed out that significant change in the illumination of pacemaker phase is restricted to the *late* subjective night once the photoperiod exceeds about 12 h, and therefore suggested that ϕ_i must lie at B rather than A. This interpretation of asymmetrical skeleton data is consistent with the *Sarcophaga* case presented in figure 6. Experimental confirmation that ϕ_i lies at B, however, was obtained in a series of experiments, based on designs used by Lees[1,24] for *Megoura viciae*, and again involving a change of T.

In the first of these (figure 10, upper panel) larvae of *S. argyrostoma* were set up in a series of asymmetrical skeletons consisting of a longer main photoperiod (P_1 = 10 h) and a shorter scanning pulse (P_2 = 1 h). The scanning pulse was placed 3 h after the end of P_1 in a position corresponding to that of point A, and the overall zeitgeber period (T) was varied systematically from T 21 to T 27 h by altering the *terminal* hours of darkness. A second series of asymmetrical skeletons (figure 10, lower panel) was then used in which the terminal hours of darkness were held longer than the critical night-length ($>9\frac{1}{2}$ h), but the hours of darkness *between* P_1 and P_2 varied to give a range of T from 24 to 32 h. In both series the proportions of pupae entering diapause were compared with the phase relationships of ϕ_i to the light, as computed from the phase response curves.

The upper panel of figure 10 shows that diapause incidence was zero or very low in cycles where ϕ_i occurred after dawn (a response equivalent to point A in figure 6) but became high when the terminal hours of darkness exceeded the critical night-length ($9\frac{1}{2}$ h) and ϕ_i fell in the dark. In the lower panel of figure 10, on the other hand, diapause incidence was high until after the phase jump, when ϕ_i coincided directly with the 1 h scanning pulse (producing a response equivalent to point B in figure 6). These results make it clear that the long-day effects at A may be reversed by a subsequent long night, but those at B are irreversible. This distinction provides experimental evidence that ϕ_i, if it is a 'reality', lies at B, not at A.

WHAT EXTERNAL COINCIDENCE FAILS TO EXPLAIN

The external coincidence model adequately accounts for the seasonal switch between continuous development and diapause, and for the critical night-length between these alternate pathways. It fails to explain, however, a number of other properties of the insect photoperiodic response, including the fall in diapause incidence in ultra-short (<6 h) daylengths, and the low and often variable

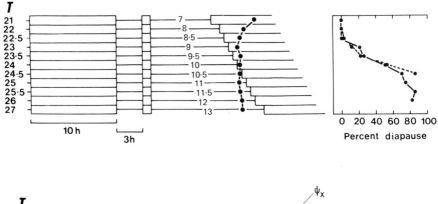

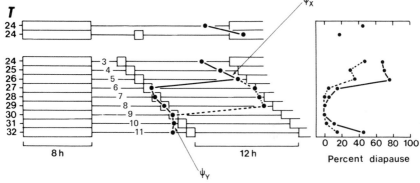

Figure 10. *Sarcophaga argyrostoma.* Computed phase relationships of ϕ_i and diapause incidence in asymmetric skeleton photoperiods whose periods are *not* equal to 24 h. *Upper panels:* The supplementary pulse (1 h) is placed 3 h into the night (at a point equivalent to **A** in figure 6); the hours of darkness following the pulse are then varied from 7 to 13 h. *Lower panels:* The supplementary pulse scans the night but is followed in all cases by a terminal dark period longer than the critical night-length. ψ_X and ψ_Y are the two phase relationships, the connecting lines (solid and dotted) the 'earliest' and 'latest' phase jumps. *Right:* The diapause responses for two replicate experiments. Note that diapause incidence is high when ϕ_i falls in the dark, but is low when it is illuminated. (Data from Saunders[14].)

incidence of diapause in continuous darkness. Strict interpretation of the model as presented in figure 1 would predict as much diapause in these régimes as in the 'strongest' short-day cycles. On the other hand, the type of *internal* coincidence, with its independently phased 'dawn' and 'dusk' oscillations, will explain these features, and more.[25]

Changes in the length of the photoperiod, or changes in the length of the zeitgeber cycle, occurred in all of the experiments described in this paper. All of them, therefore, are open to interpretation in terms of internal coincidence. However, although the probable multi-oscillator 'construction' of the *Sarcophaga* circadian system is not doubted,[26] this alternative is rejected for several reasons. First, there is no evidence for separate dawn and dusk oscillators in the photoperiodic response, or in the eclosion rhythm. Secondly, several experiments involving transfer of cultures from LD to DD, simulation of photoperiod by thermoperiod and chilling or anaesthesia in the light or the dark,[9] although still equivocal, point to a clock of the external type for *S. argyrostoma.* How, then, may those difficulties mentioned above be explained in terms of the entrainment properties of a 'single' circadian pacemaker?

The external coincidence model will account for these features of the response if we incorporate two additional ideas, both of which are related to experimental observations that a *number* of light cycles are required to elicit the response, and that these inductive cycles need to be experienced during a restricted larval 'sensitive' period.[10,23,27] The first suggestion is that the photoperiodic oscillator, *unlike* the eclosion rhythm, 'damps out' in DD within a few cycles so that an insufficient number of 'long nights' (i.e. non-illuminations of ϕ_i) occur during the sensitive period: hence the low or zero incidence of diapause in continuous darkness. This suggestion is particularly attractive because it implies a functional similarity between the circadian clock of *Sarcophaga* and the 'hour-glass' or perhaps instantly damped clock of *M. viciae*.[9] The second idea is that the photoperiodic oscillator is maintained in LD cycles, but the low incidence of diapause observed in very short (and unnatural) photoperiods is a consequence of the slow rate of approach to steady-state in trains of these 'weaker' pulses, particularly if the pacemaker is initially out-of-phase with the zeitgeber (figure 5). While the undoubtedly multi-oscillator circadian system is passing through these transients there might be a lack of internal synchrony which adversely affects the *accuracy* of night-length measurement. Preliminary experiments indicate an inverse relationship between the number of such transients and diapause incidence, as computed from eclosion phase response curves and as predicted by this hypothesis. The external coincidence model can therefore account for all experimental observations using known properties of the entrainment phenomenon provided the clock is also recognized as a complex multi-oscillator system whose internal phase relationships are of importance.

ACKNOWLEDGEMENTS

The work in this paper has been supported by a series of grants from the Science Research Council, and carried out with the expert technical assistance of Helen MacDonald and Kathleen Rothwell.

REFERENCES

1. A. D. Lees. *J. Insect Physiol.* **19**, 2279–316 (1973).
2. D. S. Saunders. *Insect Clocks.* Oxford, Pergamon (1976).
3. E. Bünning. *Cold Spring Harbor Symp. Quant. Biol.* **25**, 249–56 (1960).
4. C. S. Pittendrigh. *Z. Pflanzen.* **54**, 275–307 (1966).
5. V. P. Tyshchenko. *Zh. Obshch. Biol.* **27**, 209–22 (1966).
6. C. S. Pittendrigh. *Proc. Natl Acad. Sci. USA* **69**, 2734–37 (1972).
7. S. D. Beck. *J. Comp. Physiol.* **90**, 275–95 (1974).
8. D. S. Saunders. *J. Comp. Physiol.* **124**, 75–95 (1978).
9. D. S. Saunders. *J. Comp. Physiol.* **127**, 197–207 (1978).
10. D. S. Saunders. *J. Insect Physiol.* **17**, 801–12 (1971).
11. D. S. Saunders. *J. Insect Physiol.* **19**, 1941–54 (1973).
12. D. S. Saunders. *J. Comp. Physiol.* **97**, 97–112 (1975).
13. D. S. Saunders. *J. Comp. Physiol.* **110**, 111–33 (1976).
14. D. S. Saunders. *J. Comp. Physiol.* **132**, 179–89 (1979).
15. D. S. Saunders. In: R. Gilles (ed.). *Animals and Environmental Fitness.* (European Society for Comparative Physiology and Biochemistry, Liège, 1979.) Oxford, Pergamon, pp. 363–83 (1980).
16. C. S. Pittendrigh and D. H. Minis. *Am. Nat.* **98**, 261–94 (1964).
17. D. S. Saunders. *Physiol. Entomol.* **4**, 263–74 (1979).
18. A. T. Winfree. *J. Theor. Biol.* **28**, 327–74 (1970).

19. C. S. Pittendrigh. In: J. Aschoff (ed.). *Circadian Clocks*. Amsterdam, North-Holland, pp. 277–297 (1965).
20. P. L. Adkisson. *Am. Nat.* **98**, 357–74 (1964).
21. C. S. Pittendrigh and D. H. Minis. In: M. Menaker (ed.). *Biochronometry*. Washington DC, National Academy of Sciences, pp. 212–250 (1971).
22. D. H. Minis. In: J. Aschoff (ed.). *Circadian Clocks*. Amsterdam, North-Holland, pp. 333–343 (1965).
23. D. S. Saunders. *Physiol. Entomol.* **5**, 191–8 (1980).
24. A. D. Lees. *Insect Clocks and Timers*. Inaugural Lecture, Imperial College of Science and Technology, 1 December 1970.
25. A. S. Danilevskii, N. I. Goryshin and V. P. Tyshchenko. *Annu. Rev. Entomol.* **15**, 201–44 (1970).
26. D. S. Saunders. *Proc. Natl Acad. Sci. USA* **69**, 2738–40 (1972).
27. D. Gibbs. *J. Insect Physiol.* **21**, 1179–86 (1975).

DISCUSSION

Brown-Grant: When you were discussing the effects of single light 'pulses', you described 1 and 5 h duration pulses as 'weak' and 'strong'. At equal intensity, this is true, but by using 'weak/strong' instead of short/long are you implying that the total amount of light received rather than duration of illumination is the important factor and, if so, have you determined the effect of 1 h illumination at five time the intensity?

Saunders: All my light pulses were of 240 μW/cm^2. They are called either 'weak' or 'strong' pulses because they elicit either 'weak' (i.e. small) or 'strong' (i.e. larger) phase changes. I have not examined the reciprocity between intensity and duration, but in *Drosophila pseudoobscura* there seems to be such a reciprocity over a wide range (data of M. K. Chandrashekaran).

Lees: You have mentioned reciprocal relationships of dark period light interruptions. I might mention that in *Megoura* reciprocity is only observed over very small durations. This is especially so for monochromatic blue light interruptions introduced 1–5 h after the beginning of light/dark transition. With 0·75 h of light the highest intensities cannot induce an effect, with 1·0 h of light the response is complete and is not assisted by further extension of the illumination time.

Saunders: In the case of night interruptions the restriction in reciprocity might be due, in part, to the temporal limitations of the 'gate' at which the system is sensitive.

Elliott: Although it clearly does not affect the outcome of the elegant analysis you have shown us in *Sarcophaga*, I must point out that one should be cautious in making assumptions about the location of ϕ_i which are based on entrainment to complete photoperiods. You begin by assuming that ϕ_i is at CT 21·5 based on a critical photoperiod of 14·5 h and the observation that the oscillation starts at about CT 12 when released into DD from photoperiods longer than approximately 9 h. This ignores the likely possibility that the clock is not in fact stopped at CT 12 in these photoperiods but rather reset to about CT 12 at the time of the light/dark transition. Thus, it is possible that the pacemaker, or the rhythm of photo-inducibility, is not at CT 12 before the L/D transition and ϕ_i, therefore, would conceivably lie in the early subjective night.

Saunders: Although I recognize that the oscillation is not 'stopped' (at CT 12) in protracted light—indeed my own data show this—the oscillation is in some sense reset to about CT 12 at the light/dark transition, or is on the CT 12 isochron at that point and then winds on to the dark limit cycle within a few hours. I also concur that we have no knowledge of what 'phases' are traversed by the oscillation during the light. However, data from *Sarcophaga* and most insects are quite clear that it is *night* length being measured, and if the clock is operationally at, or close to CT 12 at the end of the light period, it is during the late subjective night that light scans the widest range of phases, and we might expect ϕ_i, if it exists, to lie there if anywhere. Making the assumption that ϕ_i lies at CT 21·5, this fits all my analyses, whereas ϕ_i at any other phase, does not.

Jones (comment): May I add a comment on whether the clock stops at CT 12. Peterson and Pittendrigh had been drawing diagrams of limit cycles in DD and at different intensities of LL. I would just like to clarify the point that, although the cycle in LL subtends only a small portion of the cycle in DD, it nevertheless goes through all possible circadian times. I have recorded the flight activity of the mosquito *Culex quinquefasciatus* in DD even though the rhythm is reset by a change to DD. Another interesting point is that the form of the cycle changes from being bimodal in DD to unimodal in LL.

INVOLVEMENT OF THE CIRCADIAN SYSTEM IN PHOTOPERIODISM AND THERMOPERIODISM IN *PIERIS BRASSICAE* (LEPIDOPTERA)

Bernard Dumortier and Jeannine Brunnarius

Institut National de la Recherche Agronomique, CNRA, Versailles, France

Summary

Overt rhythms are entrainable by zeitgeber other than the LD cycle. Temperature changes are among the most effective. In *Pieris brassicae* (and in three other insect species), diapause can be fully regulated by daily temperature changes. The significance of this so-called thermoperiodism, either as an accessory property of the photoperiodic clock or as an expression of the circadian organization, can be estimated only if the nature of the photoperiodic clock itself is ascertained. Consequently, the first part of the paper deals with a re-examination of the involvement of circadian rhythms in *Pieris* photoperiodic time measurement.

Resonance experiments of the Nanda–Hamner and Bünsow types (with and without light pulses during the protracted night) suggested that a circadian timer does operate. Asymmetric skeleton photoperiods revealed one or two light-sensitive phases (A and B) within the night, depending on its length. This feature may be accounted for by assuming that light illuminating the first photosensitive point A, which occurs early in the night, causes a phase lag of the oscillation of B, which then falls soon after dawn. How this mechanism may explain time measurement under natural conditions is discussed. Results of symmetric skeleton experiments suggest that two light pulses (90 min) are not read as the initiator and terminator of a subjective day.

Square wave heat pulses (13/21 °C) in constant darkness gave a diapause induction curve similar to that obtained with photoperiods. The length of the hot and cold sequences, as well as the amplitude of the temperature cycle and its mean level, all provide thermoperiodic information. Scanning the cold part of a 'short-day' thermoperiod with temperature pulses resulted in a distinct fall in diapause induction at one particular position (the *thermoinducible* phase). These data give some evidence for implicating the circadian organization in *Pieris* thermoperiodism.

Though rather unreliable, natural thermoperiods given in constant darkness can control diapause to a certain extent. The biological significance of thermoperiodism is discussed with regard to the general mechanisms underlying photoperiodic time measurement.

Biological Clocks in Seasonal Reproductive Cycles 83–99 (1981) (ed. B. K. and D. E. Follett: Bristol, Wright).

INTRODUCTION

It has been established for vertebrates, as well as for insects, that the light/dark cycle is not the only possible entraining agent—even if it is the most widespread—for circadian rhythms and cycles of pressure,[1] con-specific song[2,3] and artificial sound[4-6] are all capable of acting as zeitgeber in vertebrates if given in cyclic patterns. By far the most effective non-photic cue, however, is temperature, and following the review of Sweeney and Hastings,[7] its potential as an entraining agent has been shown in other invertebrates[8-15] and in one vertebrate.[16] So, where diapause control involves a clock it may be justifiable to assume that a light/dark cycle might be replaceable by one of heat and cold, so long as it is given in an identical temporal pattern. This justification has been borne out in four insects in which complete response curves have been obtained with a series of square wave temperature cycles given in constant darkness: *Nasonia vitripennis*,[17] *Diatraea grandiosella*,[18] *Pieris brassicae*[19] and *Plodia interpunctella*.[20] The results from these studies raise two questions. The first is to know whether the response to the temperature cycles is a peculiar (or accessory) property of the photoperiodic clock, or whether it is an expression of the property of circadian rhythms to entrain to thermal cues. In other words, is there an *analogy* or an *homology* between thermoperiodism and 'thermo-entrainment' of circadian rhythms? The second question relates to the biological significance of the thermoperiodic phenomenon.

A prerequisite for answering the first question is to ascertain whether the photoperiodic clock is, or is not, circadian in nature. In the four species mentioned above, resonance (Nanda–Hamner) experiments have shown the involvement of a circadian mechanism in the case of *N. vitripennis*[21,22] but not in *P. interpunctella*[20] whilst results are lacking for *D. grandiosella*. For *P. brassicae* the available data[23] require a re-examination, and the first part of this paper considers this question again.

MATERIALS AND METHODS

Experimental animals originated from a stock culture maintained since 1974 at 21 °C and under long-day conditions (LD 16 : 8). Adults were provided with a 15 per cent solution of honey and the larvae were fed on cabbage leaves. In all cases the larvae were exposed to the experimental conditions from hatching until pupation. For each treatment at least 100 were used.

For photoperiodic experiments, larvae were kept in light-tight incubators fitted with two 8 W fluorescent tubes (400 lux) controlled by a timer. The incubators were kept in a cold room (10 °C) so that the temperature could be maintained at 20 ± 1 °C at all times. Temperature was monitored potentiometrically. For the artificial thermoperiodic régimes, cooling and heating incubators were designed which could generate temperature cycles regulated with respect to both amplitude and duration. Larvae were kept in light-tight boxes within these incubators, the galvanized iron walls of the boxes avoiding any damping of the temperature cycle. The high (hot (H) = 21 °C) and low (cold (C) = 13 °C) levels of temperature were reached in about 10 min and were accurate to within 0·25 °C. All experiments were performed in constant dark (DD).

Experiments under natural conditions were conducted outdoors in screened cabinets. The animals submitted only to the daily cycle of temperature were in light-tight boxes (as described above) while the controls under natural photo- and thermoperiods were in transparent cages. Temperature was monitored continuously.

In both types of thermoperiodic experiments, the larvae never experienced light: they were fed in the dark, the faeces and scraps of uneaten leaves not being removed. Fresh air was blown through the boxes (240 l/h) to avoid fungal growth and extreme humidity.

RESULTS

General features of the photoperiodic response in *Pieris brassicae*

Involvement of a circadian mechanism

Two sets of resonance experiments were carried out. In the first (figure 1), which was carried out by J. Claret (unpublished observations) on individuals from his own culture, the night was lengthened by steps of 2 h up to a duration of 72 h, all the cycles beginning with either 8 or 10 h light. In both cases, the response was a periodic function of the length of the LD cycle and τ (the period of the rhythm) was clearly shorter than 24 h.

In the second series of experiments (figure 2), the night of a 48 h cycle (containing either 8 or 12 h of light) was scanned with 1 h light pulses. In both régimes two distinct troughs in diapause induction appear, 24 h apart in LD 8 : 40 (figure 2b), and about 19 h apart in LD 12 : 36 (figure 2d). If the responses to the light breaks in LD 8 : 40 are compared with those in LD 8 : 16 (figure 2a, b), it is clear that the two troughs in the 48 h cycle are in phase with the photosensitive point B (the single drop in diapause induction in LD 8 : 16). However, in the case of LD 12 : 36, such a close relationship with the equivalent régime of LD 12 : 12 is questionable owing to the fact that night break experiments in the cycle LD 12 : 12 give rise to two discrete points of diapause inhibition, A and B, whereas only one trough is found in the 48 h cycle.

Asymmetric skeleton photoperiods

Night interruption experiments in cycles ranging from LD 4 : 20 to LD 14 : 10 with 1 h light pulses show either one or two points where the breaks are most effective (points A and B, according to the current terminology[24]). It is clear from figure 3 that: (a) B occurs in all photoperiods, whereas the presence of A depends on the length of the night which must not exceed 14 h; (b) B is phase-set by dusk and occurs on average 9 h later (the regression line is highly significant); (c) A is preceded by about 1 h of complete insensitivity to light in the case of LD 10 : 14 and 11 : 13, but in 'longer' short-days (LD 13 : 11 and 14 : 10), it begins from very early in the dark period. All of these features can be accounted for by the theory of photoperiodic time measurement (*see* Discussion).

Symmetric skeleton photoperiods (PP$_s$)

When exposed from hatch to a régime consisting of two 90 min light pulses, the larvae give quite different responses depending on the interval between the two pulses. The diapause induction curve for such symmetrical skeleton treatments

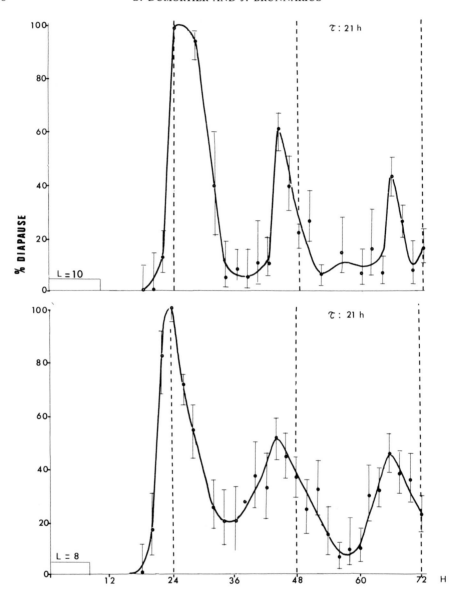

Figure 1. Resonance experiments of the Nanda–Hamner type with larvae of *Pieris brassicae*. The overall length of the cycle was varied by steps of 2 h up to a period $T = 72$ h. The cycles began with either 8 or 10 h of light. If a circadian mechanism is involved in photoperiodic time measurement (i.e. if there is a circadian rhythm of photosensitivity) this rhythm should free-run (with a period τ) in the extended dark periods. Every time the phase relationship between T and τ is such that the photosensitive point falls in light (at least once during the cycle), a long-day effect (no diapause) should follow. Thus, the response will appear cyclically as a periodic function of the overall length of the cycle. This was actually seen in these two examples, so providing evidence that a circadian process is involved in *Pieris* photoperiodism. Confidence interval: $P = 0.05$. (Data kindly provided by J. Claret.)

Figure 2 (*opposite*). Effects of 1 h light breaks (night interruptions) on diapause incidence in *Pieris brassicae* exposed to 24 h LD cycles (*a*, LD 8 : 16; *c*, LD 12 : 12) or 48 h LD cycles (*b*, LD 8 : 40; *d*, LD 12 : 36). The troughs observed in LD 8 : 40 are considered to correspond to point B in the LD 8 : 16 cycle. The relationship between the responses in LD 12 : 36 and LD 12 : 12 is less clear owing to the bimodal nature of the response (points A and B) in the 24 h cycle. Confidence interval: $P = 0.05$.

% DIAPAUSE

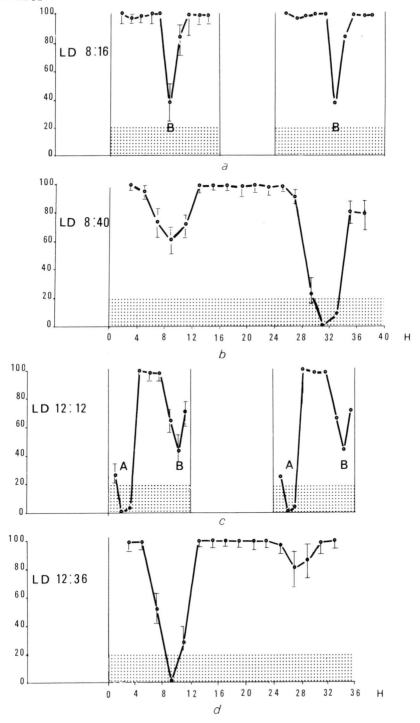

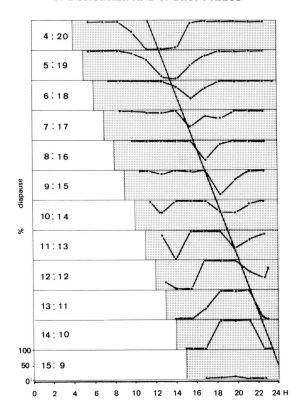

Figure 3. A scan of the night with 1 h light pulses in a series of short days (asymmetric skeleton photoperiods) results in one peak of diapause inhibition (long-day effect) when the pulse falls about 9 h after dusk (point B). A long-day effect (point A) also appears soon after dusk in the schedules with relatively short nights. The regression line joining all the B points is highly significant.

(figure 4, PP_s, solid line) is very similar to that obtained with complete photo-periods (PP_c, dotted line), but is strongly shifted along the time axis. The PP_s response curve falls within the family of curves obtained with asymmetric skeleton photoperiods (figure 4, broken lines A–F). Table 1 gives for each skeleton photoperiod the corresponding complete photoperiod (PP_c) in both the longer and the shorter possible interpretation, together with the percentage diapause expected if the animal read the two pulses as the beginning and the end of the day.

Thermoperiodic information and diapause control

The thermoperiodic response curve

Eleven régimes of alternating sequences of high (H) and low (C) temperature were used: HC 2 : 22, 4 : 20 through to 22 : 2, plus two acyclic conditions—constant high (HH) and constant low (CC) temperatures. When plotted against the thermo-period, the percentage diapause alters according to the respective lengths of the warm and cold sequences, as seen in the bottom curve of figure 5. The percentage response first increases sharply as the duration of the warm period lengthens,

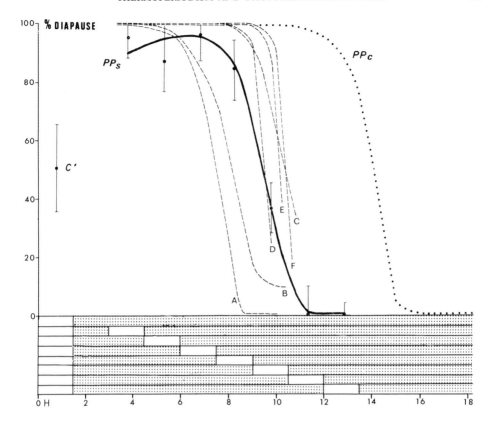

Figure 4. The incidence of diapause in *Pieris brassicae* maintained from hatching in symmetric skeleton photoperiods (PP$_s$) formed from two 90 min light pulses. Solid line, the response to symmetric skeletons (PP$_s$); dotted line, response in complete photoperiods (PP$_c$); broken lines, response curves for different asymmetric skeletons (1 h light pulse): (A) LD 4 : 20; (B) LD 5 : 19; (C) LD 6 : 18; (D) LD 7 : 17; (E) LD 8 : 16; (F) LD 9 : 15. The PP$_s$ response curve is clearly within the same family as those obtained with asymmetric skeletons. The two pulses do not simulate the corresponding complete photoperiods but act much as an asymmetric skeleton. C', diapause incidence for controls kept in a short day of LD 1·5 : 22·5. Confidence interval: $P = 0·05$. (*See also* table 1.)

reaching a maximum with HC 6 : 18. A high rate of diapause is seen up to HC 14 : 10, it then falls rapidly to zero. Figure 5, which also shows as a comparison the thermoperiodic responses of two other species, indicates that in *P. brassicae* the thermoperiodic and the photoperiodic response curves are quite similar. Constant conditions suppress diapause induction while the critical thermo- and photoperiods are about the same, respectively HC 15 : 9 and LD 14 : 10.[19]

The effect of the amplitude and the level of the temperature cycle was also tested. The highly inductive cycle HC 6 : 18, which gives 100 per cent diapause when the amplitude is 8 °C (13 to 21 °C), has only a weak 'short-day' effect when the amplitude is reduced to 5 °C. The level, i.e. the mean temperature about which the cycle oscillates, is also important, as seen in figure 6.

Extension of Bünning's hypothesis to thermoperiodism

The technique of night break experiments was transposed to thermoperiodic terms by creating asymmetrical thermoperiods. When the cold sequence (C) (13 °C) of

Table 1. Responses to symmetric skeleton photoperiods (PP_s) formed from two 90 min light pulses with the corresponding complete photoperiods (PP_c), read either in the shorter or in the longer interpretation

	PP_s			PP_c	% diapause	
L	D	L	D		Expected	Observed
1·5	1·5	1·5	19·5	{ 4·5 : 19·5 22·5 : 1·5	{ 100 0	94
1·5	3	1·5	18	{ 6 : 18 21 : 3	{ 100 0	87
1·5	4·5	1·5	16·5	{ 7·5 : 16·5 19·5 : 4·5	{ 100 0	96
1·5	6	1·5	15	{ 9 : 15 18 : 6	{ 100 0	84
1·5	7·5	1·5	13·5	{ 10·5 : 13·5 16·5 : 7·5	{ 100 0	36
1·5	9	1·5	12	{ 12 : 12 15 : 9	{ 100 5	0
1·5	10·5	1·5	10·5	{ 13·5 : 10·5 13·5 : 10·5	80	0

The percentage of *expected* diapause refers to those cases in which animals could follow either interpretation.

the 'short day' HC 8 : 16 is interrupted by a 2 h hot pulse (H) (21 °C), a significant fall of 34 per cent in the degree of diapause induction occurs if the pulse is centred 10 h after the beginning of the cold sequence. A 3 h heat pulse lowers the response to 10 per cent.[28] In figure 7 the effects of heat and light pulses are compared in the same HC 8 : 16 and LD cycle.

Thermoperiodism under natural conditions

Eleven batches of larvae were exposed from hatching to natural thermoperiods only, from the beginning of May to the end of October. During the same period, controls were exposed to both the natural photoperiod and thermoperiod. The results (table 2 and figure 8) show that the daily cycle in temperature elicits markedly different responses in the spring, summer and autumn. Until the end of August, the percentage diapause is near zero, but it then rises sharply and reaches a maximum at the beginning of October. The shape of the response curve, therefore, is roughly the same (though definitely delayed) under thermoperiodic as under photo- and thermoperiodic conditions. However, a number of individuals escape diapause when they experience only the temperature cycle, even if, with respect to the natural season, such continuous development is clearly lethal. Some of the responses may be irrelevant (*see* batches B and J in table 2).

DISCUSSION

The results from the resonance experiments lead to the conclusion that photo-periodic time measurement in *P. brassicae* involves a circadian mechanism, and so this species may be added to the small number (*N. vitripennis*,[21,22]

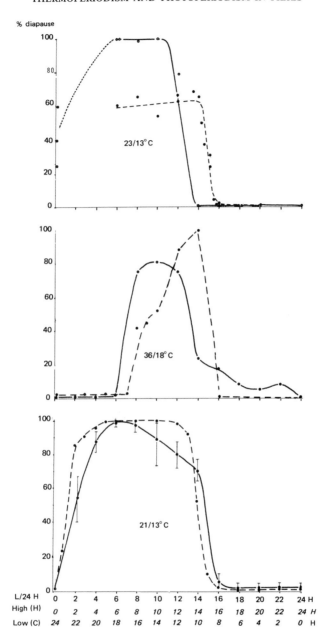

Figure 5. The effect of thermoperiod (in DD) on diapause induction in three insects. The curves show the close correspondence between the thermoperiodic (solid lines) and the photoperiodic response curves (broken lines). Numbers within the curves give the values of the hot (H) and the cold (C) sequences. On the abscissa, the duration of light in the LD cycle under constant temperature (photoperiodic curves) and the duration of the high–low thermal sequences in the thermoperiodic cycle are shown. The uppermost panel (*Nasonia vitripennis*) is redrawn from Saunders,[25,26], that in the middle (*Diatraea grandiosella*) from Chippendale and Reddy[27] and Chippendale et al.[18] The lowest panel (*Pieris brassicae*) is taken from Dumortier and Brunnarius.[19] Confidence interval: $P = 0.05$.

Table 2. Diapause-inducing effects of natural thermoperiods (*T*) as compared with those of natural photo- and thermoperiods (*TP*)

Batch	Conditions	Stages 4/5 (d.mth)	No. of pupae	% diapause	Daylength (h·min)
A	*T*	1·06	14	0*	(16·40)
	TP	5·06	100	0	16·50
B	*T*	10·06	35	43*	(16·56)
C	*T*	10·06	67	0*	16·56
	TP	9·06	76	0	16·54
D	*T*	17·07	87	2	(16·25)
	TP	16·07	194	0	16·26
E	*T*	3·08	76	1	(15·48)
	TP	2·08	143	9	15·52
F	*T*	31·08	88	12	(14·12)
	TP	26·08	64	96	14·30
G	*T*	4·09	74	55	(13·54)
	TP	4·09	134	100	13·54
H	*T*	18·09	106	76	(13·06)
	TP	12·09	136	100	13·26
I	*T*	30·09	111	84	(12·22)
	TP	1·10	63	100	12·20
J	*T*	9·10	112	41	(11·46)
	TP	10·10	13	100	11·44
K	*T*	25·10	33	73	(10·46)

The values of the daylength (actually experienced only in *TP*) are given for the time that roughly corresponds to the point of moulting between instars 4 and 5. These instars are known to be the sensitive period in *Pieris brassicae*.[29] The photoperiodic threshold of this species is about 5 lux and, thus, values of daylength are those corresponding to the time elapsed between the moments when illumination reached 5 lux at dawn and fell below it at dusk (49° Lat. N).

* These points correspond to thermoperiods accompanying increasing daylengths. They are depicted by a triangle with apex upwards in curve *d*, figure 8.

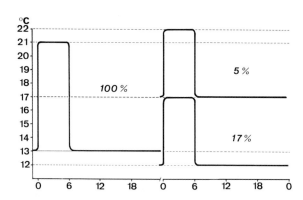

Figure 6. The effect of amplitude and level of the square wave temperature cycle on diapause induction in *Pieris brassicae*. The thermoperiodic cycle HC 6 : 18 gives 100 per cent diapause when its amplitude is 8 °C (13–21 °C). Reducing the amplitude to 5 °C lessens the response markedly, the effect depending on the level of the oscillation. The higher the mean temperature level, the lower the response. $\chi^2 = 8 \cdot 08$: significant difference ($P = 0 \cdot 01$) between percentage at high (17–22 °C) and low (12–17 °C) levels.

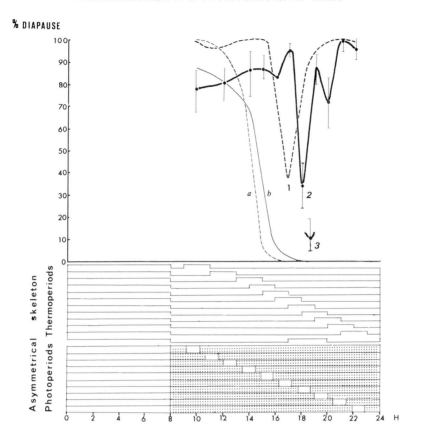

Figure 7. The diapause-inhibiting effect of asymmetrical skeleton thermoperiods (in DD). Interruption of the cold sequence (13 °C) of the HC 8 : 16 cycle by a series of 2 h warm pulses (21 °C) results in a fall of 34 per cent in diapause about 10 h after temperature step down. A 3 h pulse lowers the response to 10 per cent. Curve 1 (broken line), response to asymmetrical skeleton photoperiods (LD 8 : 16, 1 h light pulses); curve 2 (solid line), response to asymmetric skeleton thermoperiods; 3, effect of a 3 h heat pulse; *a* and *b*, part of the photoperiodic and thermoperiodic response curves. Confidence interval: $P = 0.05$. (Redrawn after Dumortier and Brunnarius.[28])

S. argyrostoma,[30,31] *Pterostichus nigra*,[32] *Aëdes atropalpus*[33]), in which such an involvement seems proved. The period of the free-running rhythm of photoperiodic sensitivity, as revealed in the Nanda–Hamner experiments, is clearly shorter than 24 h in *Pieris*. This may be uncommon but is unlikely to be unique. One methodological remark may be justified. In some instances, the resonance experiments were conducted with cycles of the type $T + n^{(T+2)}$ (where $T = 24$ h) thus giving only cycles of 36, 48, 60 and 72 h length. Under these conditions, it is plain that resonance can only be easily disclosed if τ is equal to, or near, 24 h. This may be seen in figure 1 if in the upper curve the points corresponding to abscissae values 24, 36, 48, 60 and 72 h are joined together. A new curve is generated in which periodicity is only weakly indicated.

Night break experiments in a 48 h cycle strengthen the conclusion that photoperiodism in *P. brassicae* is circadian based. However, as stressed earlier, the coincidence between phase point B in the 24 h cycle and each trough in the 48 h cycle when both cycles contain only 8 h of light, is no longer obvious when light

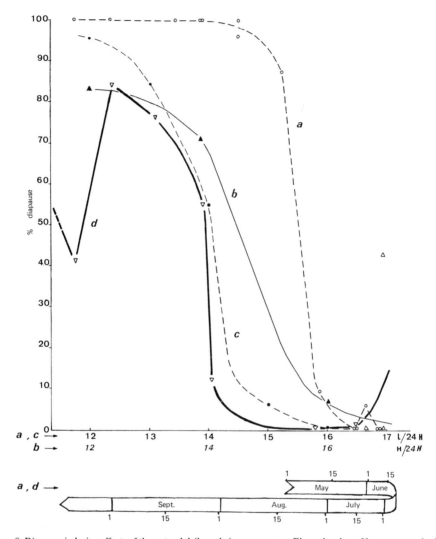

Figure 8. Diapause-inducing effects of the natural daily cycle in temperature. Eleven batches of larvae were submitted from hatching to the action of the natural thermoperiod only (May to October). The effect of this treatment is seen in curve *d* (heavy solid line). Curve *a* shows responses of the controls in natural photo- and thermoperiodic conditions. Parts of the response curves under artificial thermo- or photoperiods are respectively given in *b* and *c*. For *a*, the percentage diapause is plotted at one and the same time on a scale given both in months and in the corresponding daylength. Points in curve *b* are plotted relative to the length of the warm sequence (c.f. figure 5). Curve *c* relates to the first abscissa (L/24 h). In *d* the abscissa to be considered is the time scale. Curves *a* and *d* must be read from right (long days) to left (short days). (*See also* table 2.)

reaches 12 h in duration. Moreover, the period then decreases to 19 h and the amplitude (percentage of non-diapausing pupae) in the two troughs is clearly reversed in LD 12 : 36 by comparison with the situation in LD 8 : 40. It should be noted that the unimodal shape of the recurrent trough in diapause induction is a rather general result in experiments where an extended night is scanned with light pulses.[21,33,34] Until now the identity of the missing component (point A or point B) has at best been conjectural. A possible method of settling this question might be to use a specific property of A or B (e.g. their spectral sensitivity or threshold) as a *marker* which would then make it possible to ascribe the response to one of the two points. Such a spectral difference between early- and late-night photosensitivity has already been found.[35,36]

How A and B act, or more probably interact, is obviously the crux of the *formal* part of the problem of photoperiodic time measurement. In this connection, some first rate contributions have increased our knowledge about the black box. A thorough experimental analysis of the processes involved in the particular case of *P. brassicae* was beyond the scope of this work, since in its photoperiodic part it was restricted to assessing the circadian nature of the mechanisms. Nevertheless, purely empirical data such as those related to asymmetric photoperiods (figure 3) are quite useful. Besides showing the dynamics of the phenomenon of night photosensitivity, as well as the existence of one or two light-sensitive points depending upon the length of the day (this was already indicated in Adkisson's data[37,38]), they provide some information on how seasonal changes in daylength act as a cue for diapause control. This is developed below.

In so far as photoperiodic time measurement is concerned, *Pieris* may well behave like *P. gossypiella*. In fact, the model proposed by Adkisson in 1964,[37] together with the interpretations of Pittendrigh and Minis,[24] fits the case of *Pieris*, at least with respect to the significance of point B (the photoinducible phase). Point B is phase-set by dusk, as in *S. argyrostoma*[31] and in *A. atropalpus*,[33] in such a way that the point of maximal light sensitivity occurs about 9 h after lights-off. In any photoperiod where the night is equal to or less than 9 h, B will occur at or after dawn in the following day and so will act as the initiator of the next photoperiod (according to Pittendrigh's terminology).[24] Thus the external coincidence model, derived from Bünning's hypothesis,[24,39] can account for the photoperiodic responses in *Pieris* (apart from the responses in very short days). The significance of the point of photosensitivity which occurs early in the night (point A as revealed by asymmetric skeletons) is more equivocal at first sight. In LD 10 : 14 and 11 : 13 the first hour of darkness is completely insensitive to light, and a light pulse is fully effective only 3 h after dusk. This seems to preclude any role for A under natural conditions. However, for 'longer' short days (those actually experienced in nature), say LD 13 : 11 and 14 : 10, A occurs immediately after dusk (a pulse starting 30 min after lights-off completely abolishes diapause induction); so if the day is long enough to overlap A diapause inhibition will ensue. Point A, therefore, might be said to act as a terminator[24] of the preceding day. Yet it is plain that dusk occurs later as the day lengthens and so point B (phase-set by dusk) ends up coinciding with dawn, thus giving 'long-day' information twice. As a consequence, according to this hypothesis that two photosensitive times exist early and late in the subjective night, the insect would be provided with a redundant system of photoperiodic time measurement. However, the problem is certainly even more complex. In particular, the significance of A is open *a priori* to two interpretations. Either it is a real photosensitive point just like B (the hypothesis considered above), or responses observed at this time arise from the fact that light given early in the night restarts the clock (and creates a new dusk) so that point B is delayed $(-\Delta\phi)$ and so falls beyond dawn. Thus, the responses seen at A come, in fact, from illumination of B. Actually, when in an LD 10 : 14 cycle a pulse of light is given during the third hour of darkness (highest point of response for A), scanning the remaining night with a second pulse shows that B is phase-delayed by 3 h (unpublished observations). This recalls the behaviour of the clock controlling the eclosion of *Antheraea pernyi*[40] and, more generally, the phase shifting of overt rhythms under the action of light pulses in the subjective night. Thus the shift of B observed in the above-mentioned experiment would represent a casual point on the phase response curve of B. Very careful analysis of the significance of A and B, carried out on *S. argyrostoma*,[41] has shown that the diapause response curve in asymmetric skeletons may be interpreted as the result of the phase-shifting of B under the action of the scanning pulse. Depending upon whether B is phase-delayed or phase-advanced, it falls during the light (either the light around dawn or the light of the pulse itself) or during the dark, thus promoting diapause or not. If we assume the same dynamics for *Pieris*, it becomes possible to understand why A appears only for 'longer' short-day photoperiods: the night is then short enough for B to be pushed into the next day (stopping diapause) under the action of a light pulse given during A. When the night lengthens, an early pulse would also cause a phase delay in B, but it would still fall before the end of the night, thus giving a high percentage diapause.

In conclusion, A could be considered as a gate where the oscillation of B can be reset under the effects of light, but the true photo-inductive role would be devolved to B, as already postulated.[39] Given these data, one cannot rule out the possibility that A has a part in photoperiodic time measurement.

The results of the experiments with symmetric skeleton photoperiods (PP_s) have to be taken as an integral part of those related to asymmetric skeletons. The diapause induction curve with PP_s does not stand out from those curves obtained with asymmetric skeletons when the scanning pulse illuminates point B (curves A–F in figure 4). The two pulses do not act as the beginning and the end of a complete photoperiod (PP_c), neither in the shorter nor in the longer interpretation (table 1), but it seems that one is read as the 'main' photoperiod, the other as a scanning pulse in an asymmetric skeleton.

When transferred from LD 18 : 6 to PP_s (the first pulse starting 24 h after the beginning of the last 18 h of light), *S. argyrostoma*[31,42] behaves like *Pieris*: its response curve, shifted in comparison with the PP_c response curve, shows a range of long-day effects that cannot be accounted for by the hypothesis that the two pulses mimic a complete photoperiod. However, when cultures are transferred from LL to PP_s régimes, long-day or short-day effects are elicited depending on which interval (long or short) is 'seen' first. This example stresses the importance of the conditions prevailing before the PP_s régime is begun.

The circadian nature of the processes implied for photoperiodic time measurement in *Pieris* meets the requirements laid down in the Introduction for justifying the question: does thermoperiodism belong to the circadian organization of the individual? The analogy between thermoperiodic and photoperiodic response curves is especially striking in the case of *Pieris* (figure 5). This suggests that the analogy is attended by an homology in the nature of the underlying mechanism. The results of experiments with asymmetric skeleton thermoperiods show that this might indeed be the case, since there is a time during the cold sequence when a temperature pulse acts just as a light pulse does when it illuminates point B during the night.[28] Thus, in a formal sense, one is justified in speaking of a *thermoinducible phase*. Strictly speaking, however, this allows one only to extend the range of the analogy because it is known that a phase of night photosensitivity (our point of comparison) marks hour-glass as well as oscillatory mechanisms. Moreover, the result was obtained with only one thermoperiodic régime (HC 8 : 16). Finally, from an epistemological point of view, the formal similarity of the two phenomena does not prove them to be causally identical. Bearing in mind these reservations, one may nevertheless consider that there is at least presumptive evidence for implicating the circadian organization of *Pieris* in thermoperiodism.

It must be emphasized that, on the basis of the few available data, thermoperiodism does not seem to be connected with a very precise type of photoperiodic clock. Thermoperiodism was first found in *N. vitripennis*,[25] a species in which photoperiodic responses follow the internal coincidence model,[22,43–47] but thermoperiods in DD are apparently ineffective in *S. argyrostoma*, a species of the external coincidence type.[47] Moreover, a circadian basis for photoperiodism cannot be shown for *P. interpunctella*, which nevertheless shows thermoperiodic sensitivity.[20] Finally, the external coincidence could refer to both photo- and thermoperiodism, as in the case of *P. brassicae*.

If square wave heat pulses meet the requirements for giving unequivocal information (i.e. clearly distinct 'on' and 'off' signals), the same is far from being true with the natural daily cycle of temperature, which shows large fluctuations in shape and may even become quite indistinct because of factors such as the wind or nebulosity. Nevertheless, it must be agreed that, however ambiguous it may be, the natural thermoperiod carries information that the insect is able to use when information about photoperiod is artificially lacking (figure 8, curve *d*). Taking into consideration the high reliability of natural photoperiodic information and the fact that under experimental conditions the photoperiod entirely controls the type of development in *Pieris*, the existence of thermoperiodic sensitivity is puzzling at first sight.

Natural cycles of light and temperature, of course, are present and oscillate together, so it is physically possible for the insect to derive information from both phenomena. Thus the question is whether the animal actually uses the thermal information under natural conditions; in other words, does it make use of its thermoperiodic sensitivity or not? Figure 8 gives indirect evidence that this is indeed the case. A comparison of curves *a* and *b* (percentage diapause: *a* in natural photo- and thermoperiods; *b* in artificial photoperiods and constant temperature) shows that, under artificial conditions, the population as a whole does not produce diapausing pupae until the photoperiod becomes as short as LD 12 : 12 (the critical photoperiod is LD 14 : 10). Such a short day, which corresponds to the photoperiod at the beginning of October, is never actually experienced by larvae in nature at the latitude of Paris. Curve *a* shows that diapause is *naturally* induced much earlier, the second fortnight of August (critical photoperiod LD $15\frac{1}{2} : 8\frac{1}{2}$). So, the inducing effect of short days is markedly shifted towards longer values of daylength under natural conditions as compared with artificial ones. This may be ascribed to the action of the daily temperature cycle. The mechanism of this interaction of temperature and light in phasing the clock has been thoroughly analysed in other species.[13,20,44] Therefore, one may be justified in considering thermoperiodism, not as a simple side-effect connected with the functioning of the clock,[19] but as an ecologically important mechanism able to correct the time given by the photoperiodic clock.

ACKNOWLEDGEMENTS

We are greatly indebted to Dr Jacques Claret for kind permission to use his unpublished data relating to the resonance experiments (figure 1). It is also a pleasure to thank Mrs J. Ligeour for her technical assistance.

REFERENCES

1. P. Hayden and R. G. Lindberg. *Science* **165**, 1288–9 (1969).
2. E. Gwinner. *Experientia* **22**, 765 (1966).
3. M. Menaker and A. Eskin. *Science* **154**, 1579–81 (1966).
4. J. T. Enright. In: J. Aschoff (ed.). *Circadian Clocks.* Amsterdam, North-Holland, pp. 112–24 (1965).
5. M. Lohmann and J. T. Enright. *Comp. Biochem. Physiol.* **22**, 289–96 (1967).
6. A. Mayer. *Naturwiss.* **55**, 234–5 (1968).
7. B. M. Sweeney and J. W. Hastings. In: *Biological Clocks.* (Cold Spring Harbor Symposium No. 25.) Cold Spring Harbor, NY, pp. 87–104 (1960).
8. F. Moriarty, *J. Insect Physiol.* **3**, 357–66 (1969).
9. S. K. de F. Roberts. *J. Cell. Comp. Physiol.* **55**, 99–110 (1960).
10. S. K. de F. Roberts. *J. Cell. Comp. Physiol.* **59**, 175–86 (1962).
11. A. C. Neville. *Q. J. Microscp. Sci.* **106**, 315–25 (1965).
12. H. Kirchner. *Z. Vergl. Physiol.* **48**, 385–99 (1964).
13. W. F. Zimmerman, C. S. Pittendrigh and T. Pavlidis. *J. Insect Physiol.* **14**, 669–84 (1968).
14. D. G. Tweedy and W. P. Stephen. *Comp. Biochem. Physiol.* **38A**, 213–31 (1971).
15. B. Rence and W. Loher. *Science* **190**, 385–7 (1975).
16. R. G. Lindberg and P. Hayden. *Chronobiologia* **1**, 356–61 (1974).
17. D. S. Saunders. *Science* **181**, 358–60 (1973).
18. G. M. Chippendale, A. S. Reddy and C. L. Catt. *J. Insect Physiol.* **22**, 823–8 (1976).
19. B. Dumortier and J. Brunnarius. *CR Acad. Sci. [D] Paris* **284**, 957–60 (1977).
20. S. Masaki and S. Kikukawa. This volume, pp. 101–12.
21. D. S. Saunders. *Science* **168**, 601–3 (1970).
22. D. S. Saunders. *J. Insect Physiol.* **20**, 77–88 (1974).
23. E. Bünning. *Photochem. Photobiol.* **9**, 219–28 (1969).
24. C. S. Pittendrigh and D. H. Minis. *Am. Nat.* **98**, 261–94 (1964).
25. D. S. Saunders. *Science* **181**, 358–60 (1973).
26. D. S. Saunders. *J. Insect Physiol.* **12**, 569–81 (1966).
27. G. M. Chippendale and A. S. Reddy. *J. Insect Physiol.* **19**, 1397–408 (1973).
28. B. Dumortier and J. Brunnarius. *CR Acad. Sci. Paris [D]* **285**, 361–4 (1977).
29. J. Claret. *CR Acad. Sci. Paris [D]* **274**, 1055–8 (1972).
30. D. S. Saunders. *J. Insect Physiol.* **19**, 1941–54 (1973).
31. D. S. Saunders. *J. Comp. Physiol.* **97**, 97–112 (1975).
32. H. U. Thiele. *Oecologia* **30**, 331–48; 349–65 (1977).
33. R. F. Beach and J. B. Craig jun. *J. Insect Physiol.* **23**, 865–70 (1977).
34. W. M. Hamner. *J. Insect Physiol.* **15**, 1499–504 (1969).
35. A. D. Lees. In: M. Menaker (ed.) *Biochronometry.* Washington, DC, National Academy of Sciences, pp. 372–80 (1971).
36. D. S. Saunders. *Nature* **253**, 732–4 (1975).
37. P. L. Adkisson. *Am. Nat.* **98**, 357–74 (1964).
38. P. L. Adkisson. *Science* **154**, 234–41 (1966).
39. C. S. Pittendrigh. *Z. Pflanzen.* **54**, 275–307 (1966).
40. J. W. Truman. *Z. Vergl. Physiol.* **76**, 32–40 (1972).
41. D. S. Saunders. *J. Comp. Physiol.* **132**, 179–89 (1979).
42. D. S. Saunders. *Insect Clocks.* Oxford, Pergamon (1976).
43. V. P. Tyshchenko. *Zhr. Obshch. Biol.* **27**, 209–22 (1966).
44. C. S. Pittendrigh. In: *Biological Clocks.* (Cold Spring Harbor Symposium No. 25.) Cold Spring Harbor NY, pp. 159–84 (1960).
45. C. S. Pittendrigh. *Proc. Natl Acad. Sci. USA* **69**, 2734–7 (1972).
46. D. S. Saunders. *Science* **156**, 1126–7 (1967).
47. D. S. Saunders. *J. Comp. Physiol.* **127**, 197–207 (1978).

DISCUSSION

Turek: If insects are entrained to a short-day (LD 8 : 16) light cycle, will a temperature pulse in the middle of the night terminate diapause? If insects are entrained to a short-day temperature cycle, will a light pulse in the middle of the night terminate diapause?

Dumortier: Concerning the first part of your question, I have only one result: when giving a short-day light and temperature cycle together, an added 2 h thermal pulse in the position of its maximal efficiency (from 9 to 11 h after the end of the warm sequence) does not suppress diapause. But it is important to stress the fact that, up to now, I have not scanned the entire cold and dark sequence, so no conclusion is possible. However, and this is the answer to the second part of your question, it is certain that entrainment of the photoperiodic clock with one type of zeitgeber, namely a temperature cycle, and response to another type of information (light) during the inducible phase, is possible. Actually, with a short-day temperature cycle in DD, a 2 h light pulse completely suppresses diapause when given from 5 to 7 h after the end of the warm sequence. It must be emphasized that, with the same short-day régime made of light and dark, the efficient light pulse is 8–10 h after dusk. This may be interpreted as a phase advance (−3 h) of the photo-inducible phase connected with the step down of temperature.

Brinkmann: It is difficult to imagine that thermoperiodism acts on the basis of temperature effects on reaction rates. At least such a type of mechanism must be highly sensitive to external perturbations of rate balances. The high precision of the system you are presenting implies rather that temperature acts by changing the physical state of a critical component as it does, for instance, by lipid phase transitions in mitochondria. The feature of those mechanisms is the restriction of the actual temperature to characteristic thresholds. In your system must the thermoperiodic pulse pass a characteristic threshold temperature?

Dumortier: As I said in my introduction, little is known about thermoperiodism, and knowledge of the detailed mechanisms is entirely lacking, so it is quite impossible to answer your question precisely. Thus, in this situation all the hypotheses have to be retained *a priori*, but I agree that this one is perhaps not perfect.

Masaki: The thermoperiodic effect must be distinguished from the effect of temperature itself. The latter might be responsible for the seasonal incidence of diapause you have observed under the natural temperature conditions in DD. What do you think?

Dumortier: In DD, and in constant temperature, the response is always 'no diapause', at any rate in the range of 13–21 °C. So, if we note diapause induction from about the middle of August, it cannot be under the action of a mean low temperature, but actually because of the prevailing thermal cycle. The existence of a thermosensitive phase also militates in favour of this view.

THE DIAPAUSE CLOCK IN A MOTH: RESPONSE TO TEMPERATURE SIGNALS*

Sinzo Masaki and Shigeru Kikukawa

Laboratory of Entomology, Faculty of Agriculture, Hirosaki University, Japan

Summary

The Indian meal moth *Plodia interpunctella* responds to a short-day thermoperiod (e.g. 8 h 30 °C : 16 h 20 °C) by entering diapause and to a long-day thermoperiod (e.g. 16 h 30 °C : 8 h 20 °C) by averting it in conditions of both LL and DD. This excludes any external coincidence model for the diapause clock. The critical duration of the thermoperiod is less well defined than the critical photoperiod. Different patterns of temperature cycles exert different effects on the incidence of diapause, even though the mean temperature of the cycle (25 °C) is the same. When such temperature cycles are superimposed on short days, the incidence of diapause varies as a function of the phase angle with the light cycle, diapause being mainly prevented by a temperature rise in the scotophase. These results can be compared with night interruptions by light, and suggest that the diapause clock can derive one of the two daily signals for night measurement from lights-on or lights-off and the other from some phase point in the temperature cycle. The latter can be taken either as dawn or dusk. Therefore, the same time measuring system is sensitive to both light and temperature. However, the effect of temperature as a signal depends on the cycle pattern and the interacting light cycle.

INTRODUCTION

Insects measure the duration of day or night in order to programme their seasonal development. The physiological mechanisms underlying this function can perhaps be called a 'diapause clock'. Like the circadian clock, the diapause clock can derive time signals not only from light cycles but also from temperature cycles. This is well illustrated by the so-called thermoperiodic response in *Nasonia vitripennis*,[1] *Diatraea grandiosella*,[2] *Pieris brassicae*[3] and several other species of moth.[4,5] One of the major questions about the diapause clock is its relationship to the circadian clock. Comparisons of the two have been performed mainly by analysing their responses to light régimes[6-11] but their sensitivity to both light and temperature offers a further means of comparison.[8] Such a comparison may be particularly useful since the circadian clock has already revealed an important feature in relation to its sensitivity to light and temperature signals. It comprises at least two component oscillators—a light-sensitive driving oscillator and temperature-

* Contribution No. 84 from the Laboratory of Entomology, Hirosaki University.

Biological Clocks in Seasonal Reproductive Cycles 101–112 (1981) (ed. B. K. and D. E. Follett: Bristol, Wright).

sensitive driven oscillator, as found in *Leucophaea maderae* and *Drosophila pseudoobscura*.[12]

The dual sensitivity to light and temperature also opens an interesting way of approaching the diapause clock. There should certainly be a difference between the terminal receptor mechanisms for these two very different stimuli and, therefore, the response of the time measuring components might be distinguishable from the response of the receptor components by comparing the effects of light and temperature. What happens, for example, when the phase angle between the two cycles is systematically changed? It is not difficult to produce different patterns of temperature cycle simply by cutting a revolving plastic disc into various shapes and coupling it to a temperature regulator. Most previous studies have tested only a few phase angles between LD (light and dark) and HC (hot and cold) cycles, although Pittendrigh and Minis[8] and Dumortier and Brunnarius[13] have demonstrated that short-day effects are perturbed by concurrent temperature cycles given at particular relationships to the light cycle. They related their results to the so-called external coincidence model of photoperiodic time measurement.

We have used the Indian meal moth *Plodia interpunctella* for a further analysis of the thermoperiodic response. This insect may enter diapause at the end of larval development. The larvae that continue development pupate in the food (rice bran), but those that enter diapause come out in search of hibernation sites. The diapause response can therefore be assessed before the actual onset of diapause, saving a lot of time.

The work was started by M. Takeda of Delaware University when he was at Hirosaki; S. Kikukawa, who is now at Missouri University, followed him, and the present paper is mainly based on the data collected during the latter half of our joint study. In all the following experiments a treatment comprised five replicates each containing 25–30 insects. Mortality was usually less than 20 per cent.

THE PHOTOPERIODIC RESPONSE

At an optimal temperature of 25 °C, the larvae showed a photoperiodic response of the type commonly found among insects (figure 1). The critical photophase was

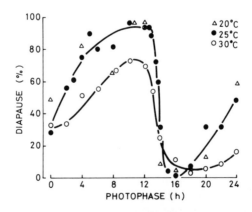

Figure 1. The diapause response to photoperiod in the Hirosaki stock of the Indian meal moth at three different temperatures.

about 13·5 h. Similar response curves were obtained at 20 °C and 30 °C. Although the incidence of diapause was generally lowered at the higher temperature, there was no clear shift in the critical photoperiod. The response changed sharply around the critical point and a difference of as small as 15 min in photoperiod caused a substantial difference in the incidence of diapause.

THE THERMOPERIODIC RESPONSE

The effects of thermoperiod were tested by exposing the insects to cycles in which there was a 10 °C step up or step down in the temperature from 20 to 30 °C or *vice versa*. The larvae were highly sensitive to these so-called thermoperiods in both the presence and the absence of light (figure 2). The response curve was similar but not

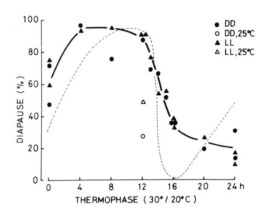

Figure 2. Diapause response to thermoperiod in the Hirosaki stock of the Indian meal moth in the presence (LL) or absence (DD) of light. The broken line shows the photoperiodic response at 25 °C redrawn for comparison.

quite identical to the photoperiodic curve. The response was less accurate to thermoperiod than to photoperiod, as indicated by the smaller slope of the curve in the critical range. In the photoperiodic response, the incidence of diapause reached its minimum at LD 16 : 8, from which it increased again as the photoperiods became very long. This rebound phase was absent in the thermoperiodic response, and the percentage of insects entering diapause continued to decrease slightly even in continuous high temperatures (HH, 30 °C). This was apparently due to the diapause-suppressing effect of the high temperature.

Although the action of photoperiod is not completely simulated by thermoperiod, it seems highly probable that a time measuring function is involved in this response. The incidence of diapause was thus much higher in the range from hot : cold (HC) 4 h : 20 h to HC 12 : 12 than at a constant temperature of 25 or 20 °C. Therefore, the temperature step up and step down may be substituting for dawn and dusk to give a short-day effect.

Another important point to be made is the effect of light on the thermoperiodic response. There was no clear difference between the results obtained in LL and DD. From this observation, any simple external coincidence model can be excluded.

Pattern effect of the temperature cycle

Several different patterns of temperature cycle were tested for their effects on diapause (figure 3). In this figure the low point or the starting point of the temperature rise is taken as hour zero, if only because this phase point tends to occur at dawn in nature. All of the cycles had the same mean temperature of 25 °C and an amplitude of 5 °C. Therefore, the times above and below the mean were both 12 h.

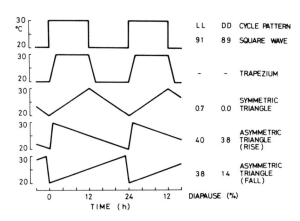

Figure 3. Patterns of temperature cycle tested in the present study, and their effects on the induction of diapause in LL or DD.

Despite this, the effects on the incidence of diapause were remarkably different in both LL and DD. A short-day effect was simulated only by the square-wave cycle. Since a long-day thermoperiod prevented diapause (figure 2), the square-wave pattern itself could not have been responsible for this effect. The interval between the twice daily changes in temperature was important. This inference was supported by the results with the asymmetric triangular cycles. These cycles gave either a sharp drop or rise in temperature of 10 °C in 1 h and then a slow return to the original level during the following 23 h. Both cycles resulted in an incidence of diapause similar to that in LL or DD. This might be due to the failure of these asymmetric cycles to give the necessary pair of time signals required to define the duration of 'day' or 'night'. On the other hand, diapause was prevented by a symmetric triangular cycle comprising 12 h rising and falling phases. Therefore, this cycle somehow generated a 'long day'. In other words, the diapause clock derived information from a certain pair of phase points separated by such an interval that a 'long day' or a 'short night' measurement was effected.

From these results, it can be pointed out that (*a*) temperature gives a time signal when it changes at a rapid rate, and (*b*) a pair of signals separated in time are required to elicit the diapause response. It should be kept in mind that these results were obtained in the absence of light cycles. A different conclusion might have been reached when there was a concurrent light cycle.

Phase angle effect

In order to measure time, the diapause clock requires a pair of stimuli signalling dawn and dusk. If there are clock components with different sensitivities to light and temperature, they might be 'dissected' by concurrent manipulation of the two cycles in various ways. In the first series of such an attempt, two diapause-inducing cycles, LD 12 : 12 and HC 12 : 12 (trapezoidal cycle), were combined at different phase angles (figure 4). The incidence of diapause clearly varied as a function of the

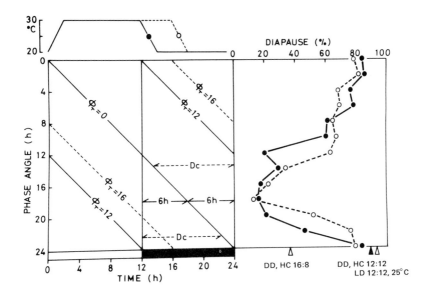

Figure 4. The effects of phase angle (*left*) between LD 12 : 12 and HC 12 : 12 (● and −) or between LD 12 : 12 and HC 16 : 8 (○ and -----) on the incidence of diapause (*right*). The temperature cycles are shown *top left* and the photoperiod *bottom left*. ϕ_τ signifies the pháse (in hours) of the temperature cycles from the start of the rising phase ($\phi_\tau = 0$). $\phi_\tau = 0$ divides the scotophase into two portions by the resetting effect, the longer one being assumed to be effective for time measurement. The minimum duration of effective night is shown to be 6 h. Dc is the critical length of dark period for the induction of diapause. Note that the suppression of diapause occurs between the phase angles forming dark periods of the critical length on either side of $\phi_\tau = 0$. DD, HC 16 : 8 (△) and DD, HC 12 : 12 (▲) show the percentages of diapausing larvae obtained with the respective square-wave thermoperiods. With trapezoidal cycles, the percentages of diapausing larvae were as follows: 54 per cent in HC 12 : 12 (DD), 52 per cent in HC 12 : 12 (LL), 15 per cent in HC 16 : 8 (DD) and 12 per cent in HC 16 : 8 (LL). In these cycles, H is the time at the temperatures above 25 °C, and C is the time at the temperatures below 25 °C. (*See* figure 3.)

phase angle between them. A temperature rise suppressed diapause remarkably when it occurred in the scotophase, but not in the photophase.

A similar tendency was observed when HC 16 : 8 was superimposed on LD 12 : 12 (figure 4). Although this thermoperiod alone exerted a diapause-preventing effect, its action in reversing the short-day effect was entirely dependent on the phase angle with the light cycle. A peak of diapause prevention appeared when a temperature rise occurred in the middle of the scotophase.

From these experiments, the following two points emerge: (*a*) any temperature version of the external coincidence model is excluded (c.f. Dumortier and Brunnarius[13]); (*b*) the temperature fall is ignored by the diapause clock. The reason for the first point is as follows. When the light and temperature cycles are completely out of phase, the scotophase overlaps with the warm phase so that the thermo-inducible phase, if it existed, should necessarily coincide with the warm phase. However, the peak of diapause prevention was not reached until the temperature-rise point came close to midnight, at least in the experiment with HC 16 : 8. The second point has been derived from the fact that the 4 h difference in time of the temperature-fall point between HC 12 : 12 and HC 16 : 8 did not produce any corresponding shift in phase of the response.

Since the diapause-preventing effect of temperature rise mainly occurs in the scotophase, it can be compared with the night-interruption effect of a light pulse.

Night interruption by light

When the scotophase of LD 8 : 16 (at 25 °C) was scanned by a 2 h light pulse, a diapause-preventing effect was produced by a midnight perturbation (figure 5). In such a régime, both portions of the scotophase separated by the pulse were shorter than the critical night length (10·5 h). When the pulse fell earlier or later in the

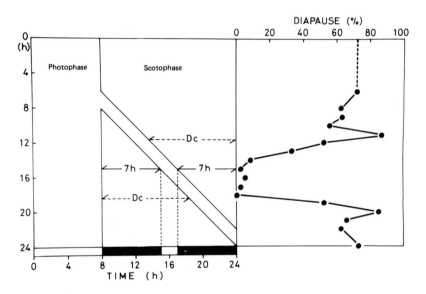

Figure 5. The effects of night interruption by light pulses (2 h) in LD 8 : 16 at 25 °C. Note that the suppression of diapause occurs when the longer of the two portions of scotophase separated by a light pulse is shorter than the critical length (Dc).

scotophase, either one of them became longer than the critical length, and diapause was induced. It seemed that diapause induction was determined by the longer of the two dark periods. However, this was true only in LD 4 : 20, 6 : 18 and 8 : 16, but not in LD 10 : 14 and LD 12 : 12, as shown in figure 6. In the latter photoperiods, the dark period after the pulse should be much shorter than the critical length to prevent diapause. This difference, caused by the background photoperiod, might be a reflection of the characteristic of the photoreceptor system but not of the clock mechanism itself. A certain duration of the dark period is probably required before the system becomes sensitive to light and thus able to reset the clock mechanism. When the night measurement is perturbed by a temperature signal, the result conforms to the prediction from the duration of the undisrupted portion of the scotophase, even in LD 12 : 12 (figures 4 and 7).

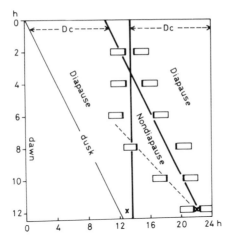

Figure 6. Summarized results of night interruption experiments in different photoperiods. The vertical axis represents the duration of photophase and the horizontal axis the time of the night interruption from lights-on. The bars show the position of light pulses (2 h) giving 50 per cent diapause. The thick end of each bar shows the lights-off or lights-on stimulus serving as a time signal. Diapause is suppressed when a light pulse falls between the two bars in each photoperiod. The two crosses near the bottom show the positions of temperature rise giving 50 per cent diapause. Diapause is suppressed by the temperature signal between them (figure 7).

Resetting model

Except for the difference due to the characteristics of the photoreceptor system, there is a similarity between the responses to light and temperature signals. In both cases, a midnight perturbation causes the largest effect on diapause suppression. One of the simplest interpretations common to both effects may be the resetting of the dusk or dawn signal, as suggested in figure 4. The diapause clock measures, in effect, the longer of the two portions of the scotophase separated by the resetting signal. Thus, a peak of diapause suppression is predicted when an effective time signal occurs at midnight of a scotophase shorter than twice the critical length (about 10·5 h). If the scotophase is much longer, such an effect would not be produced. The results of an experiment to test this prediction are given in the next section.

When the light cycle is LD 12 : 12, the two periods of night defined by the midnight rise of temperature are both 6 h—well below the critical duration and hence preventing diapause. As the temperature rise moves towards either dawn or dusk, the longer period of one of the two uninterrupted scotophases increases. When it exceeds the critical value, the diapause-preventing effect disappears.

The range of diapause prevention by the temperature rise is somewhat wider than the prediction by this model, particularly in LD 12 : 12 (figure 4). This may be due to a high-temperature effect in the scotophase.[14-16] It is not certain whether the slight difference between the results with HC 12 : 12 and HC 16 : 8 is real or not.

The above interpretation demands one apparently unlikely assumption: a temperature rise can be taken as either a dusk or a dawn signal, depending on when it occurs in the scotophase. This assumption is, nevertheless, supported by the close similarity between the responses to a temperature rise and a light pulse, which can provide either a light-on (dawn) or a light-off (dusk) stimulus.

Temperature rise and fall

In order to confirm the finding that only a temperature rise provides a zeitgeber stimulus in the presence of a light cycle, the asymmetric triangular temperature cycle with a sharp rise and a slow fall, or the reverse, was superimposed on LD 12 : 12 (figure 7). A temperature rise at midnight almost completely reversed the

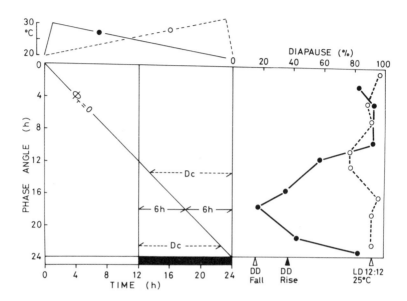

Figure 7. The effects of the phase angle between LD 12 : 12 and the temperature-rise cycle (● and −) or the temperature-fall cycle (○ and ----) on the incidence of diapause. (For key *see* figure 4.)

short-day effect. This is very similiar to the effect of the trapezoidal cycle (figure 4). Thus, a temperature rise alone is sufficient to reset the night-measuring process. In contrast, a fall in temperature almost completely failed to modify the short-day effect in any phase relationship with the light cycle (figure 7). Similar results were obtained with LD 10 : 14. An almost complete reversal of the short-day effect was again observed with a temperature rise beginning at midnight, but not with a temperature fall.

Light pulses and the rise or fall of temperature

In the next series of experiments the photophase was reduced to a single 2 h light pulse and combined with either a sharp rise or fall of temperature at different phase angles (figure 8). The primary aim of this design was to determine more clearly the sorts of information (dawn or dusk) provided by a rise or a fall in temperature. A short light pulse may be taken as either a dawn or a dusk signal, but not as both in each cycle. Therefore, it would form an effective time interval with a single temperature stimulus on either side of it. Which interval is measured as night is determined by the kind of information given by the temperature signal.

The temperature-rise cycle gave a clear diapause-inducing effect, when the low point, i.e. the starting point of the rise, occurred in the middle portion of the

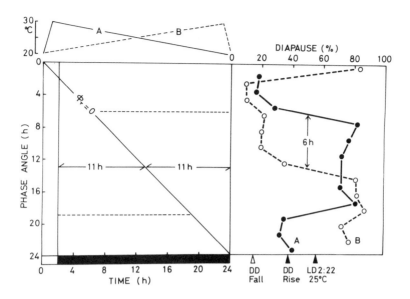

Figure 8. The effects of the phase angle between LD 2 : 22 and the temperature-rise cycle (● and −) or the temperature-fall cycle (○ and ----) on the incidence of diapause. The horizontal broken lines indicate the possible limits of phase angle for the effective time measurement. (For key *see* figure 4.)

scotophase. Since LD 2 : 22 alone gives only a moderate level of diapause incidence (figure 1), like arrhythmic conditions (LL and DD), a long night must have somehow been formed by the light and temperature signals. If it is assumed as before that a new night is formed by resetting, the time is measured either between lights-off and the low point in temperature, or between the low point and lights-on. The effective duration of the night should be 11 h, when the low point falls at midnight. This is longer than the critical duration and so induces diapause. As the low point moves towards lights-off or lights-on, however, there is an abrupt drop in the incidence of diapause. This lowered level of diapause is similar to that seen using this temperature cycle in DD. It seems likely, therefore, that the diapause clock fails to effect time measurement when three signals—lights-on, lights-off and temperature rise—come in close succession over a short period of time.

A temperature-rise stimulus thus seems to serve as either a dawn or a dusk signal. If it comes before midnight, it is taken as dusk, if after midnight, it is read as dawn. In view of the asymmetric structure of the temperature cycle, it is rather surprising to find such a 'symmetric interpretation' by the insect.

Although the temperature-fall cycle was quite ineffective in modifying the short-day effect of LD 12 : 12 or LD 10 : 14, the incidence of diapause varied conspicuously as a function of the phase angle with LD 2 : 22 (figure 8). The response curve is, however, very different from that obtained with the cycle involving the temperature-rise. This difference may be ascribed to a phase shift in the response. The phase of the response to the temperature-fall cycle is about 6 h behind that to the temperature-rise cycle. A close similarity between them emerges if the percentage diapause is plotted against the phase angle between hour 6, instead of hour 0, of the temperature-fall cycle and lights-on. With this adjustment, the resetting model still holds. The phase point that gives a time signal may not be the sharp drop in temperature itself, but may actually occur in the rising phase 6 h after the low point.

These results, together with preceding ones, suggest that the input of temperature signals to the diapause clock is not simply determined by the intensity of the physical stimulus. It depends not only on the rate of the temperature change, but also on the cycle pattern, availability and sequence of other possible signals. For example, the temperature cycle with a sharp fall and a slow rise failed to reverse the short-day effect of LD 12 : 12, but strongly modified the effect of LD 2 : 22. Both the temperature step-up and step-down may be 'read' in the absence of light cycles, but only the rise in temperature is effective if there is a concurrent light cycle.

Symmetrical triangular cycles

A further complication was encountered in experiments with a symmetrical triangular cycle. When this cycle was superimposed on LD 12 : 12, a clearly bimodal response appeared (figure 9). If the resetting model is not to be rejected, it

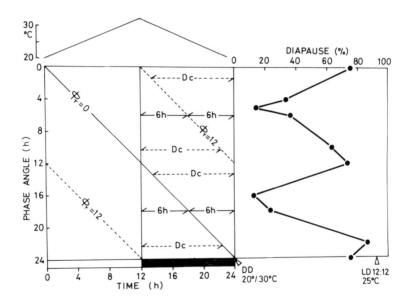

Figure 9. The effects of the phase angle between LD 12 : 12 and the symmetric triangle cycle of temperature. (For key *see* figure 4.) Note the two peaks of diapause suppression both occurring when the longer of the two portions of scotophase separated by the low point ($\phi_\tau = 0$) or the high point ($\phi_\tau = 12$) is shorter than the critical duration (Dc).

has to be assumed that both the low and the high points of the temperature cycle exert resetting effects. In fact, both occurred at midnight, so indeed effectively reversing the short-day effect. This temperature cycle was also tested on a background of LD 10 : 14, and an essentially similar result was obtained. If the concurrent light cycle was a long day (LD 16 : 8), however, no phase angle effect occurred, the incidence of diapause being very low at any phase angle. This result again supports the view that a night perturbation is responsible for the phase angle effect.

OSCILLATOR OR HOUR-GLASS?

It was totally unexpected that the diapause clock could take the same temperature change as either dusk or dawn despite the fact that the low point in temperature usually occurs at dawn under natural conditions. The resetting model simply states that a temperature rise gives a dusk signal before midnight and a dawn signal after midnight. A further elaboration on the model is clearly required for this switch from dawn to dusk. The 'subjective interpretation' of the temperature signal might be related to the course of time measurement. If, however, the time-measuring process simply starts at each lights-off, it goes on as a function of time, and the system cannot 'know' when midnight is reached without any entrainment by the previous cycles. Entrainment is of course a function of a self-sustaining oscillation. However, the present investigation has failed to reveal any light-sensitive and temperature-sensitive components comparable to those found in the circadian clock of certain insects.[12] Both dawn and dusk were signalled by temperature as easily as by light. It seems probable, therefore, that the same system is sensitive to both light and temperature signals.

An alternative interpretation comes from an hour-glass mechanism similar to that postulated for *Megoura viciae*.[7] The switch from the dusk to dawn effect of the temperature signal might occur when the critical duration for time measurement has passed. Before this, the system is in a reversible state and easily set back to hour zero (dusk) by a signal, but thereafter an irreversible stage is entered and a signal terminates the night (dawn).

The available data are not sufficient to determine which alternative is the more likely. From other sources of information, it is also difficult to determine whether the diapause clock of this insect is an oscillator or an hour-glass. A long series of resonance experiments was carried out, using cycle lengths from 18 to 108 h[17], but a clear prevention of diapause occurred only when the scotophase was shorter than the critical duration of about 10 h, and no circadian fluctation was found in the incidence of diapause.

Although the meal moth larvae responded to skeleton photoperiods formed by a pair of 2 h pulses, they could not 'read' the shorter dark interval. When they were exposed to complete photoperiods every three days in otherwise dark conditions, their response was obscure. However, when these two régimes were combined, that is, one day of a complete photoperiod was followed by two days of skeleton photoperiod, they showed a response closer to the normal pattern (S. Kikukawa, unpublished data). The possibility cannot be ruled out that the entrainment to a complete photoperiod facilitates the response to the following 2 days of skeleton photoperiods.

Thus, the various pieces of available information do not fit a single picture, and a clear basic plan of the diapause clock of this moth cannot be drawn as yet.

REFERENCES

1. D. S. Saunders. *Science* **181**, 385–60 (1973).
2. G. M. Chippendale, A. S. Reddy and C. L. Catt. *J. Insect Physiol.* **22**, 823–8 (1976).
3. B. Dumortier and J. Brunnarius. *CR Acad. Sci. Ser. D [Paris]* **284**, 957–60 (1977).
4. N. I. Goryshin and R. N. Kozlova. *Zh. Obshch. Biol.* **28**, 278–88 (1967).
5. M. Menaker and G. Gross. *J. Insect Physiol.* **11**, 911–14 (1965).
6. A. D. Lees. *Circadian Rhythmicity*. Central Agricultural Publishing House, University of Wageningen, pp. 87–110 (1971).

7. A. D. Lees. *J. Insect Physiol.* **19,** 2279–316 (1973).
8. C. S. Pittendrigh and D. H. Minis. *Biochronometry.* Washington, National Academy of Sciences, pp. 212–250.
9. D. S. Saunders. *J. Comp. Physiol.* **110,** 111–33 (1976).
10. D. S. Saunders. *Insect Clocks.* Oxford, Pergamon (1976).
11. D. S. Saunders. *J. Comp. Physiol.* **127,** 197–207 (1978).
12. C. S. Pittendrigh. *Cold Spring Harb. Symp. Quant. Biol.* **25,** 159–84 (1960).
13. B. Dumortier and J. Brunnarius. *CR Acad. Sci. Ser. D [Paris]* **285,** 361–4 (1977).
14. S. D. Beck. *Biol. Bull.* **122,** 1–12 (1962).
15. N. I. Goryshin. *Entomol. Obozr.* **43,** 43–6 (1964).
16. R. Thurston. *Environ. Entomol.* **5,** 626–7 (1976).
17. M. Takeda and S. Masaki. Proceedings of US–Japan Seminar on Stored Product Insects, Manhattan, pp. 186–201 (1976).

LOCALIZATION OF CIRCADIAN PACEMAKERS IN INSECTS

Terry L. Page*

Department of Biological Sciences, Hopkins Marine Station
of Stanford University, Pacific Grove, California, USA

Summary

Circadian pacemakers that control a variety of behavioural rhythms have been localized in the brain (supra-oesophageal ganglion) in a number of insects. Eclosion behaviour in silkmoths (*Antheraea pernyi* and *Hyalophora cecropia*) and adult activity in *Drosophila melanogaster* have been shown in tissue transplantation experiments to be controlled by brain-centred pacemakers. Locomotor activity rhythms in cockroaches and locomotor and stridulatory activity rhythms in crickets are abolished by bilateral ablation of the optic lobes of the protocerebrum. In cockroaches the optic lobe cells involved in sustaining rhythmicity have been localized to a small region ventral to and near the lobula neuropile. Further studies utilizing both lesions and localized low-temperature pulses have provided evidence that these cells are part of a bilaterally redundant pacemaking system composed of two mutually coupled oscillators.

Other experiments on insects have also pointed to the multi-oscillator nature of the circadian system. In the beetle, *Blaps gigas*, the rhythms of electro-retinogram amplitude in the two eyes are controlled by apparently independent circadian pacemakers. In the cockroach, *Leucophaea maderae*, a rhythm of cuticle deposition in newly moulted adults does not require the optic lobes. Finally, in cockroaches and crickets there is some evidence that a strongly damped oscillator outside the optic lobes is also involved in the activity and stridulatory rhythms.

The results are consistent with the generally accepted view that circadian organization in insects, and in multicellular organisms in general, is derived from a population of anatomically discrete circadian oscillators within the individual.

INTRODUCTION

Because insects can generate varied and complex behavioural patterns with a relatively simple nervous system they have increasingly become favourite subjects for the neurobiologist interested in the cellular mechanisms of behaviour. For the same reason they are particularly suitable material for the search for general

* Present address: Department of General Biology, Vanderbilt University, Nashville, Tennessee, USA.

principles underlying circadian organization in the metazoa. While the 'circadian repertoire' of insects includes a variety of well-defined circadian rhythms, precise photoperiodic time measurement, and time-compensated celestial navigation, the organization of the neural and neuroendocrine systems that are ultimately responsible for maintaining this temporal organization is 'simple' enough that an explanation of its cellular basis is a realistic goal.

The realization of this goal is dependent on two general approaches. The first is the anatomical localization of the components of the circadian system and the identification of the specific cells involved. The second is the elucidation of the physiological mechanisms by which these components interact to form an organized regulatory system. In the past 15 years there has been substantial progress on both fronts. Invariably, the investigations begin with the model, initially put forward by Pittendrigh and Bruce,[1] which is illustrated in figure 1. Four functionally defined

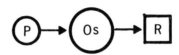

Figure 1. Oscillator model of the circadian system. Os, oscillator; P, photoreceptor; R, overt rhythm. Coupling pathways are designated by arrows.

components are shown—an oscillator, a photoreceptor for entrainment and two 'coupling' pathways, one that mediates the flow of information from the photo-receptor to the oscillator and a second between the oscillator and the function that it controls.

Substantial progress has been made in insects in identifying the anatomical correlates of these functionally defined component parts of the circadian system. Two general points of interest have emerged from these studies. Although in no instance have circadian oscillators or photoreceptors been linked to particular, identified cells, it is clear that restricted regions of the nervous and neuroendocrine systems can fulfil the functions of the various components in figure 1. Oscillators, photoreceptors and steps in the coupling pathways have been localized in various insects. On the other hand, there is much evidence that the model is incomplete as a representation of the circadian system. In any one individual there may be several circadian oscillators—in some cases independent in function, in other cases interactive. The emerging picture of circadian organization in insects is similar to that for other metazoan groups, temporal order within the individual appearing to be derived from a population of anatomically discrete oscillators and depending not only on the properties of the individual pacemakers but also on the coupling relationships between them.[2–4]

The following is a summary of efforts to localize circadian pacemakers in several insects. The first section recounts data from transplantation and lesion experiments in which circadian pacemakers have been localized to the brain. In the second section the anatomical and physiological characterization of the cockroach pace-making system is considered in some detail. Finally, the evidence for the multi-osciliator organization of insect circadian systems is discussed.

LOCALIZATION OF CIRCADIAN PACEMAKERS IN THE NERVOUS SYSTEM

Circadian pacemakers that control a variety of behavioural rhythms have been localized in the brain in a number of insects (table 1). The most direct evidence has come from experiments utilizing tissue transplantation techniques that have been successful in cases where there is a humoral link in the output pathway of the pacemaker. In silkmoths, for example, the time of emergence of the pharate adult

Table 1. Insect circadian pacemakers localized to the supra-oesophageal ganglion

Organism	Function	Reference no.
Silkmoth	Eclosion	5, 6, 7
(*Antheraea pernyi*,	Flight activity	11
Hyalophora cecropia)		
Fruitfly	Locomotor activity	9
(*Drosophila melanogaster*)		
Cricket	Stridulation	12
(*Teleogryllus commodus*)	Spermatophore production	13
	Locomotor activity	14
Cockroach	Locomotor activity	15, 21, 22, 27
(*Leucophaea maderae*,		
Periplaneta americana)		

from its pupal case is gated by a circadian pacemaker that times the release of an eclosion hormone that in turn triggers the emergence behaviour.[5] When the brain is surgically removed, eclosion of individuals of a population occurs at random times of day; but the population rhythm is restored if the brains are transplanted to the abdomens.[6] It has also been shown that the transplanted brain not only restores rhythmicity, but imposes its characteristic phase on the time of eclosion. If brains are cross-transplanted between two species of moth (*Hyalophora cecropia* and *Antheraea pernyi*) that emerge at different times of day, the time of eclosion is determined by the donor brain rather than the host moth.[7] The fact that the transplanted brains not only restore rhythmicity, but in addition, dictate the rhythm's phase, leaves little doubt that the pacemaker that controls the time of emergence is located in the brain. Further localization was attempted by subdividing the brain prior to transplantation. The optic lobes are unnecessary since the intact cerebral lobes adequately gate eclosion. When the cerebral lobes were subdivided, however, by cuts lateral to the medial neurosecretory cells, where the hormone that initiates eclosion is produced,[8] the isolated median piece did not gate emergence.[6] These data suggested that the pacemaker may reside in the lateral portions of the cerebral lobes.

The transplantation technique has also been used in adult *Drosophila melanogaster*.[9] The experiments made use of two X-chromosome-linked mutant strains. One of the strains (per°) invariably exhibits aperiodic locomotor activity. The other strain (per^s) exhibits an activity rhythm with a period in constant darkness of about 19–20 h, markedly shorter than the wild-type period of about 24 h.[10] Brains of short period mutants were transplanted into the abdomens of aperiodic mutants.[9] Four of the 55 animals that survived the surgery exhibited three or more cycles of periodic activity in constant conditions, and the periods of these rhythms were between 16 and 20 h. Although the success rate was (understandably) low the fact that some per° hosts expressed a rhythm with a period similar to that of the per^s donors indicates that the brain is the site of the pacemaker that controls activity; and that its effects can be mediated *via* a humoral pathway.

In other insects putative pacemaker sites have been localized in the brain with surgical or electrolytic lesions. In the silkmoth, removal of the brain or severing its neural connections with the thorax abolished the flight-activity rhythm of the adult.[11] Ablation of the optic lobes did not disrupt rhythmicity, which suggested that the pacemaker for the activity rhythm resides in the cerebral lobes.

In crickets, ablation of the optic lobes abolishes the circadian rhythms of stridulation,[12] spermatophore production[13] and locomotor activity.[14] Similarly, bilateral ablation of the optic lobes in cockroaches causes persistent arrhythmicity of locomotor activity.[15–17] Lesions to other regions of the brain will also cause arrhythmicity in both crickets and cockroaches, but other work, particularly on the cockroach, has suggested that it is the optic lobes that function as the driving oscillator for the system.

ORGANIZATION OF THE CIRCADIAN PACEMAKING SYSTEM OF THE COCKROACH

Photoreceptors for entrainment

Several studies have provided unequivocal evidence that the compound eyes of cockroaches (*Leucophaea maderae* and *Periplaneta americana*) are the sole source of photoreceptive information for entrainment of the circadian rhythm of loco-motor activity. Painting over the compound eye[18] or surgical transection of the optic nerves[19,20] abolishes entrainment by light. The ocelli are neither necessary nor sufficient for entrainment,[18,19] even under very high light intensities.[20]

The optic lobes—locus of the driving oscillation?

Nishiitsutsuji-Uwo and Pittendrigh[15] first suggested that the optic lobes of the protocerebrum might be the site of the pacemaker that controls the cockroach activity rhythm when they found that bilateral ablation of the optic lobes or section of the optic tracts causes persistent arrhythmicity (figure 2). They also found that unilateral optic lobe ablation or optic tract section did not abolish the rhythm. This suggested the possibility that each of the optic lobes might contain an oscillator sufficient to drive the locomotor activity rhythm.

Mutual entrainment of bilaterally redundant oscillators

These ideas prompted a systematic investigation of the effects of unilateral optic lobe ablation in which two questions were raised.[21] First, if there are two pacemakers are they functionally equivalent? Second, does ablation of one optic lobe have any effect on the rhythm that would suggest an interaction between optic lobe pacemakers? In these experiments one optic lobe was removed (or the optic tract cut) in 39 animals that were free-running in constant darkness (23 right lobe, 16 left lobe). The results indicated that, at least as measured by ability to maintain rhythmicity and average free-running period (τ), the optic lobes were functionally

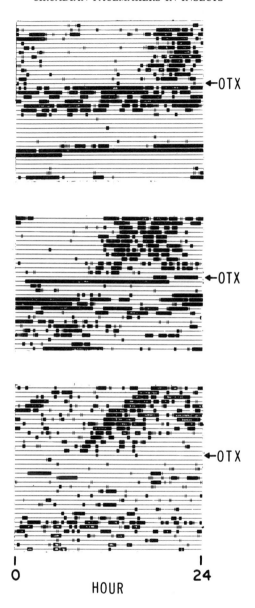

HOUR

Figure 2. Three examples of activity records showing the disruption of the locomotor rhythm in the cockroach (*Leucophaea maderae*) following bilateral section of the optic tracts (OTX).

redundant. However, it was found that ablation of either the right or left optic lobe led to a small, but consistent and significant, increase in τ. In the 39 animals studied, $\bar{\tau}$ before surgery was about 23·7 h ($23{\cdot}73 \pm 0{\cdot}26$), whereas rhythms driven by either the right or left lobe alone had average periods of about 24·0 hours ($\bar{\tau}_{left} = 23{\cdot}95 \pm 0{\cdot}28$; $\bar{\tau}_{right} = 23{\cdot}96 \pm 0{\cdot}24$). The surgical control of sectioning the optic nerve had no effect on $\bar{\tau}$.

These results were important in two respects. First, the demonstration that the optic lobes were not only necessary to sustain the activity rhythm, but also were involved in determining its free-running period supported the idea that these structures were part of the pacemaking system. Second, the data suggested that the bilaterally redundant optic lobe oscillators were mutually coupled to form a compound pacemaker. This suggestion was reinforced by the finding that either one of the compound eyes was sufficient to entrain both oscillators (figure 3). The

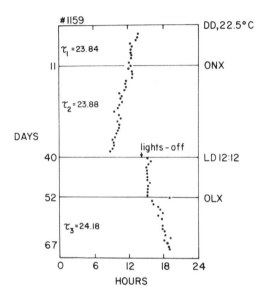

Figure 3. Record of the activity rhythm in a cockroach, *Leucophaea maderae*, illustrating the sufficiency of either compound eye (here the left eye) to mediate entrainment of the contralateral pacemaker, and showing the effects of optic lobe ablation on the free-running period. Onsets of daily activity peaks are shown by filled circles. The record begins with the animal free-running in constant darkness (DD). Section of the right optic nerve (ONX) has no significant effect on the period. On day 40 the animal was placed in an LD cycle which imposed a large phase delay on the rhythm. On day 52 the left optic lobe was removed (OLX). The cockroach began to free-run in the light cycle, driven by the right optic lobe pacemaker, with an initial phase determined by the light cycle. The left eye was sufficient to entrain the right pacemaker. The free-running period after removal of one lobe was significantly longer than the period in the earlier DD free run.[21]

experiments involved animals in which one optic lobe was isolated from the input from its own compound eye by optic nerve section. After maintaining the animals for several days in a light cycle which imposed a major phase shift on the activity rhythm, the unoperated optic lobe was removed. The animals began to free-run in the light cycle being driven by the optic lobe which had been isolated from its own eye but with a phase determined by the light cycle.[21] These data suggested the model illustrated schematically in figure 4. Each oscillator (Os), which independently has a free-running period of 23·95 h, receives input from photoreceptors (P) of the ipsilateral compound eye. In the intact animal the oscillators are mutually entrained with a free-running period (τ) of the coupled system being 23·72 h. Mutual coupling could also provide a pathway by which either compound eye can entrain both pacemakers.

The hypothesis of mutually coupled optic lobe oscillators has been recently tested in a series of experiments utilizing localized low temperature pulses.[22] The

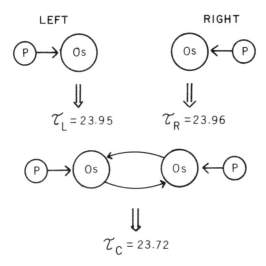

Figure 4. Schematic illustration of the hypothesis of mutually coupled circadian pacemakers in the cockroach. Each optic lobe (left and right) contains an oscillator (Os) and receives input from photoreceptors (P) of the ipsilateral compound eyes. Each oscillator has a free-running period (τ) of about 23·95 h. Mutual coupling causes a reduction in τ to 23·72 h and provides a pathway by which either compound eye can entrain both oscillators.[21,22,27]

methodology was based on the observation that the phase of the activity rhythm can be reset by pulses of low temperature ($<12°$).[23,24] There is reason to believe these pulses are acting directly on the pacemaker and not on the light entrainment pathway since it has been shown that the phase response curve for short duration (6 or 8 h), low temperature pulses is composed entirely of delaying phase shifts[24] (S. K. Roberts, personal communication) while the phase response curve for light includes both delays and advances.[23,25] Furthermore, low temperature pulses generate large phase shifts at phases where the pacemaker is relatively insensitive to light.

Localized cooling was accomplished by positioning an insect pin attached to a Peltier block just ventral to the optic lobe. The temperature of the contralateral optic lobe was controlled by another insect pin attached to a 560 ohm resistor. By separately adjusting the current through the Peltier block and the resistor, one optic lobe could be cooled to about 7·5 °C while the other was maintained near 25 °C. Animals free-running in constant darkness were removed from their activity cages, treated with low temperature pulses (for 6 h beginning at activity onset) and returned to the cages.

The results of several experiments with localized low temperature pulses are summarized in figure 5. In animals in which one optic tract had been cut, cooling the intact optic lobe consistently caused a phase delay of several hours (figures 5A, 6A). In contrast, cooling either the neurally isolated optic lobe (figures 5B, 6B) or the midbrain[22] had little or no effect on the phase of the rhythm. The demonstration that the optic lobes are involved in control of phase as well as period leaves virtually no doubt that they are involved in generating the driving oscillation for the activity rhythm.

These results also suggested a method for the direct demonstration of a coupling pathway between the optic lobes. In these experiments low temperature pulses were given to one optic lobe of intact animals. When the treated lobe was removed

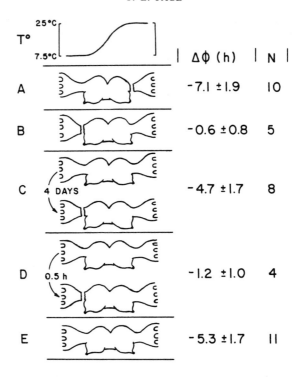

Figure 5. Summary of protocols and results of low temperature pulses. The spatial temperature gradient across the brain (shown below) is schematically represented at the top of the figure. In each of the five treatments shown the left optic lobe was cooled to about 7·5 °C while the right lobe was maintained at about 25 °C. For each treatment the average phase shift ± standard deviation (Δφ) and the sample size (N) are given to the right.

four days after the pulse, the subsequent rhythm (driven by the untreated contralateral lobe) was phase-delayed by several hours (figures 5C, 6C); but, if the treated lobe was removed only 0·5 h after the pulse, the phase shift in the contralateral optic lobe pacemaker was prevented (figures 5D, 6D)—the small phase delay was essentially the same as that caused by optic tract section alone.[22] The results indicated that the low temperature pulse caused a phase shift in the pacemaker of the treated optic lobe (without *directly* affecting the phase of the contralateral pacemaker) that was subsequently transmitted to the contralateral pacemaker. Mutual coupling also appeared to reduce the amplitude of the phase shift generated by localized low temperature pulses. Cooling one optic lobe of an intact animal resulted in an average steady-state phase shift nearly 2 h less than the phase shift obtained when the optic lobe contralateral to the pulse was neurally isolated (figure 5E). This observation suggests that following desynchronization of the two optic lobe pacemakers the motion of the coupled system to steady-state involves a phase advance in the treated pacemaker as well as the phase delay in the untreated pacemaker.

Localization of pacemaker function within the optic lobe

The region within the optic lobe that contains the cells involved in controlling the activity rhythm has been further localized. Following removal of one optic lobe,

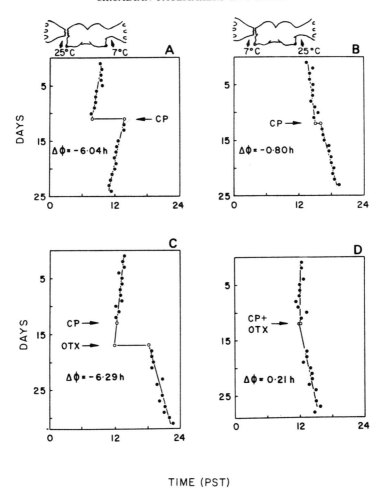

TIME (PST)

Figure 6. Examples of data from animals that were treated with localized low temperature pulses. ●, time of activity onset for each day; ○, projected phases of the rhythms before and after the pulse; lines are linear regressions. Pulses were 6 h in duration and began at activity onset. In A, the intact optic lobe of an animal in which one optic tract was cut was cooled (CP) while the neurally isolated optic lobe was maintained at 25 °C; in B, the neurally isolated optic lobe was cooled while the intact lobe was maintained at 25 °C. Cooling the intact lobe caused a large phase delay (Δφ) while cooling the neurally isolated lobe had little effect. C and D illustrate the effects of a low temperature pulse to one lobe on the rhythm driven by the contralateral lobe. In C the optic tract of the treated lobe was cut (OTX) 4 days after the pulse (CP). The subsequent rhythm, driven by the untreated lobe, is phase-delayed by several hours. In D the optic tract of the treated lobe was sectioned 0·5 h after the pulse. Optic tract section shortly after the pulse prevents the phase shift in the rhythm.

surgical or electrolytic lesions distal to the second optic chiasm or dorsal to the lobula have no effect on rhythmicity, while lesions on the ventral half of the lobe near the lobula frequently abolish the activity rhythm.[16,26,27] Furthermore, lesions to this region of the optic lobe prevent entrainment of the contralateral pacemaker via photoreceptors of the eye of the lesioned lobe, and concurrently increase $\bar{\tau}$ of the rhythm.[27] These results suggest that the cells necessary to pacemaker function have their soma and/or crucial processes in the region of the lobula.

MULTIPLE OSCILLATORS IN INSECTS

Several other lines of evidence suggest that circadian organization in insects involves a number of oscillatory components. For example, in the beetle *B. gigas* each of the two compound eyes exhibits a circadian rhythm in the amplitude of its electroretinogram (ERG). In an individual the two ERG rhythms can free run with quite different periods, and the phase and period of the rhythm in one eye can be controlled by local illumination of the eye without having any detectable effect on the rhythm in the contralateral eye.[28] The data suggest that the ERG rhythms are controlled by a bilaterally distributed pair of circadian pacemakers that, in contrast to the situation in the cockroach, are independent.

In cockroaches and crickets there is some evidence that a strongly damped oscillator survives ablation of the optic lobes. Rence and Loher[29] reported that following optic lobe ablation, crickets in a 24 h temperature cycle restricted their stridulatory activity to the first few hours after the high temperature (H) to low temperature (C) transition. The conclusion that the rhythmicity was not simply forced by the change in temperature was prompted by the observations that animals did not exhibit a rhythm when subjected to a temperature cycle outside the circadian range (HC 15 : 15) and that there was evidence for one or two transient cycles of rhythmicity following a single 12 h low temperature pulse. Lobeless cockroaches will also exhibit a rhythm of locomotor activity when placed in a 24 h temperature cycle[17] (T. L. Page, unpublished), although, unlike crickets, the rhythm is also evident in a 30 h temperature cycle.[17] Interestingly, in *Leucophaea* the phase angle between the onset of activity and the H to C transition varies as a function of the period of the temperature cycle (T. L. Page, unpublished). These observations on crickets and cockroaches are difficult to explain as a simple reflex response to a change in ambient temperature, and are at least consistent with the idea that the temperature cycle is acting through a strongly damped oscillator.

The best evidence for circadian oscillators existing outside the optic lobes of cockroaches comes from a recent study on the rhythm of cuticular deposition in cockroaches. In many insects the formation of the endocuticle early in adult life involves the daily deposition of alternately lamellate and non-lamellate chitin.[30] This results in the formation of growth layers which are visible under polarized light. It was found in the cockroach *Blaberus fuscus* that after imaginal ecdysis, the circadian rhythm in the formation of the endocuticle was still evident in animals whose optic lobes had been removed in the last larval instar or following decapitation of the adult shortly after imaginal ecdysis.[31] The results demonstrate the existence of an extra-cephalic pacemaker and raise the possibility that the cells of the epidermis are autonomously rhythmic.[30]

CONCLUSION

The emerging descriptions of the anatomical and physiological organization of insect circadian pacemaking systems parallels those from other groups of multi-cellular organisms. Circadian organization is derived from a system of anatomically distributed oscillators and is dependent not only on the properties of the individual pacemakers, but also on the coupling relationships between them. In no instance is the description of this system complete, but substantial progress has been made in localizing pacemakers in several insects; and in a few cases we are beginning to uncover the 'internal' structure of the timing system. The results suggest that within the individual organism there may be independent pacemaking systems, as appears to be the case in the control of the ERG rhythm in the two eyes of the beetle *B. gigas* or in the control of locomotor activity *v.* cuticular deposition in the cockroach. In other cases a single pacemaking system (e.g. optic lobes in crickets) may control several rhythms. Finally, the control of a single rhythm may involve multiple oscillators. In the cockroach the driving oscillation is produced by a pair of mutually coupled, bilaterally redundant pacemakers. Moreover, in both the cockroach and cricket the possibility has been raised that the pacemaking system of the optic lobes may act through yet another damped slave oscillator.

Clearly much work remains to be done before we can claim a complete understanding of the organization of the insect's circadian system. Ultimately, future progress will depend on the identification of the specific cells that are responsible for the generation of circadian oscillations and for the transmission of information between oscillators and photoreceptors, among oscillators, and between the pacemaking system and the effectors under its control. Progress made thus far in localizing components of the pacemaking system and in elucidating their role in circadian organization suggests that this is a realizable goal.

REFERENCES

1. C. S. Pittendrigh and V. G. Bruce. In: D. Rudnick (ed.) *Rhythmic and Synthetic Processes in Growth.* Princeton, NJ, Princeton University Press, pp. 75–108 (1957).
2. C. S. Pittendrigh. In: F. O. Schmitt and F. G. Worden (ed.) *The Neurosciences: Third Study Program.* Cambridge, Mass., MIT Press, pp. 437–458 (1974).
3. G. D. Block and T. L. Page. *Ann. Rev. Neurosci.* **1**, 19–34 (1978).
4. T. L. Page. In: J. Aschoff (ed.) *Handbook of Behavioral Neurobiology*, IV. *Biological Rhythms.* New York, Plenum (in the press) (1981).
5. J. W. Truman. In: F. O. Schmitt and F. G. Worden (ed.) *The Neurosciences: Third Study Program.* Cambridge, Mass., MIT Press, pp. 525–529 (1974).
6. J. W. Truman. *J. Comp. Physiol.* **81**, 99–114 (1972).
7. J. W. Truman and L. M. Riddiford. *Science* **167**, 1624–6 (1970).
8. J. W. Truman. *Biol. Bull.* **114**, 200–11 (1973).
9. A. M. Handler and R. J. Konopka. *Nature* **279**, 236–8 (1979).
10. R. J. Konopka and S. Benzer. *Proc. Natl Acad. Sci. USA* **68**, 2112–16.
11. J. W. Truman. *J. Comp. Physiol.* **95**, 281–96 (1974).
12. W. Loher. *J. Comp. Physiol.* **79**, 173–90 (1972).
13. W. Loher. *J. Insect Physiol.* **20**, 1155–72 (1974).
14. P. G. Sokolove and W. Loher. *J. Insect Physiol.* **21**, 785–99 (1975).
15. J. Nishiitsutsuji-Uwo and C. S. Pittendrigh. *Z. Vergl. Physiol.* **58**, 14–46 (1968).
16. S. K. Roberts. *J. Comp. Physiol.* **88**, 21–30 (1974).
17. R. Lukat and F. Weber. *Experientia* **35**, 38–9 (1979).
18. S. K. Roberts. *Science* **148**, 958–9 (1965).
19. J. Nishiitsutsuji-Uwo and C. S. Pittendrigh. *Z. Vergl. Physiol.* **58**, 1–13 (1968).
20. R. J. Driskell. Master's Thesis, University of Delaware (1974).
21. T. L. Page, P. C. Caldarola and C. S. Pittendrigh. *Proc. Natl Acad. Sci. USA* **74**, 1277–81 (1977).
22. T. L. Page. *Am. J. Physiol.* (in the press) (1981).
23. S. K. Roberts. *J. Cell. Comp. Physiol.* **59**, 175–86 (1962).
24. G. Wiedenmann. *J. Interdiscipl. Cycle Res.* **8**, 378–83 (1977).
25. G. Wiedenmann. *Z. Naturforsch.* **32**, 464–5 (1977).
26. P. G. Sokolove. *Brain Res.* **87**, 13–21 (1975).
27. T. L. Page. *J. Comp. Physiol.* **124**, 225–36 (1978).
28. W. K. Koehler and G. Fleissner. *Nature (Lond.)* **274**, 708–10 (1978).
29. B. Rence and W. Loher. *Science* **190**, 385–7 (1975).
30. A. C. Neville. *Biology of the Arthropod Cuticle.* New York, Springer-Verlag (1975).
31. R. Lukat. *Experientia* **34**, 377 (1978).

DISCUSSION

Jacklet: Is it possible to do a chronic low temperature experiment? That is, lower the temperature of a specific brain area for days while recording the locomotory activity and then remove the temperature block and observe the alteration in the rhythm caused by the interacting bilateral clocks?

Page: It would be very difficult in a freely moving animal, but it might be possible to do the experiment in a restrained animal by monitoring leg movements or muscle potentials.

Bittman: Since unilateral cooling causes only phase delays, it would be desirable to examine coupling in a situation more representative of entrainment, i.e. where one oscillator is phase-advanced rather than slowed or stopped. Have you performed unilateral illumination experiments to phase-advance one optic lobe? Does unilateral cooling produce a larger phase delay if the contralateral (uncooled) lobe is removed after the manipulation than if both optic lobes are left intact? Is there any evidence that the optic lobes can adopt any stable phase relationships to one another other than 0° (e.g. antiphase)?

Page: In answer to the first question, unilateral illumination experiments have been done, and the results have been published. As to the second question, that is certainly what you would predict, but the experiment has not been done. We have no evidence for any other stable phase relationships.

Rusak: Is there a circadian constraint on temperature entrainment? Did you release optic lobe-lesioned animals entrained to temperature cycles into DD and did they show any transient persistence of rhythmicity? What happens to activity during the four days between the time of cooling and the optic lobe excision?

Page: To answer your questions serially. We have not done the studies in cockroaches. Rence and Loher say that bilobectomized crickets cannot entrain to 30 h days. We saw no convincing evidence that rhythmicity persisted. On the first day there is little activity. By the second or third day activity has already shifted 4–5 h and appears to be in a steady-state.

THE PHOTOPERIODIC CONTROL OF POLYMORPHISM IN APHIDS: NEUROENDOCRINE AND ENDOCRINE COMPONENTS

A. D. Lees and J. Hardie

ARC Insect Physiology Group, Imperial College at Silwood Park, Ascot, Berkshire

Summary

The mode of production of the female sexual morphs in aphid species shows some striking differences. In the non-host-alternating (monoecious) vetch aphid *Megoura viciae* the egg-laying oviparae are produced directly by parthenogenetic females (virginoparae) which can be either apterous or alate. The maternal photoperiodic response induces the embryos to develop as virginoparae under long-day conditions or as oviparae in short days. In the heteroecious (host-alternating) black bean aphid, *Aphis fabae*, the production of oviparae requires the interpolation of an additional morph, the winged gynopara. In this instance an alate morph is obligatory in the production of the ovipara although this alata can arise either from short-day exposure or from first instar crowding.

In both species there is evidence that the photoperiodic response is mediated by the endocrine system. Cautery and microillumination experiments, using *Megoura*, have led to the suggestion that the light receptor and photoperiodic clock are located in those areas of the brain (protocerebrum) lateral to a prominent medial group of neurosecretory (group I) cells. These group I cells may be connected synaptically to the neural clock, receiving and accumulating its daily output. Cautery of the group I cells eliminates the long-day response, all embryos subsequently developing into oviparae. The neurosecretory product therefore appears to act, either directly or indirectly, as a virginopara-inducing agent.

Juvenile hormone (JH) is also involved in aphid polymorphism, although its role varies. Topical application of JH I or kinoprene causes juvenilizing effects, the third instar larva being particularly sensitive; disturbances in metamorphosis, however, are less marked and may indeed be absent if first or fourth instar larvae are treated, so that JH could, in theory, be utilized as a morph-controlling agent. In *Megoura*, topical applications of JH have virtually no effect on long-day virginoparae whether they are crowded alata producers or isolated aptera producers. However, when fourth instar or adult short-day ovipara producers are so treated, groups of embryos develop as near-perfect alate virginoparae or as ovipara/alate virginopara intermediates, often with mixed ovaries. Nevertheless, it seems doubtful whether these events mirror the natural control of polymorphism, as the induced alatae originate from somewhat younger embryos than those normally responsive to a maternal crowding stimulus. In addition, no typical apterae are produced. Applica-

Biological Clocks in Seasonal Reproductive Cycles 125–135 (1981) (ed. B. K. and D. E. Follett: Bristol, Wright).

tions of JH to first instar larvae cause some maternal juvenilization without affecting embryonic morph differentiation. It seems, therefore, that the neuroendocrine component and not JH may have a direct effect on virginopara determination, although it is likely that JH has a role in the promotion of embryogenesis. *A. fabae* differs from *Megoura* in that the presumptive gynopara responds directly to photoperiod during late embryonic development and this sensitivity extends into the post-natal period. The action of topically applied JH I on first instar presumptive gynoparae precisely mimics the action of LD in that the gynopara develops into a normal embryo-producing apterous vivapara. In this species therefore the terminal effector is probably JH.

INTRODUCTION

The sexual morphs, the oviparae and males, of most aphid species from temperate climates are produced in late summer and autumn and are preceded during the summer months by a series of parthenogenetic generations consisting of apterous or alate viviparous females (virginoparae). The fertilized eggs laid by the oviparae overwinter in diapause. This transition from parthenogenetic to sexual reproduction is usually, although not invariably, controlled by length of day. Such a response requires a photoreceptor and clock to measure the length of day or night and an endocrine link that communicates this signal to the target organs—in this case the oocytes and embryos developing in the abdomen of the parthenogenetic virginoparae.

Evidence relating to the mediating role of the endocrine system in aphid photoperiodic responses is still scarce and depends on a limited range of techniques. The problem is also complicated by the variety of routes by which different aphid species can achieve sexuality, sometimes involving a succession of morphs each with its own intrinsic capacities to respond to the photoperiodic cue. We wish to refer particularly to two species with very different life cycles. The first, the vetch aphid *Megoura viciae*, can complete its life cycle on a single perennial host plant (*Lathyrus pratensis*). In conformity with this simple pattern (monoecy) the oviparae are produced directly by the virginoparae under short-day conditions. The parental virginoparae can be either apterous or alate, although the latter require lower temperatures to produce a full complement of oviparous daughters. The apterous parent has been studied more thoroughly. In the second species, the black bean aphid *Aphis fabae*, short-day exposure during the development of the aptera leads to the production of an additional morph, the alate gynopara, which in turn gives birth to the oviparae after migrating to the primary host (the spindle *Euonymus europaeus*) in early autumn (heteroecy). This more elaborate train of events can, however, be considerably simplified and shortened if the insects are crowded together under short-day conditions. Whereas uncrowded short-day insects develop into apterous gynopara producers, crowded short-day aphids become winged gynoparae. Unlike *Megoura*, an alate morph is obligatory in the production of the ovipara but it is immaterial whether the alata is produced by short-day conditions or by crowding.

The normal crowding response, which can be studied under long-day conditions, has no obvious physiological connection with the photoperiodic system in *Megoura* and depends on the tactile stimulation received by the mother when jostled by other aphids. *A. fabae* differs from this condition in that sensitivity to the crowding effect is centred on the first instar larva and, as already mentioned, there is a more intimate connection with the photoperiodic response.

Male production is highly dependent on photoperiod in most species, although *Megoura* is an exception. In many monoecious species such as *Acyrthosiphon pisum*, the pea aphid, males are derived from the same virginopara that produces oviparae under short-day conditions but are born towards the end of her reproductive life. In the host-alternating species *A. fabae* and *Myzus persicae* late-born males appear in the same generation as the gynoparae. Male production is probably under hormonal control but the evidence on this point has only recently begun to accrue (*see below*).

TIME COURSE OF EMBRYONIC DEVELOPMENT

Since photoperiod eventually influences the egg-producing germaria or the developing embryos, some reference to the chronology of embryogenesis is desirable. In the apterous virginopara of *Megoura* the abdomen of the adult is packed with 16–18 ovarioles, each containing 6 or 7 embryos of gradually increasing maturity. At this stage the production of young embryos by the germarium is almost complete. Proceeding backwards in time, the fourth instar aphid has ovarioles with 4 or 5 embryos while the third has 4 and the neonate first instar, 2. Embryogenesis therefore begins during the prenatal period when the mother is herself an embryo within the abdomen of the grandparent. An inspection of third or early fourth instar grandparental aphids shows that under long-day conditions embryogenesis is just beginning in the largest embryos present at this time in the five-membered embryo chain. In the adult this growth stage is attained by the penultimate embryo in the chain. This is depicted semi-diagrammatically in figure 1.

This long period of gestation suggests that embryos may be particularly accessible to maternal environmental cues and may indeed be influenced by a sequence of stimuli, hormonal or otherwise, which direct development towards the appropriate developmental pathway. The chromosomal sex-determining mechanism, which is of the XX/XO type in aphids, provides the first point of control. In *Acyrthosiphon*, male production can first be 'switched on' by prenatal short-day treatments when the grandmother is in the third or fourth instar. According to the chronology outlined above, the largest embryos (representing the future maternal generation) will already have one or two embryos in each ovariole, but these are obligatorily female and cannot be influenced by earlier short photoperiods.[1] The remaining oocytes can develop either as XX or XO parthenogenetic eggs. Many are ovulated post-natally and male production can in fact be 'turned on' up to the time the mother is in her third larval instar, although the number of females progressively increases at the expense of the male progeny. During the post-natal period of development, female embryos become determined either as oviparae or as virginoparae, depending on the photoperiod. Switching experiments made by transferring aphids from long days to short days and *vice versa* provide an approximate guide to the embryonic stage when this occurs in *Megoura*.[2] Thus it can be shown that the two embryos already formed in the first instar are not yet determined as either morph. The progeny records further show that the two largest embryos in the chain of 7 in the adult are irrevocably committed to differentiating as oviparae. It seems, therefore, that medium-sized embryos occupying positions

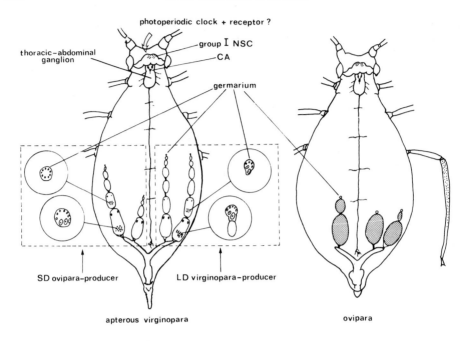

Figure 1. Representation of the adult apterous virginopara of *Megoura viciae* (*left*) and the ovipara (*right*), showing some of the internal structures. The ovarioles of the virginopara, many of which have been removed, contain chains of developing embryos. When reared in long-day conditions (*right*) the largest embryos contain functional germaria which have already produced one or two embryos (encircled). Reared in short-day conditions (*left*) the embryos are destined to become oviparae, the germarium producing yolky haploid eggs. The latter are featured in the adult ovipara on the right. Note also the numerous pheromone-producing scent plaques on the hind tibia (NSC, neurosecretory cells; CA, corpus allatum).

two to five must be the target of the photoperiodic control system. The mother herself develops as an externally normal virginopara in short-day conditions. It is noticeable, however, that in adult 'ovipara producers' the production of the yolky haploid eggs by the ovipara-generation embryos is considerably delayed in comparison with embryogenesis in the corresponding long-day 'virginopara producer' (figure 1). In *Megoura* and *Acyrthosiphon* fully grown embryos in position seven in the ovarioles of long-day virginopara-producers can respond to crowding or isolation by developing into alatae or apterae. This response appears to be unconnected with the chronologically earlier photoperiodic response.

By making use of microilluminators of various kinds it is possible to expose the head and abdomen of the aphid to different photoperiods. This has proved helpful in deciding whether the photoperiodic response is controlled maternally or is the result of light reaching the embryos through the translucent cuticle of the maternal abdomen. Since aphid generations overlap, this method can be used for charting the duration of the photoperiodic response within the life of the individual. In *Megoura*, for example, long-day stimulation of the head induces uncommitted embryos to develop as virginoparae whereas direct stimulation of the embryos in the abdomen does not. On the other hand, pre-natal photostimulation of the abdomen can be shown to influence the initiation of the photoperiodic response in late embryos. It seems from such experiments that the photoperiodic mechanism in *Megoura* is continuously active from 2 to 4 days before the birth of the mother to at least half-way through adult reproductive life, by which time all the embryos are determined. During the adult stage the photoperiodic systems of the advanced embryos and of the adult are functioning simultaneously and presumably independently.

In *A. fabae* the photoperiodic sensitivities of the mother and her embryonic progeny are patterned differently in relation to the generation succession and the individual's own ontogeny. When reared uncrowded in short-day conditions *A. fabae* develops as a 'gynopara producer' externally indistinguishable from a long-day aptera. However, a maternal photoperiodic effect presumably biases the embryos in favour of the gynoparous condition. At a later point in development advanced embryos in the abdomen of the gynopara producer can be prevented from differentiating into gynoparae by long-day stimulation of the abdomen of the gynopara producer. This is certainly a direct, non-maternal effect of photoperiod on the embryos ,as in this species photoperiodic sensitivity extends into the post-natal period when first instar larvae (presumptive gynoparae) can be completely apterized by long-day exposure.[3] In this instance photoperiod influences both the external morphology and the embryos which proceed to develop as virginoparae rather than as oviparae. Later on in the middle instars the gynopara can no longer be changed externally by long days but can still be switched over to the production of virginoparous progeny.[4]

ROLE OF THE NEUROENDOCRINE SYSTEM IN *MEGOURA*

Maternal photoperiodic control of virginopara and ovipara production has been examined in adult apterae of *Megoura* in experiments designed to locate the photoreceptors and any possible neuroendocrine effectors. Two techniques were used: first, localized long-day stimulation was applied to small areas of the head; secondly, lesions were made in the brain and elsewhere by microcautery. Observation of the progeny sequences then showed whether the aphid reared to the adult stage in short days could still respond to long days by switching over from ovipara to virginopara production. Results showed that long-day stimulation of the protocerebrum was indeed essential for initiating this switch. Although accurate localization was prevented by light scattering in the cuticle and brain tissue, photoperiodic sensitivity was found to be centred in the mid-dorsal and antero-lateral areas of the protocerebrum. Light microscopy revealed five groups of neurosecretory cells (NSC) in this region, of which one, the group I NSC, was particularly prominent (figure 2).[5] Cautery of the group I cells in the pars intercerebralis area resulted in complete failure to switch to virginoparae in response to the change in photoperiod, although cytological examination of the sectioned brains showed that as long as some group I cells remained the aphids retained the capacity to respond.[6] The response remained intact after elimination of other NSC cell groups. However, failure to respond was noted if the lesions invaded the areas lateral to the group I cells, even though the latter were judged by cytological standards to be undamaged. From this evidence we concluded that the light receptor (and probably the photoperiodic clock) was located in this lateral area but that the group I cells were the endocrine effectors. In other experiments aphids were reared without change of photoperiod. It was then found that after elimination of group I cells by cautery, short-day insects continued to produce oviparae, whereas long-day insects reverted spontaneously to ovipara production.

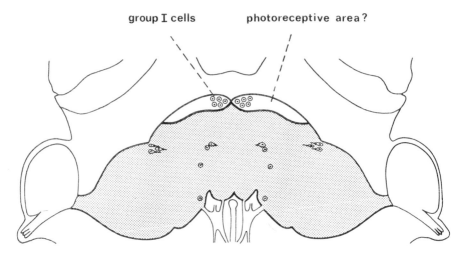

Figure 2. Brain of *Megoura viciae*, viewed from the dorsal aspect, showing the arrangement of the different groups of neurosecretory cells.

There seemed little doubt, therefore, that embryos differentiate as virginoparae in the presence of this endocrine factor and as oviparae in its absence.

Further evidence on the photoperiodic receptor/effector system is unfortunately meagre. Extensive studies have shown that the clock can be regarded as an hour-glass which 'measures' night length.[7] Although day length is not critical in the same sense, the scotophase is, of course, defined by light which therefore performs an essential role in the functioning of the clock. The action spectra with a predominant blue sensitivity indicate that a yellow absorbing pigment, possibly a caroteno-protein,[8] is involved in photoreception. However, no pigmentation has yet been detected in the putative neuronal receptor area of the brain, either by visual or by microspectrophotometric means. A very low pigment concentration is therefore presumed. Electron microscopy has not yet revealed any organized photoreceptor structure or indeed any 'unusual' neurones in the suspected area.

Examination of the group I cells by PAF staining has shown that the neurosecretory material is stored principally in the swollen axonal reservoirs, with very little neurosecretory material remaining in the perikarya in either long-day or short-day conditions. The axons of the group I cells run ventrally through the brain and suboesophageal ganglion and project to the fused thoracic–abdominal ganglion. The release points of the neurosecretion have not yet been found. The group I axons also branch near the perikarya, these collaterals apparently ending in the neuropile near the central body. Such branches may represent the points of synaptic contact between the group I cells and the probable neuronal area of the photoperiodic clock.

These anatomical considerations suggest that the group I cells receive the daily output of the clock and may indeed be responsible for the process of cycle summation since in aphids, as in other insects, the effect of long-day stimulation is highly dependent on the number of daily cycles (scotophases) experienced. The precise role of the neurosecretory cells has not yet been resolved. If the neurosecretory product of the group I cells is acting directly as a virginopara-determinant, it might reach the embryos by some form of 'directed delivery', via the thoracic ganglion[5] or it might be released into the blood as a neurohormone. Alternatively, there is the possibility that the other components of the endocrine system, such as the corpus allatum, are implicated. The possible involvement of juvenile hormone (JH) in photoperiodic control is discussed below.

MORPHOGENETIC EFFECTS OF JUVENILE HORMONE

Studies in which juvenile hormone (JH) or JH analogues were applied topically to the cuticle of the developing aphid have shown that JH is concerned both in the control of metamorphosis, as in other insects, and in the regulation of polymorphism, including morph changes induced by photoperiod.

When applied to the middle (second and third) instars JH causes strong juvenilizing effects which are recognizable in the fifth (normally adult) instar. Larval features, which are most conspicuous in the alate morphs, are dose-dependent and can serve as an assay.[9] Slight effects are shown mainly by the abnormal crumpling of the wings, by the weakened marginal pigment spots in *Megoura* and by the retention of abdominal wax patches in *A. fabae*. In more severely affected insects the thorax is less sclerotized and the sutures partially obliterated. In highly affected insects the pterothorax is completely unsclerotized, the wings take the form of larval wing pads and the dorsal ocelli and genital pore are lacking. These individuals invariably undergo a supernumerary ecdysis. Some characters, particularly those based on negative features such as the absence of abdominal pigment spots or of ocelli, are typical both of larvae and of the adult aptera. However, a comparison of the wing characters in the complete series indicates clearly that these features are indeed larval and are not the result of an apterizing process.

Dose–response curves for alatae developing in long-day conditions show that the third (penultimate) larval instar is the most sensitive to juvenilization, approximately 0·2 ng JH I or the mimic kinoprene (ZR–777), eliciting a response of about half the maximum 'morphological effect' in *Megoura*. Responsiveness is first apparent during the pre-natal period and increases rapidly during the second post-natal instar (figure 3). Aphids are unusual, however, in that the last larval instar (the fourth) is almost totally insensitive to the juvenilizing effect of JH. Evidently, adult characters are determined by the absence of endogenous JH during the third instar. It is tempting to infer also that this exceptional state of affairs is an adaptation which permits JH to be utilized in the fourth instar as a morph-controlling agent without disturbing the metamorphosis of the mother.

Morph-controlling effects of juvenile hormone

Mittler et al.[10] have recently reported that kinoprene considerably reduces the production of males by apterous virginoparae of *Myzus persicae* under short-day conditions. It seems that the analogue either directly or indirectly interferes with the chromosomal sex-determining mechanism, increasing the ratio of XX to XO eggs. Applied JH, however, also has the ability to affect the post-natal differentiation of female aphids at points in their development when there is a diminished sensitivity to the juvenilizing effects of JH. This is notably the case in the fourth instar.

Applied JH has little or no influence on the progeny of fourth instar alate or

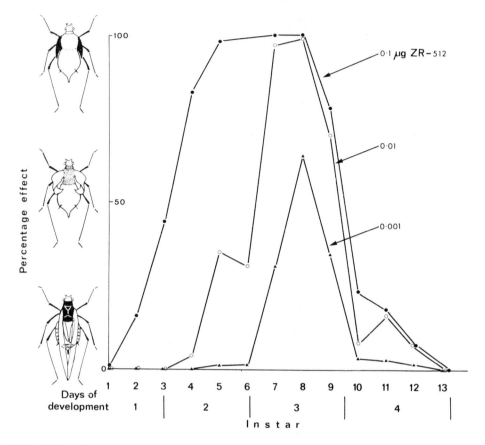

Figure 3. Curves illustrating the development of sensitivity in *Megoura* alatae to the juvenilizing effect of the juvenile hormone (JH) analogue hydroprene (ZR–512) when applied topically at three different dosages. In these experiments ZR–512 was some 30 per cent less active than the analogue kinoprene (ZR–777) or JH I. The juvenilized forms scoring 50 and 100 per cent 'morphological effect' are indicated alongside the ordinate.

apterous progeny of *Megoura* reared in long-day conditions. The maternal response to crowding in long-day conditions (production of alatae) is also unaffected as is the response to isolation (production of apterae). The effect in question is seen in short-day ovipara-producing aphids which in *Myzus persicae* and *A. fabae* are fourth instar alate gynoparae and in *Megoura* are apterous virginoparae reared from birth in short-day conditions.[11,12] In both instances a small group of embryos (about 10–20) is induced to develop as alate virginoparae or as intermediates between oviparae and alatae. These intermorphs can be arranged in a graded series (figure 4). The least affected aphids are wingless and ovipara-like but may show some thoracic sclerotization. Intermediate grades are characterized by small rudimentary wings or larger wings that are severely twisted and blistered. The most extreme forms are near-perfect alatae with fully formed but sometimes foreshortened wings and a complete abdominal pattern. These alatae may, however, still retain a few pheromone-producing scent plaques on the hind tibiae, an oviparous characteristic. No typical apterous virginoparae are produced by this treatment. This has appeared as something of a paradox since JH has frequently

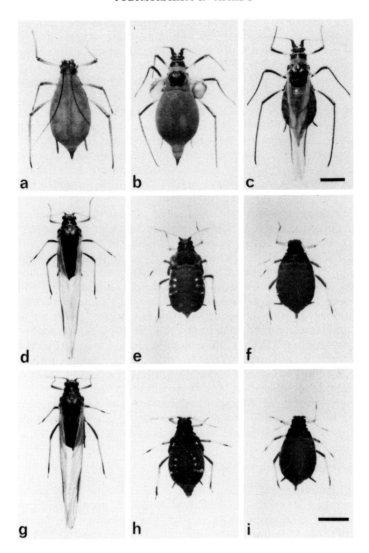

Figure 4. *a–c*, Adults of *Megoura viciae* showing the effect on their progeny of applying 0·1 μg JH I to fourth instar ovipara producers reared throughout in short days. *a*, Normal ovipara; *b*, ovipara/alate virginopara intermediate with rudimentary wings and partially sclerotized pterothorax; *c*, near-perfect alate virginopara exhibiting maximum effect. *d–f, Aphis fabae*. Series of adults showing apterizing effect produced by exposing first instar gynoparae to long-day treatment during development. *d*, Normal gynopara control reared throughout in short days; *e*, intermediate adult with small wing pads; *f*, fully apterized virginopara. *g–i*, A comparable series of *A. fabae* showing an identical apterizing effect produced under short-day conditions by applying 0·1 μg JH I to first instar gynoparae. *g*, Normal control gynopara, no JH; *h*, intermediate gynopara/apterous vivapara; *i*, fully transformed aptera. Scale bars 1·0 mm.

been regarded as an apterizing agent, irrespective of whether the alate morph has arisen from a crowding stimulus or short-day exposure (*see below*).

Juvenile hormone also has a dramatic influence on the maturation of the female eggs. In low-grade intermediates ovipara-like mixed ovaries are common, some ovarioles containing haploid yolky eggs, as in the normal ovipara, while others contain a small number of embryos. In still other instances the production of viable eggs is so disrupted that the germarium fails to generate either eggs or embryos and the maternal abdomen is virtually empty. Insects which are near-perfect alatae usually possess ovarioles with a full complement of embryos. The pregnant females are, however, invariably infertile. Although the embryos continue to grow and distend the abdomen, parturition fails to take place.

If the induction of alatae by JH were an integral part of the natural photoperiodic response to long-day conditions, it would be necessary to assume that an apterizing mechanism is present which would facultatively convert the incipient alata into an aptera just prior to birth. However, none of the treatments that induce aptera-formation in *Megoura* (isolation, CO_2 anaesthesia) prevents JH-treated fourth instar ovipara producers from yielding alatae. Therefore, either the embryos or the morph-control system is defective.

A further possible reason for the absence of apterae may be that the dosages of JH have been inappropriate. However, repeated smaller doses applied during the fourth instar were no more successful than single large doses. Experiments involving short-day/long-day switching also indicate that embryos become committed to develop as virginoparae when the mother is somewhat younger than the fourth instar. However, the third instar cannot tolerate JH without severe juvenilization and the treatment of earlier instars with moderate JH doses does not initiate virginopara production. It seems from the present evidence that JH is not the sole terminal hormone inducing embryos to develop as virginoparae. If this is a function of the neuroendocrine system and the group I cells, JH secretion may still be required at a slightly later stage to promote parthenogenesis in the target embryos. According to this view the alata-promoting effect of JH may result from the non-specific action of a potent insect growth regulator acting at the wrong stage of development. Finally, it should be emphasized that once the embryos are fully established in long-day conditions as virginoparae, their apterous or alate character does not appear to involve differential corpus allatum activity.

The role of juvenile hormone in *Aphis fabae*

In this heteroecious species the fourth instar ovipara producers (gynoparae) respond to applied JH in the same way as *Megoura*, producing a group of alatae and ovipara/alata intermediates. Developing gynoparae also exhibit the same response profile as *Megoura* alatae with regard to the juvenilizing effects of JH. However, there is evidence that JH plays a much more significant role in the mediation of photoperiodic response. We have seen that post-natal photoperiodic sensitivity is virtually confined to the first instar. When first instar gynoparae receive a sufficient dose of JH I in short-day conditions they develop into normal parthenogenetic apterae, thus mimicking the action of long days[13] (figure 4 *d–i*). Gynopara/aptera intermediates are also found which are identical to those induced by long-day exposure (figure 4 *e,h*). Their distinctive morphology differs entirely from that of juvenilized larval/adult intermediates. This evidence strongly suggests that in the gynopara of *A. fabae* the photoperiodic clock controls larval development through the agency of the corpus allatum, influencing both the external form of the larva and the type of gonads. After the first instar, the external form is no longer responsive to long days, although the germaria can still be induced to form parthenogenetic embryos rather than haploid eggs.[4] This process can again be imitated by topical applications of JH I.[14]

It is also of great interest that crowded presumptive gynoparae reared in short-day conditions do not respond like the isolated first instar gynopara to the apterizing effect of topically applied JH.[13] It seems that there must be an additional unknown hormone responsible for controlling alata or aptera differentiation in long-day conditions.

We conclude that the endocrine systems mediating the photoperiodic responses of aphids have unexpected complexities which differ radically in the two species we have studied.

REFERENCES

1. R. L. Blackman. *Int. J. Morphol. Embryol.* **7,** 33–44 (1978).
2. A. D. Lees. *J. Insect Physiol.* **3,** 93–117 (1959).
3. A. D. Lees. *Nature (Lond.)* **267,** 46–8 (1977).
4. J. Hardie. *Physiol. Ent.* **5,** 385–96 (1980).
5. C. G. H. Steel. *Gen. Comp. Endocrinol.* **31,** 307–22 (1977).
6. C. G. H. Steel and A. D. Lees. *J. Exp. Biol.* **67,** 117–35 (1977).
7. A. D. Lees. *J. Insect Physiol.* **19,** 2279–316 (1973).
8. A. Veerman and W. Helle. *Nature (Lond.)* **275,** 234 (1978).
9. A. D. Lees. *J. Insect Physiol.* **26,** 143–51 (1980).
10. T. E. Mittler, J. Eisenbach, J. B. Searle, M. Matsuka and S. G. Nassar. *J. Insect Physiol.* **25,** 219–26 (1979).
11. T. E. Mittler, S. G. Nassar and G. B. Staal. *J. Insect Physiol.* **22,** 1717–25 (1976).
12. A. D. Lees. In: P. J. Gaillard and H. H. Boer (ed.) *Comparative Endocrinology.* Amsterdam: Elsevier/North-Holland, pp. 165–8 (1978).
13. J. Hardie. *Nature (Lond.)* **286,** 602–4 (1980).
14. J. Hardie. *J. Insect. Physiol.* (in the press) (1981).

CIRCADIAN CLOCKS IN LIZARDS: PHOTORECEPTION, PHYSIOLOGY AND PHOTOPERIODIC TIME MEASUREMENT

Herbert Underwood

Department of Zoology, North Carolina State University, Raleigh, USA

Summary

Extra-retinal photoreceptors located in the brain are involved both in entrainment of circadian activity rhythms to 24 hour light/dark cycles and in photoperiodic photoreception in lizards. The lateral eyes participate in entrainment but not in photoperiodic photoreception.

Removal of the pineal organs of iguanid lizards free-running in continuous light (LL) or continuous darkness (DD) causes splitting of the rhythm, changes in its period (τ) or arrhythmicity. Exogenous administration of melatonin to intact, blinded, pinealectomized, or blinded–pinealectomized lizards causes a significant lengthening of the period of the activity rhythm. Blinding can also produce changes in τ or arrhythmicity. The data show that the lizard's circadian system is a multi-oscillator system and the pineal (and eyes) act to preserve the integrity of this system, possibly through the release of melatonin.

In contrast to the higher vertebrates, the photoperiodic time-measuring system in the male lizard *Anolis carolinensis* seems to rely on an hour-glass timer which lacks endogenous rhythmicity. Studies on the nature of the time-measuring system in *Anolis* have shown:

1. Resonance (Nanda–Hamner) and *T* lighting régimes fail to demonstrate a photo-inducible phase characteristic of organisms utilizing a circadian rhythm of photosensitivity to measure photoperiodic time.

2. Comparison of entrained activity patterns with testicular responses to various light cycles also argues against the participation of a circadian clock in photoperiodic time measurement.

3. Exposure to as few as six long days can elicit a maximal testicular response.

4. The hour-glass timer measures the length of the light portion of LD cycles.

5. The timer does not have to be 'reset' daily by a dark period.

6. The dark period of a LD cycle may be involved in a complex fashion in reversing a light-initiated reaction.

INTRODUCTION

Within the past few decades a considerable amount of data has been accumulated on the existence of daily rhythms and annual reproductive cycles in poikilotherms,

Biological Clocks in Seasonal Reproductive Cycles 137–152 (1981) (ed. B. K. and D. E. Follett: Bristol, Wright).

but only recently have advances been made into the physiological mechanisms involved.[1] The present discussion will focus primarily on the physiology of circadian pacemakers in lizards and on the nature of their photoperiodic time-measuring system. Since the pineal organ is involved in both the circadian and photoperiodic systems of lizards a brief discussion of the structure, evolution and physiology of the pineal system is included.

All pineal organs are derived embryologically as evaginations from the roof of the diencephalon.[2] Although birds and mammals possess only a pineal organ, the pineal 'system' of lower vertebrates is more complex, often consisting of two distinct parts: an intracranial pineal organ which remains attached to the roof of the diencephalon, and a more superficial parapineal organ (called the parietal eye in lizards) which originates either as an outpouching from the pineal organ or as a separate diverticulum from the diencephalon. The pineal organs of birds and mammals are glandular in appearance whereas the pineal organs of fish, amphibians and reptiles possess sensory cells with a characteristic photoreceptive appearance.[2] The parapineal organs of lower vertebrates also are photoreceptive: the lizard's parietal eye is highly eyelike in morphology and contains well-organized photoreceptors organized in a retina, as well as a cornea and lens.[3] Nerve cells are present which show synaptic contacts with the photoreceptors and send axons to the rest of the brain. The pineal and parapineal organs of lower vertebrates have also been shown to be photosensitive on the basis of neurophysiological evidence.[4,5] In general, pineal organs show achromatic responses to light whereas parapineal organs show chromatic responses. Major differences also exist in the innervation of pineal organs—in lower vertebrates pinealo-fugal (afferent) nerves are present which project centrally whereas in the higher vertebrates pinealo-fugal nerves are less common. In the lower vertebrates the presence of pinealo-petal (efferent) innervation has not been definitely established but pinealo-petal innervation is common (if not universal) in higher vertebrates. The general tendency for pineal evolution, therefore, seems to be a shift from that of a photosensory organ which sends photic information to the brain to a glandular organ in higher vertebrates. However, this distinction is greatly blurred in that the pineal organs of lower vertebrates clearly possess a great deal of secretory potential while the pineal organs of birds and juvenile mammals show evidence of photoreceptive abilities.[6-9]

In 1958 Lerner and his co-workers isolated a substance from bovine pineals which aggregates pigment granules in amphibian melanophores and identified it as 5-methoxy-N-acetyltryptamine (melatonin).[10] Subsequently, a number of studies elucidated the biosynthetic pathways for melatonin and related indoleamines.[11,12] The basic pathway is as follows: tryptophan → 5-hydroxytryptophan → serotonin → N-acetylserotonin → melatonin. Melatonin, in particular, excited much interest because it was thought to be unique to the pineal organ and could be detected in the plasma, although subsequently several other areas have been shown to possess melatonin-synthesizing capacities, such as the retina and the Harderian gland.[13] A notable feature of pineal biochemistry is the daily variation in the amounts and activities of various substrates and enzymes.[11,12] While most investigations into pineal biochemical rhythmicity have used mammals or birds, a number have demonstrated such rhythms in lower vertebrates as well.[14-16] Although daily rhythms in mammalian pineals are dependent upon intact sympathetic innervation, several laboratories have shown that the rhythm in pineal N-acetyltransferase activity in avian pineals can persist for at least a few cycles in organ-cultured pineals.[17,18]

PHOTORECEPTION

All eukaryotic organisms examined to date exhibit entrainment of circadian rhythms to 24 hour light/dark cycles. Among non-mammalian vertebrates the perception of entraining light cycles does not necessarily involve the eyes, entrainment persisting in blinded animals.[9] Among lizards, eight different species (*Sceloporus olivaceus, S. magister, S. clarkii, Anolis carolinensis, Lacerta sicula, Coleonyx variegatus, Hemidactylus turcicus, Xantusia vigilis*) representing four different families (*Iguanidae, Lacertiidae, Gekkonidae, Xantusiidae*) have been assayed for their ability to entrain to LD 12 : 12 fluorescent light cycles (of 30–50 lux intensity) after blinding.[19,20] All individuals tested to date readily entrained to the light cycle after blinding, although, in some cases, blinding altered the pattern of the entrained activity rhythm. Removal of the pineal system of blinded lizards does not abolish entrainment.[19] These data clearly show that lizards possess extra-retinal receptors which are fully capable of mediating entrainment and, further, that the light-sensitive pineal system is not necessary for entrainment. It must be emphasized, however, that the data do not exclude the possibility that the pineal system (parietal eye and/or pineal) may be an alternative route of photoreception. Localization experiments have shown that the brain is the site of this extra-retinal photoreception but the exact loci of the photoreceptors within the brain is, as yet, unknown.[20] The extra-retinal receptors are remarkably sensitive; for example, approximately half (8 of 15) of the *S. olivaceus* tested could be entrained to LD 12 : 12 (1·0 lux : 0) electroluminescent light cycles even after removal of the lateral eyes, parietal eye and pineal organ.[19] The lateral eyes can also participate in entrainment since removal of the lateral eyes of *S. olivaceus* entrained to a dim green electroluminescent LD 12 : 12 (0·05 lux : 0) light cycle causes them to free-run (exhibit their endogenous circadian periodicity).[19]

Extra-retinal photoreceptors are involved in photoperiodic photoreception as well. Blinded and blinded-parietalectomized male lizards, *A. carolinensis*, will respond to stimulatory long-day photoperiods by exhibiting testicular growth and maturation.[21] Blocking light penetration to the brains of sighted anoles by painting the heads (but leaving the eyes exposed) abolishes the stimulatory effects of long days (figure 1). These results show that extra-retinal receptors located in the brain mediate the photoperiodic response in *Anolis* and that the eyes are not involved. At present the relationship between the extra-retinal receptors mediating entrainment and those mediating photoperiodic photoreception is unknown.

PHYSIOLOGY OF THE CIRCADIAN SYSTEM

It is becoming increasingly apparent that the circadian system of vertebrates is a 'multi-oscillator' system. The most compelling evidence of the existence of multiple circadian clocks in vertebrates is the 'splitting' of free-running activity rhythms into two distinct components which sometimes occurs when animals are maintained under conditions of continuous illumination for a number of days. Typically, the split components free-run with different periods until they are approximately 180° antiphase to one another and then lock on to a stable condition. Such behaviour has been noted in hamsters, tree shrews and starlings.[22-24] There is also compelling evidence that the lizard's circadian system is a multi-oscillator system.[25] Figure 2 shows the effects of removing the pineal system on the activity rhythm of an iguanid

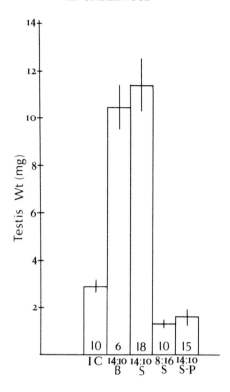

Figure 1. Average testis weight of the left testis (±1 s.e.) of anoles exposed to several experimental treatments for 21 days in the autumn. I.C. denotes initial controls killed at the beginning of the experiment, B (blinded), S (sighted), S-P (heads painted with a mixture of collodion and carbon black leaving the eyes exposed). The numbers within the bars show sample size.

lizard *Sceloporus olivaceus* free-running in LL. The rhythm clearly splits into two circadian components which free-run with different frequencies. Interestingly, the components do not lock on at 180° antiphase but continue to free-run and cross several times. Removal of the pineal organs of a number of *S. olivaceus* and *S. occidentalis* under conditions of both LL and DD produces a variety of effects—splitting, changes in τ and arrhythmicity. Removal of the parietal eye alone is without effect.[25]

These data strongly support the notion that the lizard is a multi-oscillator system and that the pineal organ acts to keep together the component oscillators comprising the system. The variety of effects seen after pinealectomy represent the varying degrees of integrity of the remaining component oscillators; in some cases the component oscillators break apart and free-run with their own frequencies, producing splitting or arrhythmicity, whereas in other cases the remaining component oscillators still remain coupled together but the strength of the coupling between them is altered so producing a change in the period of the coupled system. The exact function of the pineal organ in such a system is still uncertain. The pineal could be acting as a master oscillator in a hierarchical system which unilaterally entrains other subordinate oscillators located elsewhere, or the pineal may lack endogenous rhythmicity itself but act as a coupling device between mutually interacting oscillators located elsewhere.

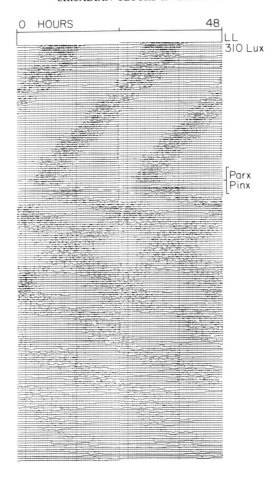

0 HOURS   48

LL
310 Lux

[Parx
[Pinx

Figure 2. Splitting of the activity rhythm of a *S. olivaceus* into two circadian components after removal of the pineal system.

The lateral eyes are also an important component of the circadian system in lizards. They clearly have a role in mediating entrainment of the circadian system to 24 hour light/dark cycles but they also may play an additional role.[19,20] Removal of the lateral eyes of *S. olivaceus* free-running in LL can produce a shortening in τ or arrhythmicity.[20] These results are not interpretable on the basis that removal of the eyes has decreased the amount of light reaching the circadian clock(s) and suggest that the eyes may function as coupling devices or, perhaps, as the loci of circadian oscillators. Potentially, either neural or hormonal routes may be involved in transmitting information between the pineal or eyes and the rest of the circadian system. Recently, a number of studies have demonstrated a direct retino-hypothalamic neural pathway which terminates in the suprachiasmatic nuclei (SCN) in a variety of vertebrates, including lizards.[26] Ablation of the SCN in birds and rodents can abolish circadian rhythmicity, suggesting that the SCN may be the location of a circadian clock(s).[1] Significantly, in the rat, a circadian rhythm in multiple unit activity persists in the SCN when it is isolated from the rest of the brain, strongly suggesting that the SCN is an autonomous circadian pacemaker.[27] In view of these data, it is possible that the SCN is the location of a circadian clock or clocks in lizards as well and the retino-hypothalamic pathway may be used to transmit photic information to the SCN.

The pineal or the eyes may have an endocrine function within the circadian system. The indoleamine synthesizing ability of both the pineal and eyes has excited a great deal of interest. In the lizard *Lacerta muralis* the pineal exhibits a circadian rhythm in serotonin content and the pineal of the turtle *Testudo hermani* shows a daily rhythm in melatonin content, with peak melatonin levels at night.[15,28] In the scincid lizard *Trachydosaurus rugosus* plasma melatonin levels show a daily rhythm, with peak levels occurring at night.[16] Both the pineal organ and lateral eyes of the scincid lizard *Lampropholas guichenoti* exhibit a daily rhythm, under LD cycles, in hydroxyindole-O-methyltransferase activity.[14] Ocular rhythms have been noted in other vertebrates as well; ocular N-acetyltransferase activity shows a higher activity during the dark (on LD 12 : 12) in the chick, the house sparrow and the rat.[29]

Exogenous administration of melatonin to both birds and lizards, *via* capsules which release melatonin at constant rates, causes changes in the period of activity rhythms or arrhythmicity in animals maintained under constant conditions.[30,31] Also, daily injections of melatonin can entrain the activity rhythms of pinealectomized starlings.[32] In the lizard *S. occidentalis* exogenous melatonin consistently lengthens the period of the activity rhythm in intact, as well as in pinealectomized and/or blinded, lizards showing that melatonin exerts its action at extra-pineal and extra-ocular sites. In view of these data it seems likely that melatonin may be a hormone which links the pineal (or eyes) to the rest of the circadian system.

PHOTOPERIODIC SYSTEMS

Many vertebrates use the annual change in daylength to time such important behavioural and physiological events as migration, fattening, moulting and reproduction. The adaptive significance of such photoperiodic responses is obvious— animals can anticipate and prepare for adverse conditions and can confine reproduction to the time of year most conducive to the survival of the individual and its offspring. The use of the annual change in daylength to time these events is not surprising since it is the most consistent and 'noise-free' stimulus available in the environment.

Relatively little is known about the concrete mechanisms involved in photoperiodic time measurement. Most of our knowledge concerning the physiology of vertebrate photoperiodic systems has been obtained from studies with mammals, mainly the golden hamster.[11,12] These studies have shown that the pineal and the suprachiasmatic nuclei are importantly involved. The mammalian pineal organ appears to secrete antigonadal substances since pinealectomy tends to produce progonadal effects. The SCN, which receives photic information from the eyes *via* the retinohypothalamic pathway, controls daily rhythmicity in a host of functions including pineal indoleamine metabolism.[1] Melatonin is a probable antigonadal substance secreted by the pineal although melatonin can have progonadal effects under some conditions.[12] Partially purified pineal peptides have also been implicated as possible antigonadal substances.[11,12] In birds reports on the endocrine function of the pineal are not extensive and are sometimes contradictory. However, effects of pinealectomy and melatonin have been observed in some species.[12,33] Only a few studies have examined the role of the pineal or melatonin in reproduction in reptiles. Levey[34] demonstrated that pinealectomy at certain times of year can cause accelerated ovarian maturation in *A. carolinensis* maintained on LD 6 : 18 and this effect can be reversed by melatonin injections. Haldar and Thapliyal[35] showed that pinealectomy affected the annual testicular cycle of the lizard *Calotes versicolor*; pinealectomy resulted in an earlier recrudescence and a later regression of the testes of lizards exposed to natural conditions. Also, pinealectomy during early summer inhibited the decrease in testicular size of *C. versicolor* exposed to short photoperiods whereas during early winter pinealectomy accelerated testicular growth of lizards subjected to long photoperiods.[36] A preliminary study by Packard and Packard[37] showed that melatonin injections elicited testicular regression in the lizard *Callisaurus draconoides*. These few studies with lizards, therefore, suggest a role of the pineal in reproduction similar to that seen in mammals.

The most thoroughly studied North American lizard with respect to the environmental control of reproductive cycles is the iguanid lizard *Anolis carolinensis*.[38] In the field the testes of *Anolis* regress in late summer and recrudescence begins in October, proceeds gradually throughout the winter and is completed in early spring. Both photoperiod and temperature are involved in the control of this cycle but the importance of the two stimuli varies with the phase of the cycle.[38] Testicular recrudescence (between late autumn and early spring) is controlled mainly by temperature whereas the maintenance and subsequent regression of the testes in late summer are primarily influenced by the photoperiod. However, in the laboratory the testicular response can be manipulated by photoperiod between late June and mid-October. Summer animals brought into the laboratory will maintain testicular size and spermatogenesis when exposed to artificial long days (above about 13·5 h) but will exhibit testicular regression on short days. In animals brought into the laboratory in early autumn (September to October) long days will stimulate increases in testicular size and spermatogenesis.

Recently investigations have been conducted on the formal properties of the photoperiodic time measuring system in *Anolis*. There are two major theories concerning the mechanisms of photoperiodic time measurement. One theory, first formulated by Bünning,[39] suggests that an endogenous circadian rhythm of responsiveness to light lies at the basis of photoperiodic time measurement. If light falls on a photosensitive phase of the rhythm photoperiodic induction occurs, whereas the organism is insensitive to light during the rest of the cycle. Pittendrigh and co-workers[40,41] have emphasized that light must have a dual role in this kind of photoperiodic system. First, the light cycle acts as an entraining agent for the organism's many circadian rhythms, including the circadian rhythm of photoperiodic photosensitivity (CRPP) and, secondly, light may be photoperiodically inductive but only if the organism's CRPP is entrained in such a way that the photosensitive portion of the CRPP is illuminated. This model, which assumes a temporal coincidence between the photosensitive phase of a CRPP and light, has been termed the external coincidence model.[41] A second theory states that photoperiodic time measurement is dependent upon an hour-glass or interval timer which lacks endogenous rhythmicity. According to this hypothesis a reaction product accumulates during the light (or dark) which is inactivated during the other part of the light/dark cycle. If enough reaction product accumulates (i.e., if the light, or dark, periods are of sufficient duration) a photoperiodic response is initiated. A number of studies indicate that higher vertebrates (birds and mammals) use a CRPP to measure photoperiodic time.[42,43] However, similar studies on the lizard *A. carolinensis* suggest that this species uses an hour-glass timer.[44]

To determine whether or not the circadian system is involved in photoperiodic time measurement in *Anolis*, lizards were subjected to various experimental lighting régimes and the testicular response subsequently assayed.[44] The first experiment involved exposing groups of male anoles to one of four different light cycles, consisting of an 8 h photoperiod coupled with durations of darkness ranging from 18 to 54 h (figure 3a). This protocol was first used by Nanda and Hamner[45] on plants and will be referred to as a 'resonance' experiment. If a circadian system is involved, induction should rise and fall as a function of the period of the light cycle—animals exposed to LD 8 : 28 or LD 8 : 52 should show induction since the light would fall on the sensitive portion of a CRPP whereas LD 8 : 16 and LD 8 : 40 should be ineffective. However, none of these cycles was effective in maintaining testicular size and spermatogenesis in *Anolis*. The negative result suggests that an hour-glass mechanism is in operation since, in each cycle, the duration of light (8 h) is less than the critical length (13·5 h) in this species and the duration of darkness is longer than the critical length (10·5 h) below which induction occurs.

A second class of experiments also yielded results compatible with an hour-glass timer in *Anolis*. This protocol exploited the action of light as an entraining agent for the circadian system in *Anolis*. Varying the period (T) of a light cycle varies the position of the light relative to the circadian system; light controls both the period and the phase of endogenous circadian rhythms.[40,41] Therefore, a light pulse can, potentially, act both as an entraining agent and a photoperiodic stimulus. A major assumption in such 'T-experiments' is that all of an organism's many circadian rhythms (including the CRPP) should bear a fixed phase relationship with each other. Therefore, it should be possible to assay one of an organism's overt circadian rhythms, such as locomotor activity, and to assume that the light cycle is affecting the CRPP in the same way as it affects the activity rhythm. Physiologically this must mean that all of an organism's many circadian rhythms are driven by one clock or, more likely, if more than one clock is involved, they are all coupled together or share the same response to light.[25] Even if the above assumption does not hold and the phase relationship between the CRPP and the activity rhythm varies as a function of T the use of a range of T values which encompasses the range of entrainment should yield at least one light cycle that illuminates the sensitive part of the putative CRPP. Although the range of entrainment of the putative CRPP is not known with certainty, it should lie within the maximal ranges observed for most circadian rhythms ($T = 18$–30 h). Figure 3b shows the results of a T-experiment utilizing cycles from $T = 18$ to $T = 28$.[44] None of the T cycles was photoperiodically inductive. Again these results suggest an hour-glass mechanism.

It has been well established that the critical photoperiod for a standard ($T = 24$) light cycle is about 13·5 h (LD 13·5 : 10·5).[38] Accordingly, if an hour-glass timer is involved, and if it is measuring the length of the light period, only cycles with photoperiods $\geqslant 13\cdot5$ h should be inductive. If the dark period is being measured, cycles with dark periods $\leqslant 10\cdot5$ h should be inductive and, if the light/dark ratio is critical, cycles with L/D ratios of 1·29 (13·5/10·5) or greater should be inductive. Which component of the light/dark cycle was being measured was determined by exposing male anoles to LD 16 : 20 (L/D = 0·8).[44] This cycle was inductive indicating that the light component was being measured, if the L/D ratio or the

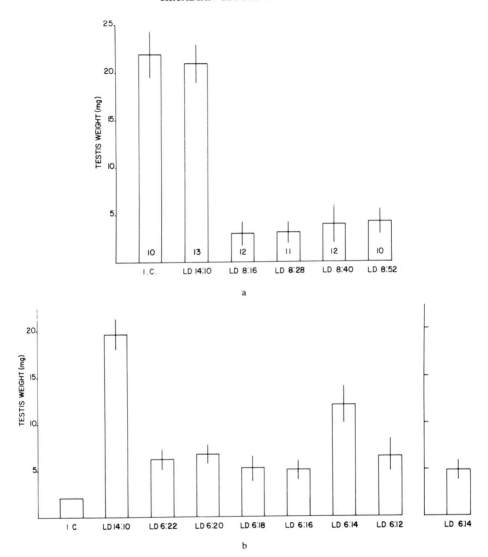

Figure 3. Evidence that a circadian rhythm of photosensitivity is not involved in photoperiodic time measurement in *Anolis*. *a*, Average testis weights (left testis only) of anoles exposed to various 'resonance' light cycles for 25 days in the summer. *b*, Average testis weights (left testis only) of anoles exposed to various *T* cycles for 24 days in the autumn. The apparent testicular growth which occurs on LD 6 : 14 is not believed to be the result of photoperiodic induction[44] and is not a repeatable observation (figure on the far right shows the response to LD 6 : 14 the following autumn).

duration of dark was being measured, no induction should have occurred.[44]

The responsiveness of the *Anolis* photoperiodic system was determined by exposing groups of lizards to only a few long (16 h) photoperiods over a period of several weeks (figure 4). These experiments showed that as few as six long days could stimulate a maximal photoperiodic response and significant testicular growth could be seen in response to as few as three long days.

Additional properties of the photoperiodic time-measuring system in *Anolis* were explored in a further series of experiments. The rationale behind these experiments

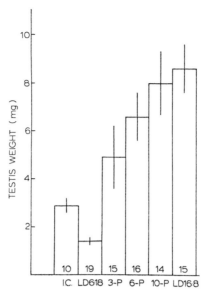

Figure 4. Sensitivity of the testicular response to several long-day exposures. The lizards were otherwise maintained on LD 6 : 18 and given 3 (3-P), 6 (6-P) or 10 (10-P) long days (LD 16 : 8) at evenly spaced intervals over a total period of 21 days in the autumn.

relied heavily on results previously obtained by others on the nature of photo-periodic systems in certain insects, notably the aphid *Megoura viciae*.[46,47] Certain species of insects, including *Megoura*, are the only other organisms known to date that utilize an hour-glass type timer. As is typical in these insects, the *Megoura* hour-glass measures the length of the dark period; when the daily dark period is less than 9·5 h, parthenogenic female offspring (virginopara) are produced, whereas oviparous offspring (ovipara) are produced when the daily dark period exceeds 9·5 h.[46] The dark reaction is initiated by dark onset and reaches completion after 9·5 h (the critical dark period). However, the dark timer does not function accurately unless preceded by at least 6 h of light—light has a 'priming' action which 'resets' the hour-glass and permits the dark reaction to proceed once again. For example, in DD there is no light component to serve as a primer so the dark reaction can go to completion only once, resulting in only partial virginopara inhibition. Light can also function as a terminator of the dark reaction if it intercepts the dark reaction before completion. Using a light cycle (LD 13·5 : 10·5) containing a night just longer than critical, Lees demonstrated that night breaks (1 h light pulses) produced two peaks of short-night effects (or virginopara production): one about 2 h after dark onset, the other during the last 6 h of the night. Light pulses applied during the middle of the night had no effect in terminating the dark reaction. Lees[46] concluded that the *Megoura* dark reaction was actually a complex of at least four stages, defined by their sensitivities to light.

As a first approximation, the *Anolis* system was considered to be analogous to the insect system except that the critical reaction occurs during the light. The first experiment examined the photoperiodic response to LL; if the *Anolis* timer needed to be 'primed' or 'reset' daily by a dark pulse the LL would be predicted to be non-inductive. Figure 5 shows that the testes are maintained on LL; therefore, the system does not have to be reset daily.

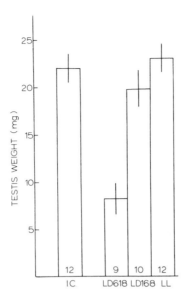

Figure 5. Maintenance of testicular weight in *Anolis* by exposure to continuous illumination (LL) for 21 days in the summer.

The simplest model for an hour-glass timer would involve a light reaction which is reversed by dark. For example, light may cause an accumulation of photoproduct which, if it exceeds a critical amount, initiates the chain of events involved in gonadal growth or maintenance. In such a system dark would dissipate the reaction product. Accordingly, in this simple model, provided that dark-induced dissipation is not instantaneous, a certain duration of darkness would be necessary to dissipate photoproduct below critical amounts. Therefore, even short photoperiods may be inductive if the accompanying dark period is of insufficient length to dissipate enough photoproduct to prevent accumulation to critical levels.

Two experiments were conducted to determine whether or not the dark period played a role in the dissipation of a putative photoproduct in *Anolis*. The first experiment assayed the testicular response of *Anolis* to light/dark cycles consisting of 6 h light periods coupled to dark periods ranging from 1 to 18 h (table 1A). These data fit a simple model involving accumulation/dissipation of photoproduct: LD 6 : 1, LD 6 : 3 and LD 6 : 6 are inductive, suggesting that when coupled to a 6 h photoperiod, dark periods ranging from 1 to 6 h are of insufficient duration to prevent accumulation of critical amounts of photoproduct; however, dark periods of 8 or 10 h are sufficiently long to dissipate photoproduct below critical levels.

A second experiment involved coupling a 6 h dark period to light periods ranging from 2 to 14 h (table 1B). These data show that the photoperiodic response is more complex than can be accounted for by a simple model. For example, LD 4 : 6 and LD 6 : 6 are inductive whereas cycles with the same dark period (6 h) but even longer photoperiods (LD 8 : 6, LD 10 : 6, LD 12 : 6) are not. In order to incorporate these results into a model involving accumulation/dissipation of a putative photoproduct it is necessary to invoke a varying sensitivity of the light reaction to dark interruptions. That is, 6 h dark interruptions beginning 4 or 6 h after light onset are ineffective in reversing the light reaction (therefore, LD 4 : 6

and LD 6 : 6 are inductive) whereas 6 h dark interruptions beginning 8–12 h after light onset can reverse the light reaction (therefore, LD 8 : 6, LD 10 : 6 and LD 12 : 6 are not inductive). Lees has proposed a similar kind of mechanism to explain the varying sensitivity of the aphid's hour-glass to night breaks.[46]

In addition to measuring the testicular response, the circadian rhythm of locomotor activity was assayed under a number of different light cycles. If a CRPP exists in *Anolis*, and if it bears a fixed phase relationship to the activity rhythm, some strong predictions can be made about the photoperiodic effectiveness of different lighting régimes. Figure 6 shows the phase relationships between activity

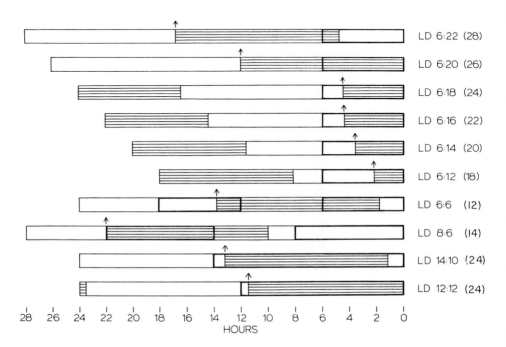

Figure 6. Phase relationships between entraining light cycles and the activity rhythm in *Anolis*. The heavily outlined rectangles represent the light phase and the thinly outlined rectangles represent the dark phase of the light cycles. Activity onsets are marked by vertical arrows. The active portion of the activity rhythm is assumed to be 12 h in duration and is shown by horizontal ruling. The period (T) of each light/dark cycle ($L + D = T$) is shown in parentheses.

and light under a number of different light cycles. For the purposes of discussion the lizard's activity cycle is divided into two equal portions—the subjective day which begins with activity onset and extends for 12 h and the subjective night which occupies the remaining part of the cycle. An LD 12 : 12 cycle is non-inductive in *Anolis* whereas LD 14 : 10 is strongly inductive.[38] If a photo-inducible phase (ϕ_i) of a putative CRPP is being illuminated on LD 14 : 10 but not on LD 12 : 12, ϕ_i must lie early in the lizard's subjective night or late in the subjective night since these are the only portions of the lizard's circadian cycle which are illuminated by LD 14 : 10 but not by LD 12 : 12. Examination of the activity rhythm as well as the testicular response to various light cycles, however, fails to conform with the hypothesis that a photo-inducible phase exists early or late in the lizard's subjective night. For example, some cycles which are non-inductive illuminate either the lizard's early subjective night (LD 6 : 22) or late subjective night (LD 6 : 18, LD 6 : 16, LD 6 : 14, LD 6 : 12) (figures 3b, 6).

The results obtained so far with the *Anolis* photoperiodic system are clearly not compatible with the external coincidence model in which light has a dual role (entrainment and induction). An alternative model, the internal coincidence model, states that the phase relationship between two circadian oscillators (or groups of oscillators) determines whether or not induction will occur.[41] One phase is set by the dawn transition and the other by dusk, so that the mutual phase relationship between the two oscillators changes as the photoperiod shortens or lengthens. In internal coincidence light has only one role—that of entraining the constituent oscillators. If an internal coincidence model is operating in *Anolis* several assumptions would have to hold:

1. None of the *T* cycles, or resonance cycles, utilized so far placed the two oscillators in the correct phase relationship for induction (figure 3).

2. Since LL is inductive (figure 5), LL must hold the oscillators in the correct phase for induction.

3. Both LD 4 : 6 and LD 6 : 6 hold the oscillators in the inductive phase whereas the other non-standard ($T \neq 24$ h) light cycles examined to date do not (table 1).

Table 1. Effectiveness of various lighting régimes in stimulating testicular growth in *Anolis*

	A			B	
	Light cycle	Induction		Light cycle	Induction
	LD 6 : 1	+		LD 2 : 6	−
	LD 6 : 3	+		LD 3 : 6	−
	LD 6 : 6	+		LD 4 : 6	+
	LD 6 : 8	−		LD 6 : 6	+
	LD 6 : 10	−		LD 8 : 6	−
Controls {	LD 6 : 18	−		LD 10 : 6	−
	LD 16 : 8	+		LD 12 : 6	−
				LD 14 : 6	+
			Controls {	LD 6 : 18	−
				LD 16 : 8	+

+, inductive; −, non-inductive.

Saunders[48] has demonstrated that the level of ambient temperature is critical to the expression of 'resonance' in diapause incidence in the flesh-fly *Sarcophaga argyrostoma*. Resonance experiments conducted at constant low temperatures produce a response of the '*Megoura*-type' lacking circadian peaks of diapause incidence, whereas at higher temperatures resonance is observed. Therefore, the possibility can be entertained that the *Anolis* system failed to show a circadian component because the temperature was not 'optimal'. All the experiments with *Anolis* were conducted at a constant 32 °C which is close to the mean preferred temperature of this species. The level of constant temperature required to reveal a photoperiodic response in *Anolis* is relatively narrow; constant temperatures of 35 °C or higher cause somatic damage to the reproductive system whereas photoperiodic responsiveness is abolished at temperatures of 25 °C or lower.[49,50] Although the experiments with *Anolis* were conducted at a temperature which is behaviourally 'preferred', the possibility that a circadian component to photoperiodic time measurement in *Anolis* exists but requires either a slightly different level (or cycle) of temperature to be revealed obviously warrants further investigation.

Very few studies have examined the nature of photoperiodic time measurement in other kinds of lower vertebrates. Night break experiments have been conducted on the female medaka *Oryzias latipes*[51] and on the stickleback *Gasterosteus aculeatus*.[52] In female medaka exposed to LD 7 : 17 with the dark period being interrupted by 1 h of light, the 1 h light pulse stimulated gonadal development when it fell 9 h after dark onset. A similar study in sticklebacks (exposed to LD 6 : 18 with 2 h night breaks) showed a maximal sensitivity 8–10 h after dark onset. Although these results are compatible with the hypothesis that a CRPP is involved in these fish, further experimentation (i.e., using resonance or T cycles) is required to confirm the existence of a CRPP. Positive night break results, alone, are not sufficient to prove the existence of a CRPP. For example, in the aphid, which utilizes an hour-glass timer, long-day effects can also be produced by night breaks.[46]

CONCLUSION

Reptiles occupied a central position in the evolution of the higher vertebrates. By studying the similarities, and perhaps even more importantly, the differences between present-day reptiles and the higher vertebrates, it should be possible to gain insights into the evolution of circadian and photoperiodic systems. It seems likely that many of the details of circadian systems in vertebrates are a product of convergent evolution, but there may well be a central core to the structure of circadian systems, which is ancient. For example, many differences exist between the detailed effects of pinealectomy on the activity rhythms of fish, reptiles and birds, yet it seems likely that the pineal is playing a central role, that of organizing circadian oscillators so that they all share a common frequency. A major difference does seem to exist between the mechanisms of photoperiodic time measurement of lizards and those of the higher vertebrates. Birds and mammals utilize a circadian rhythm of photosensitivity to measure photoperiodic time whereas the *Anolis* time-measuring system seems to lack endogenous rhythmicity. In insects, Saunders[53] suggests that the difference between an insect which utilizes a CRPP (*S. argyro-stoma*) and one which does not (*M. viciae*) is simply that the *Megoura* clock is so rapidly damped out in DD that it requires a daily resetting. However, the possibility that the *Anolis* timer is a rapidly damped 'circadian' clock is unlikely since the *Anolis* photoperiodic system fails to respond to any T cycles in which a lighting stimulus is presented daily (figure 3b). Obviously, further investigations into the biochemistry and physiology of time-measuring systems are needed to gauge the actual magnitude of the differences between the time-measuring systems of *Anolis* and the higher vertebrates.

REFERENCES

1. B. Rusak and I. Zucker. *Physiol. Rev.* **59**, 449–526 (1979).
2. R. J. Wurtman, J. Axelrod and D. E. Kelly. *The Pineal.* New York, Academic Press (1968).
3. R. M. Eakin. *The Third Eye.* Berkeley, University of California Press (1973).
4. D. I. Hamasaki. *Vision Res.* **9**, 515–23 (1969).
5. D. I. Hamasaki and E. Dodt. *Pfluegers Arch.* **313**, 19–29 (1969).
6. J. P. Collin. *Progr. Brain Res.* **52**, 271–96 (1979).

7. B. L. Zimmerman and M. O. M. Tso. *J. Cell Biol.* **66**, 60–75 (1975).
8. C. R. S. Machado, L. E. Wragg and A. B. M. Machado. *Science* **164**, 442–3 (1969).
9. H. Underwood. In: E. H. Burtt jun. (ed.) *The Behavioral Significance of Color.* New York, Garland Press, pp. 127–78 (1979).
10. A. B. Lerner, J. D. Case, Y. Takahashi, T. H. Lee and W. Mori. *J. Am. Chem. Soc.* **80**, 2587 (1958).
11. W. B. Quay. *Pineal Chemistry.* Springfield, Ill., Thomas (1974).
12. R. J. Reiter. *The Pineal.* Montreal, Eden Press (1978).
13. G. A. Bubenik, R. A. Purtill, G. M. Brown and L. J. Grota. *Exp. Eye Res.* **27**, 323–33 (1978).
14. J. M. P. Joss. *Gen. Comp. Endocrinol.* **36**, 521–5 (1978).
15. B. Vivien-Roels and J. Arendt. *Ann. Endocrinol. (Paris)* **40**, 93–4 (1979).
16. B. T. Firth, D. J. Kennaway and M. A. M. Rozenbilds. *Gen. Comp. Endocrinol.* **37**, 493–500 (1979).
17. C. A. Kasal, M. Menaker and J. R. Perez-Polo. *Science* **203**, 656–8 (1979).
18. S. A. Binkley, J. B. Riebman and K. B. Reilly. *Science* **202**, 1198–201 (1978).
19. H. Underwood. *J. Comp. Physiol.* **83**, 187–222 (1973).
20. H. Underwood and M. Menaker. *Photochem. Photobiol.* **23**, 227–43 (1976).
21. H. Underwood. *J. Comp. Physiol.* **99**, 71–8 (1975).
22. C. S. Pittendrigh and S. Daan. *J. Comp. Physiol.* **106**, 333–55 (1976).
23. K. Hoffmann. In: M. Menaker (ed.) *Biochronometry.* Washington, DC, National Academy of Sciences, pp. 134–50 (1971).
24. E. Gwinner. *Science* **185**, 72–4 (1974).
25. H. Underwood. *Science* **195**, 587–9 (1977).
26. J. Repérant, J. P. Rio, D. Miceli and M. Lemire. *Brain Res.* **142**, 401–11 (1978).
27. S. T. Inouye and H. Kawamura. *Proc. Natl Acad. Sci. USA* **76**, 5962–6 (1979).
28. A. Petit and B. Vivien-Roels. *Arch. Biol.* **88**, 217–34 (1977).
29. S. Binkley, M. Hryshchshyn and K. Reilly. *Nature (Lond.)* **281**, 479–81 (1979).
30. F. W. Turek, J. P. McMillan and M. Menaker. *Science* **194**, 1441–3 (1976).
31. H. Underwood. *J. Comp. Physiol.* **130**, 317–23 (1979).
32. E. Gwinner and I. Benzinger. *J. Comp. Physiol.* **127**, 209–13 (1978).
33. C. L. Ralph. In: R. J. Reiter (ed.) *The Pineal and Reproduction.* New York, Karger, pp. 30–50 (1978).
34. I. L. Levey. *J. Exp. Zool.* **185**, 169–74 (1973).
35. C. Haldar and J. P. Thapliyal. *Gen. Comp. Endocrinol.* **32**, 395–9 (1977).
36. J. P. Thapliyal and C. Haldar. *Gen. Comp. Endocrinol.* **39**, 79–86 (1979).
37. M. J. Packard and G. C. Packard. *Experientia* **33**, 1665 (1977).
38. P. Licht. *Ecology* **52**, 240–52 (1971).
39. E. Bünning. *Berl. Dtsch. Bot. Ges.* **54**, 590–607 (1936).
40. C. S. Pittendrigh and D. H. Minis. *Am. Nat.* **98**, 261–94 (1964).
41. C. S. Pittendrigh. *Proc. Natl Acad. Sci. USA* **69**, 2734–7 (1972).
42. B. K. Follett. *J. Reprod. Fertil.* Suppl. **19**, 15–18 (1973).
43. J. A. Elliott. *Fed. Proc.* **35**, 2339–46 (1976).
44. H. Underwood. *J. Comp. Physiol.* **125**, 143–50 (1978).
45. K. K. Nanda and K. C. Hamner. *Bot. Gaz. (Chicago)* **120**, 14–25 (1958).
46. A. D. Lees, *J. Insect Physiol.* **19**, 2279–316 (1973).
47. D. S. Saunders. *Insect Clocks.* Oxford, Pergamon (1976).
48. D. S. Saunders. *J. Insect Physiol.* **19**, 1941–54 (1973).
49. P. Licht and S. L. Basu. *Nature (Lond.)* **213**, 672–4 (1967).
50. P. Licht. *J. Exp. Zool.* **172**, 311–22 (1969).
51. K. Chan. *Can. J. Zool.* **54**, 852–6 (1976).
52. B. Baggerman. *Gen. Comp. Endocrinol.* Suppl. **3**, 466–76 (1972).
53. D. S. Saunders. *J. Comp. Physiol.* **132**, 179–89 (1979).

DISCUSSION

Nicholls: The 'critical photoperiod' for testicular growth in *Anolis* is about 13 h, at least when the animals are photosensitive. Why does the minimum testicular weight

under natural conditions in Louisiana occur in September/October, and then the testes grow in November to January when daylengths are decreasing to minimal?
Underwood: The photoperiod in Louisiana in the autumn is below the critical effective photoperiod. *Anolis* is insensitive to photoperiod between November and May and the recrudescence observed is mainly under the control of temperature. Other relative roles of temperature and light on different phases of the annual cycle have been investigated by Paul Licht. According to him, photoperiod only plays a role in *termination* of the breeding season in natural populations.
Bünning: Can we be sure that only hour-glass timers are involved? Experiments with plants make me wonder. Until about 1955 all the experiments intended to check the possible involvement of circadian rhythmicity resulted in the conclusion that only hour-glass timers participate. After introducing one change in the experimental conditions, suddenly the opposite conclusion was drawn. In the case of plants, this changed factor was no longer working with incandescent light but with fluorescence tubes (which avoided an immediate damping of the circadian rhythmicity). Might it be that in the case of lizards a certain other factor is responsible for an immediate damping?
Underwood: At present I have failed to demonstrate that a circadian rhythm of photoperiodic time measurement is involved in *Anolis* using the classic experimental protocols that were specifically designed to test such an involvement. Consequently I am forced to entertain the possibility that some type of hour-glass timer is operating. Clearly, additional experimentation may lead me to revise this explanation.
Short: Could you please describe the anatomical connections between the parietal eye, the pineal organ and the brain? Is any melatonin detectable in lizards from which the eyes, the parietal eye and the pineal organ has been removed?
Underwood: The parietal eye nerve is generally considered to terminate in the habenular nuclei and may innervate the pineal as it courses over the surface of the pineal on its way to the rest of the brain. The termination of the pineal nerve is uncertain. Melatonin levels have not been examined in pinealectomized–blinded or pinealectomized lizards.

CIRCANNUAL RHYTHMS: THEIR DEPENDENCE ON THE CIRCADIAN SYSTEM

Eberhard Gwinner*

Max-Planck-Institut für Verhaltensphysiologie, Andechs,
Federal Republic of Germany

Summary

Circannual rhythms persisting for several cycles under seasonally constant conditions have been reported in a number of animal species. The formal properties of these rhythms have been relatively well described but little is known about the physiological processes by which circannual rhythms are controlled. In trying to develop hypotheses about possible physiological mechanisms the most troublesome features of circannual rhythms are the long time constants of the processes that must be involved. In an attempt to overcome these difficulties it has been suggested, therefore, that circannual rhythms might be generated by rhythms with higher frequencies, e.g. circadian rhythms. An obvious way in which this may be achieved is that the organism transforms circadian frequencies into circannual cycles by a process called 'frequency demultiplication' in oscillator theory. In effect, such a mechanism would amount to the organism counting subjective circadian days with the 'knowledge' that a year has 365 days. This frequency demultiplication hypothesis (FDH) predicts that the period of a circannual rhythm could be altered by changing the period of the underlying circadian rhythm. We have tested this prediction in experiments with garden warblers and starlings which were exposed for up to 4 years to light/dark cycles with periods of 22, 24 and 26 h. In both species the circannual rhythms of testicular size were measured and also the circannual rhythm of moult in the starlings. The results are, on the whole, not consistent with the FDH and rather suggest that circannual rhythms are not generated by circadian rhythms. This conclusion is supported by several other findings on the behaviour of circannual rhythms under certain experimental conditions, i.e. under photoperiodic cycles with periods different from 1 year and after phase shifts of the environmental photoperiodic cycle.

INTRODUCTION

It has been demonstrated in a variety of animal species that annual fluctuations of physiological and behavioural functions may persist for several years under

* Present address: Max-Planck-Institut für Verhaltensphysiologie, Vogelwarte Radolfzell, Rodolfzell/Möggingen, FDR.

Biological Clocks in Seasonal Reproductive Cycles 153–169 (1981) (ed. B. K. and D. E. Follett: Bristol, Wright).

experimental conditions devoid of seasonal variations. Such circannual rhythms lasting for 2 years or more have now been observed in at least 1 coelenterate, 1 mollusc, 2 arthropods, 1 reptile, 11 birds and 15 mammals.[1-3] Circannual rhythms behave in many respects like circadian rhythms in their formal properties. Under constant seasonal conditions they free-run with periods different from 12 months.[3] These free-running rhythms can be synchronized by annual environmental rhythms such as the annual cycle in photoperiod.[4-7] There is good evidence in some species that circannual rhythms are innate [8-10] and some evidence that they are relatively temperature-independent.[11,12] In other respects, however, the behaviour of circ-annual rhythms is different (at least quantitatively) from circadian rhythms. For instance, the period of the free-running rhythm is often much more variable, and the relative range of zeitgeber periods to which circannual rhythms can be entrained may be much greater than that of circadian rhythms.[4-6] In addition, large species differences are found in several characteristics of circannual rhythms such as period length and persistence of the rhythms, as well as the conditions under which they are expressed.[13] This diversity is in sharp contrast with the impressive uniformity of circadian rhythms in literally all groups of living beings.[14] Nevertheless, it has proved useful in the past to treat circannual rhythms, like circadian rhythms, as endogenous oscillators and to use the terminology of oscillator theory for their description.[3,7]

The formal properties of circannual rhythms have been relatively well described for several animal species, but still very little is known about the physiological processes controlling them. In trying to develop hypotheses about possible physiological mechanisms, the most troublesome feature of circannual rhythms is the long term nature of the processes that must be involved. These long time constants are often difficult to reconcile with those of the physiological processes known to be involved in seasonal phenomena. In an attempt to overcome these difficulties it has been suggested, therefore, that circannual rhythms might be generated by rhythms with higher frequencies, e.g. by circadian rhythms.[15-20] The possibility that circadian rhythms are somehow involved in the production of circannual rhythms is appealing for several reasons. First, circadian rhythms are ubiquitous and almost certainly phylogenetically older than circannual rhythms,[21] so that they were available to any organism needing to evolve a circannual clock. Secondly, circadian rhythms are known to be involved in the control of at least some seasonal phenomena, e.g. in photoperiodic time measurement.[22] Finally, a few of the properties of circannual rhythms of animals under different constant photoperiods can be related to peculiarities of the circadian system which measures photoperiod, e.g. the persistence of a circannual rhythmicity in some bird species in a 12 h photoperiod but not under other photoperiodic conditions.[23]

There are several ways in which circadian rhythms might be involved in generating free-running circannual rhythms,[8,17] but perhaps the most interesting possibility is that the organism transforms circadian frequencies into circannual rhythms by a process called frequency demultiplication or frequency division in oscillator technology. In effect, such a mechanism would amount to the organism counting subjective circadian days 'knowing' that a year has 365 days. A familiar technical analogue of such a device is the electrical clock which generates a 24 h rhythm out of the commercial periodicity of 50 cycles per second. Because circannual rhythms free-run under seasonally constant conditions, a separate synchronizing mechanism would have to be assumed in addition to such a counting mechanism (*see* Discussion).

This frequency demultiplication hypothesis (FDH) has been proposed more or less explicitly several times.[16,18,24] It can be tested easily because of its unambiguous prediction that the period of the circannual rhythmicity depends on the period of the circadian rhythmicity. The longer the circadian period the longer is the resulting period of the circannual cycle. The periods of the two rhythms should be directly proportional to one another. This prediction has been tested in a previous study with European starlings.[18] In this experiment, 9 male birds were held for 15 months in continuous dim light and their circadian rhythm of locomotor activity as well as their circannual rhythm of moult were investigated. The mean circadian periods of these birds varied from 22·5 to 24·5 h and the circannual periods from 10·6 to 13·8 months. A weak ($P < 0.05$) positive correlation was found between these two rhythms. These findings are consistent with the FDH although they could also simply represent a common dependency of the periods of the measured rhythms on external or internal factors.

A more powerful way of testing this hypothesis consists of exposing animals to light/dark cycles with periods different (but not too different) from 24 h, thereby altering the period of the entrained circadian rhythms to precisely defined values. Such a procedure allows accurate predictions about the period of the circannual rhythm because, according to the model, any change in the circadian period should alter the circannual period by the same percentage. We have carried out experiments of this type with garden warblers (*Sylvia borin*) and European starlings (*Sturnus vulgaris*), species in which circannual rhythms of various functions have previously been described.[6,23,25,26] This paper concentrates on the results of these experiments. In the Discussion, a more general evaluation of the FDH will be attempted.

METHODS

Maintenance and experimental conditions

Garden warblers

Sixteen male garden warblers were captured in the autumn of 1971 and subsequently kept for about 10 days in individual cages under the natural photoperiodic conditions of Andechs, Germany. On 26 October they were transferred to the experimental chambers in which they were subsequently kept for up to 30 months. The birds were subdivided into three groups: group I, 5 birds on LD 12 : 12, group II, 6 birds on LD 11 : 11; group III, 5 birds on LD 13 : 13.

Light was provided by incandescent bulbs which produced a light intensity of about 200 lux during periods of light and 0·01 lux during periods of darkness. Temperature was held constant at 20 ± 2 °C. The birds were kept in individual 48 × 23 × 28 cm wire mesh cages in which their perch-hopping activity was continuously recorded. These recordings indicated that the circadian locomotor activity rhythms of all birds were synchronized with the light/dark cycles at all times. The birds were provided with food and water as described elsewhere.[27]

Starlings

Twenty-four male and 6 female birds were caught in the winter of 1971/72 and kept in large outdoor aviaries at Andechs. According to the criteria of Berthold[28] and

Svensson[29] they were all first-year birds. On 21 April they were moved to the experimental chambers in which they were subsequently kept for up to 43 months. The birds were subdivided into four groups: group I_1 (LD 12 : 12, males), 10 males (I_{11}–I_{20}, in figure 2) were housed together as a group in a 58 × 58 × 50 cm wire mesh cage; group I_2 (LD 12 : 12, males with females), 6 males (I_{21}–I_{25} in figure 3) were housed together with 6 females in a wire mesh cage of the same size as that of group I_1; group I_3 (LD 12 : 12, single males), 8 males (I_{31}–I_{38} in figure 4) were housed singly in 48 × 23 × 28 cm wire mesh cages, equipped with perch contacts which allowed recording of locomotor activity; group II (LD 11 : 11, males), 10 males (II_1–II_{10} in figure 5) were housed together as a group in a 58 × 58 × 50 cm wire mesh cage.

The birds of groups I_1, I_2 and I_3 were kept together in the same 310 × 210 × 240 cm room. The cages of group I_3 were placed closely together on a board so that each bird could see several of its neighbours. The cages of groups I_1 and I_2 were placed on another board underneath so that the birds of these two groups also had optical contact with each other but not with the birds of group I_3. Light and temperature conditions for all groups were the same as for the garden warblers.

In the birds of group I_3 locomotor activity was recorded during three periods of 2 weeks in 1972, 1973, and 1974. In addition, 4 randomly selected birds of group II also were kept for 4–5 days each year in individual registration cages of the type used for the birds of group I_3 in order to register their locomotor activity. The recordings indicated that the circadian locomotor activity rhythms of the birds were synchronized in every case with the light/dark cycles to which the birds were exposed. They were provided with food and water as described elsewhere.[30] Three birds of group I_1 (I_{18}–I_{20}) and 1 male of group I_2 (I_{26}) died within a few months of the beginning of the experiment. The data of these birds as well as those from the females of group I_2 are not included in the present analysis.

Experimental procedure

Garden warblers

At the beginning of the experiment and subsequently at intervals of about 4–6 weeks the birds were laparotomized and the length of the left testis was measured with compasses to the nearest 0·1 mm. Circannual period length (τ_{TO}) was defined as the interval between dates (established graphically) at which testicular length during a phase of testicular growth exceeded 2·0 mm.

Starlings

At the beginning of the experiment and subsequently at intervals of about 4–6 weeks the birds were investigated for testicular size and bill coloration. Testicular size was established by laparotomy as in the garden warblers except that testicular width rather than length was measured.[30]

The bill colour of the starling is black in autumn and winter. Beginning with gonadal recrudescence in early spring, it gradually changes to yellow, starting at the base. During testicular regression the bill turns black again.[31] To classify the states of bill coloration the following scores were used: 1, bill entirely black; 2, bill up to one-third yellow; 3, bill half to two-thirds yellow; 4, bill more than two-thirds, but not entirely yellow; 5, bill entirely yellow.

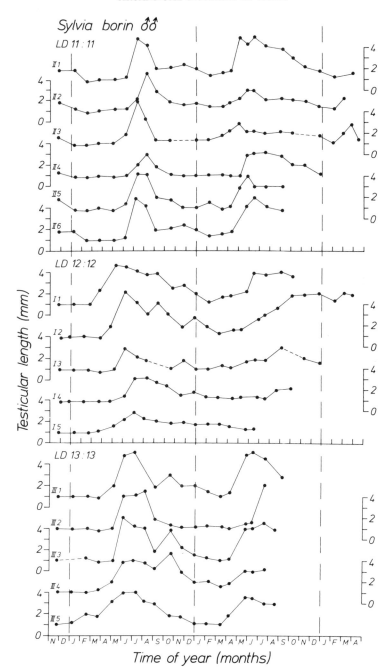

Figure 1. Variations in testicular length in male garden warblers (*Sylvia borin*) kept under three different constant light/dark cycles.

Moult protocols were made separately for body moult and moult of primaries and secondaries at intervals of 4–6 days so that the time of onset of moult could be estimated with an accuracy of about ±2 days.

Two different estimates of circannual period lengths were made using the following parameters. Moult: τ_M is the time interval between two successive onsets of moult. Onset of testes growth: τ_{TO} is the time interval between two successive dates (established graphically) when testicular width exceeded 4·0 mm during a phase of testicular growth.

RESULTS

Garden warblers

General

The results obtained from the experiment with the garden warblers are summarized in figure 1. Within 8–14 months following the transfer to experimental conditions all birds had gone through a complete cycle of testicular growth and regression. Subsequently all but one bird (I_5 which died early) started a second testicular cycle, about 1 year later. In 2 birds (II_2 and II_3) the beginning of even a third cycle was observed, the interval between successive onsets of testicular growth clearly being shorter than 1 year. In effect, these birds demonstrated circannual rhythms as had been previously found in this species under three different constant photoperiods (LD 10 : 14, LD 12 : 12 and LD 16 : 8),[26] and indicated again that the annual testicular cycle is primarily independent of exogenous seasonal variations but controlled rather by an endogenous circannual rhythmicity.

Test of hypothesis

The FDH predicts that the birds of group II would have the shortest circannual period and group III the longest, with those of group I being intermediate. Inspection of figure 1 reveals, in contrast with this prediction, that the longest

Table 1. Mean (± s.d.) duration (in days) of circannual period τ_{TO} of the three groups of garden warblers maintained under three different light/dark cycles

Group	τ_{TO}
II (LD 11 : 11)	297 ± 11 (6)
I (LD 12 : 12)	410 ± 19 (5)
III (LD 13 : 13)	362 ± 23 (5)

Numbers in parentheses refer to number of birds investigated.
Arrows indicate significant ($P < 0.01$, Mann–Whitney U test) deviations of actual differences from the differences predicted by the frequency demultiplication hypothesis. For further explanations *see* text.

periods were found among the birds of group I, the shortest ones among the birds of group II and those of group III were intermediate. This conjecture is confirmed by the quantitative data provided in table 1*.

To test the hypothesis more thoroughly, the circannual periods (τ_{TO}) of the first testicular cycle of all individual birds were expressed as the number of light/dark cycles rather than as the number of objective days which had passed between the first and second onset of testicular growth (as defined above). If the circannual cycle of the birds resulted from the counting of circadian days, the null-hypothesis is that τ_{TO} should be the same in all three groups. Table 1 shows that this was only true for the comparison of groups II and III but not for the comparison of groups I and II or groups I and III.

Starlings

General

The results obtained from the starling experiments are summarized in figures 2–5. In all birds the testes regressed within a few months of the beginning of the experiments. Associated with testicular regression, the bills gradually turned black. Subsequently all birds carried out a complete moult. Later in the experiment both testicular size and bill coloration showed pronounced and in many instances regular variations, and most birds performed at least one additional moult. In most, although not all, instances the curves for the index of bill coloration followed the curves for testis size and, as a rule, moults took place when the testes were regressing or were small.

These results, in general, are in agreement with previous findings that a circannual rhythmicity in testicular size and moult may persist in starlings under certain constant environmental conditions (e.g. LD 12 : 12). However, a substantial inter-individual variability was found in the behaviour of all three parameters measured. Particularly obvious was the occurrence in some birds of exceptionally long intervals between successive corresponding events. If a frequency distribution is plotted of all individual periods (τ_{TO} and τ_M) there is a suggestion of bimodality with τ values of about 550 days, i.e. 1·5 years, best separating the short- and the long-period part of the distribution. This may justify looking separately at rhythms with periods shorter than 550 days ('typical' circannual cycles) and rhythms with periods longer than 550 days ('atypical' circannual cycles). On the basis of this distinction, the main types of behaviour shown in the experimental birds can be classified roughly into the following groups.

1. *'Typical' circannual cycles of the three parameters, testicular size, bill coloration and moult:* In 10 out of the 23 birds that survived for at least 3 years and 4 out of the 7 birds that survived for a shorter time, relatively clear and persistent circannual rhythms with periods shorter than 1·5 years could be detected (e.g. I_{12} in figure 2). In these birds, which were found among all four experimental groups, the three parameters were in a more or less normal temporal relationship with one another; i.e. the curves for testis size and bill coloration ran parallel and moult occurred when these curves went through a trough. However, in contrast with the birds living in natural photoperiodic conditions, the time during which testes were

---

* For bird I_5, who died before it had initiated a second testicular cycle, it was assumed that its testicular length would have passed the 2 mm threshold at the date of laparotomy following its death. With regard to the hypothesis, this was a conservative assumption.

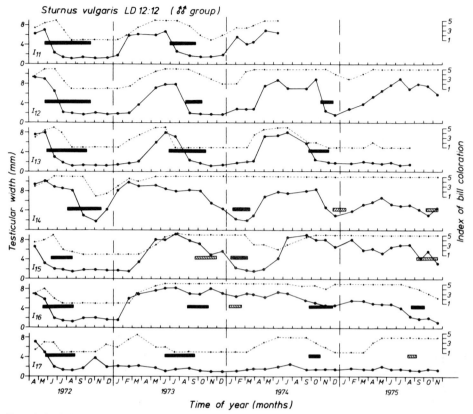

Figure 2. Variations in testicular width (solid lines) and in the index of bill coloration (dotted lines) as well as occurrence of moult. Moult of wing and tail feathers and body plumage (solid bars), moult of body plumage only (hatched bars) in the starlings (*Sturnus vulgaris*) of group I_1 (LD 12 : 12, males).

small was usually relatively short whereas the time during which testes were enlarged was sometimes as long as 12 months. This is especially clear in the birds of group II during their first testicular cycle in 1972/73 (figure 5).

2. *'Typical' circannual cycles of only one or two parameters (other parameters being apparently arrhythmic):* In several of the birds kept in LD 12 : 12 the rhythm in testis size waned after about 1 year in constant conditions whereas the rhythm in moult persisted over the whole experimental period. This was particularly obvious in two birds of group I_2 (I_{21} and I_{25} in figure 3), in which, starting in late 1973, testes size remained more or less constant at a half-developed state while the rhythm in moult continued. Similarly, in 2 birds of group I_1 (I_{17}) and I_3 (I_{38} in figure 4) testicular size remained small for at least 2 years but the rhythm in moult persisted. The converse situation of a persisting rhythm in testicular size and the absence of a rhythm in moult is suggested by the behaviour of bird I_{37} (figure 4).

Particularly interesting are a few cases in which either testes size or bill coloration showed a clear circannual rhythm whereas the other parameter did not. In bird I_{17} (figure 2) bill coloration went repeatedly through a clear circannual cycle but testis size remained small, even though in both 1973 and 1974 a minute increase in testicular width was at least suggested. Conversely, the testis size of bird I_{14} (figure

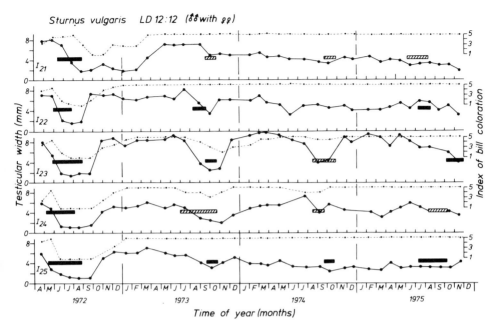

Figure 3. Cycles in the starlings of group I₂ (LD 12 : 12, males with females). For details *see* figure 2.

2) showed a clear circannual rhythmicity but the bill remained completely yellow for almost 3 years. That testis size and bill coloration may be independent parameters was supported by the fact that in some birds whose testis size and bill coloration were cyclic, the increase in the index of bill coloration preceded the increase in size of the testes (e.g. I_{37}, 1973, 1974; I_{33}, 1973; figure 4).

3. *'Atypical' circannual cycles of at least one parameter:* 'Atypical' circannual cycles with periods exceeding 550 days were found in a few birds of all four experimental groups. For instance, in bird II_8 (figure 5) the second cycle of moult and testis size was considerably longer than the first one and was about twice as long as that of the other birds in this group, suggesting a 2 : 1 relationship with the 'typical' circannual cycles. Similarly, in four birds of group I_3 (I_{31}–I_{34}, figure 4) the second cycles of moult and testis size had periods drastically exceeding the first ones and longer than 1·5 years. Finally, in bird I_{16} (figure 2) the first and only period of increased testis size lasted more than 2·5 years.

In two birds kept in LD 12 : 12 (I_{35}, I_{36} in figure 4) the cycles of testis size were unusually short. The circannual period (τ_{TO}) of these birds was in the range of 200–300 days.

Test of hypothesis

Because of the large inter-individual variability in the behaviour of the birds, figures 2–5 are inadequate for a direct comparison of the circannual periods of birds of different experimental groups. To facilitate evaluation of the results, the pertinent data for testes size and moult are presented in a different way in figure 6. The lines in the four diagrams connect the normalized mean dates at which in the different groups onset of testis growth (a,c) or moult (b,d) occurred. The curves in

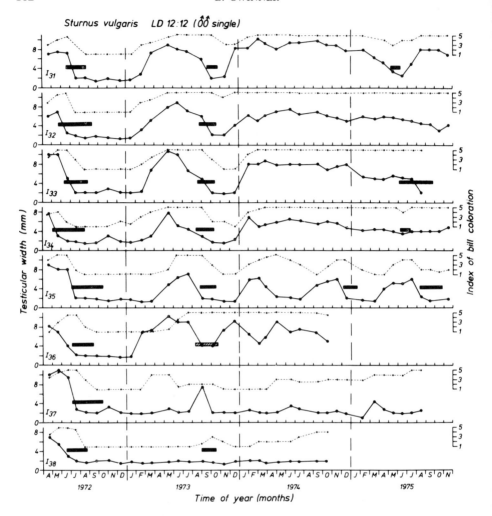

Figure 4. Cycles in the starlings of group I$_3$ (LD 12 : 12, single males). For details *see* figure 2.

diagrams *a* and *b* are based on all data whereas those in diagrams *c* and *d* are based only on the 'typical' circannual periods with values shorter than 550 days. Such a separate calculation seemed desirable because of the large effects of the atypically long periods on the mean values and variances and because of the suspicion that in some cases these periods represent 'circabiannual' rhythms.

As in the warblers, the data obtained from the starlings were in conflict with the FDH. This hypothesis predicts shorter periods in the group held in LD 11 : 11 than in the groups kept in LD 12 : 12. The results shown in figure 6 instead suggest longer periods in the birds kept in LD 11 : 11. For a quantitative evaluation of these results the same procedure was employed as in the warblers, i.e. the circannual periods were expressed in number of light/dark cycles and then the null hypothesis was tested that the birds in LD 11 : 11 had the same τ values as the birds in LD 12 : 12. This has been done separately for the first and second circannual period and the τ values of the birds in LD 11 : 11 (group II) were compared with the τ

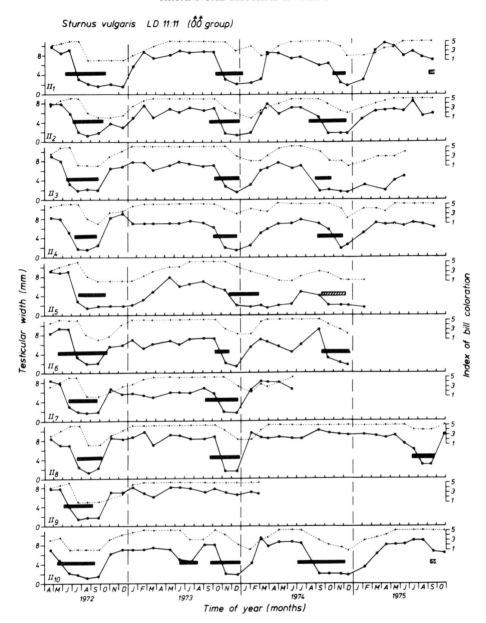

Figure 5. Cycles in the starlings of group II (LD 11 : 11, males). For details *see* figure 2.

values of all three groups held in LD 12 : 12 (group I_1, I_2, I_3).

Table 2 shows that for the first cycle in four out of six possible comparisons the null hypothesis must be rejected ($P < 0.05$), regardless of whether all (upper part of table 2) or only the 'typical' circannual periods were considered. In all these cases τ of the birds in LD 11 : 11 was longer than predicted by the hypothesis. Because of low sample size, for the second cycle only two (or one, if only the typical

Table 2. Mean duration (in days ± s.d. if $n > 4$) of circannual periods τ_{TO} and τ_M of the four groups of starlings maintained under four different conditions

Group	First cycle τ_{TO}	τ_M	Second cycle τ_{TO}	τ_M	Third cycle τ_{TO}	τ_M
II (LD 11 : 11, males)	468 ± 34 (9)	462 ± 36 (9)	395 ± 112 (6)	395 ± 130 (8)	—	372 (2)
I$_1$ (LD 12 : 12, males)	402 ± 60 (5)	443 ± 51 (7)	328 (2)	465 ± 134 (6)	262 (1)	318 (3)
I$_2$ (LD 12 : 12, males with females)	443 ± 93 (5)	459 ± 26 (5)	301 (1)	443 ± 165 (5)	—	330 (4)
I$_3$ (LD 12 : 12, males single)	318 ± 119 (7)	440 ± 23 (7)	540 (4)	592 (4)	226 (1)	250 (1)
II (LD 11 : 11, males)	468 ± 34 (9)	462 ± 36 (9)	349 ± 22 (5)	351 ± 36 (7)	—	372 (2)
I$_1$ (LD 12 : 12, males)	402 ± 60 (5)	443 ± 51 (7)	328 (2)	414 ± 58 (5)	262 (1)	318 (3)
I$_2$ (LD 12 : 12, males with females)	403 (4)	459 ± 26 (5)	301 (1)	370 (4)	—	330 (4)
I$_3$ (LD 12 : 12, males single)	274 ± 21 (6)	440 ± 23 (7)	229 (1)	464 (1)	226 (1)	250 (1)

Numbers in parentheses refer to number of birds in which the respective τ values could be determined. Data are given separately for all three successive circannual cycles. In the upper four rows all data were considered whereas in the lower four rows only the 'typical' values were taken into account. Arrows indicate significant ($\longleftrightarrow P < 0.01$, $\leftarrow - \rightarrow P < 0.05$, Mann–Whitney U test) deviations of actual differences from the differences predicted by the frequency demultiplication hypothesis. For further explanations *see* text.

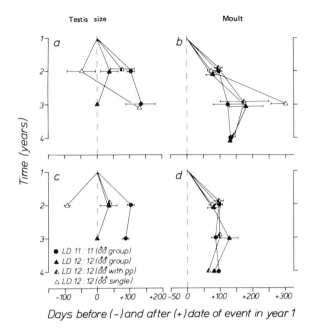

Figure 6. Circannual rhythms of testicular size (*a* and *c*) and moult (*b* and *d*) in the four groups of starlings. Mean dates of successive onsets of testicular growth and moult have been calculated for each experimental group and data were then normalized relative to the onset of the first event (= day 0 on abscissa). Delays in the occurrence of an event are plotted to the right of 0 on the abscissa and advances to the left. Horizontal lines at the symbols represent s.d. (given only when *n* > 4). The curves in (*a*) and (*b*) are based on all data whereas those in (*c*) and (*d*) are based only on the 'typical' circannual cycles as defined in the text.

periods are taken into account) comparisons are possible. They revealed no significant differences.

If the three groups held in LD 12 : 12 were compared, significant differences were found for the first testicular cycle. They indicated a shorter period in the males housed singly as compared with those housed together with other males or with females.

DISCUSSION

The present data do not support the frequency demultiplication hypothesis. As indicated by the activity recordings, the circadian rhythms of locomotor activity were synchronized with all experimental light/dark cycles, and there are good reasons to believe that under the strong zeitgeber conditions used, the same would also be true for other circadian functions. Therefore, the hypothesis would have predicted that the circannual period length is directly proportional to the period length of the environmental light/dark cycles. This, however, was not the case in either the warblers or the starlings. In the warblers, birds in a 24 h zeitgeber period had the longest and those in the 22 h zeitgeber period had the shortest circannual period; birds in the 26 h zeitgeber period were intermediate. Similarly, the first circannual period of moult and increased gonadal size measured in the starlings was longer in the birds under the 22 h photoperiod than predicted by the hypothesis. Hence, these data as a whole would rather suggest that a frequency demultiplication mechanism is not involved in the production of circannual cycles. However, at least two objections against such a proposition could perhaps be made.

First, it might be argued that essentially all the conclusions of the present paper have been derived from data on the behaviour of circannual rhythms during the first cycle in constant conditions when the system is probably not yet in steady-state; and that in steady-state conditions the predicted relationship between circadian and circannual rhythmicity might still be observed. Such an objection would not be very relevant, however, because any reasonable frequency transformation hypothesis would have to explain the transient behaviour of circannual rhythms as well. In fact, as will be pointed out below, the existence of transients is one of the major arguments against such a mechanism.

Secondly, in the present experiments the period of the environmental light/dark cycles has been altered by altering the light and the dark fraction of the cycle proportionally. Therefore, a change in cycle length also results in a change of the absolute light and dark times. The bird might interpret this as change in photoperiod. Such photoperiodic effects could be augmented by the well-known phenomenon that the phase relationship between circadian rhythms and the entraining light/dark cycles changes as a function of the zeitgeber period.[7] It could be argued, therefore, that circannual rhythms still result from frequency demultiplication but that the effects of this mechanism are secondarily obscured or superimposed by photoperiodic effects. Again, such a possibility, together with the fact that circannual cycles can indeed be modified by photoperiod, should be interpreted as a principal argument against the FDH rather than as an argument against its rejection (see below).

Apart from the experimental evidence against the FDH discussed so far, there are several features of circannual rhythms which are very hard to reconcile with such a mechanism. They include the following phenomena:

1. Circannual rhythms free-running under seasonally constant conditions can be synchronized by sinusoidal changes in photoperiod. This was shown in experiments with sika deer,[4] starlings [5,6] and garden warblers.[32] By shortening the period of the photoperiodic cycles it was possible to compress the circannual cycles of gonadal size and moult of the starling up to one-eighth of its spontaneous value, i.e. to about 1·5 months (E. Gwinner, unpublished observations). Since in these experiments the light/dark cycle synchronizing the circadian rhythms always had a period of 24 h, it would have to be assumed that the 'counting' mechanism of the animal is modified according to the rate at which photoperiod oscillates. Although such a mechanism is theoretically possible, it is unlikely in animals.

2. Entrained circannual rhythms can be phase-shifted by phase-shifting the zeitgeber conditions. An excellent example was given by Davis and Finnie[33] who displaced woodchucks (Marmota monax) from the northern to the southern hemisphere. It took the animals' circannual body weight rhythms[34] at least three cycles to resynchronize with the phase-shifted zeitgeber conditions, and during this resynchronization time the length of successive periods changed systematically. Similar transients can also be observed during the first cycles after transfer to constant conditions (see figure 6). Such changes in the circannual period length could only be explained on the basis of the FDH if it is assumed that the frequency demultiplication mechanism changes its properties continuously during the successive transient cycles.

3. In some animals the period of circannual rhythms depends, at least to some extent, on light intensity and temperature.[3] Moreover, maintenance conditions and/or social relationships may affect the circannual period as well, as suggested by the differences found between the three groups of starlings maintained under LD 12 : 12 (table 2). Lastly, large inter-individual differences exist in the period values of different individuals kept under the same set of environmental conditions. Once again a complicated weighting system for the relative duration of the day would have to be assumed to reconcile such data with the FDH.

These theoretical difficulties, together with the negative experimental evidence presented in this paper, render any frequency demultiplication mechanism highly unlikely as a possible basis of circannual rhythms. The weak positive correlation found previously between circadian and circannual cycles of starlings kept in a 12 h photoperiod (*see* Introduction) is more likely the result of common factors influencing both classes of rhythms than of a causal dependency of the circannual on the circadian system.

Frequency division is only one possible way by which circadian rhythms may be involved in generating circannual cycles, but so far all alternative mechanisms proposed have received just as little experimental support.[7,17] On the other hand, it is clear for several species including the starling[35] that a circadian rhythmicity is involved in photoperiodic time measurement and, hence, in the process of synchronization of circannual cycles. At the present time it seems not unlikely that this is the only level at which the circadian system plays a significant role in the control of annual cycles of organisms equipped with a circannual clock.

ACKNOWLEDGEMENT

This work was supported in part by the Deutsche Forschungsgemeinschaft (SPP Biologie der Zeitmessung).

REFERENCES

1. P. Berthold. *Endogene Jahresperiodik. Innere Jahreskalender als Grundlage der jahreszeitlichen Orientierung bei Tieren und Pflanzen.* Konstanzer Universitätsreden 69, Konstanz, Universitäts verlag (1974).
2. E. Gwinner. In: D. S. Farner and J. A. King (ed.) *Avian Biology*, vol. 5. New York, Academic Press, pp. 221–85 (1975).
3. E. Gwinner. In: J. Aschoff (ed.) *Handbook of Behavioral Neurobiology*, vol. 4: *Biological Rhythms.* New York, Plenum Press, pp. 391–410 (1981).
4. R. J. Goss. *J. Exp. Zool.* **170**, 311–24 (1969).
5. E. Gwinner. *Vogelwarte* **29**, 16–25 (1977).
6. E. Gwinner. *Naturwissenschaften* **64**, 44 (1977).
7. J. Aschoff. In: *Proceedings of the 17th International Ornithology Congress.* New York, Plenum Press.
8. E. T. Pengelley and S. M. Asmundson. *Comp. Biochem. Physiol.* **32**, 155–60 (1970).
9. H. C. Heller and T. L. Poulson. *Comp. Biochem. Physiol.* **33**, 357–83 (1970).
10. M. J. Ducker, J. C. Bowman and A. Temple. *J. Reprod. Fertil.* **19**, 143–50 (1973).
11. E. T. Pengelley and S. M. Asmundson. *Comp. Biochem. Physiol.* **30**, 177–83 (1969).
12. E. T. Pengelley and K. H. Kelly. *Comp. Biochem. Physiol.* **19**, 603–17 (1966).
13. E. Gwinner. In: M. Menaker (ed.) *Biochronometry.* Washington, DC, National Academy of Sciences, pp. 405–27 (1971).
14. J. Aschoff (ed.): *Handbook of Behavioral Neurobiology*, vol. 4: *Biological Rhythms.* (in the press) (1981).
15. J. King. *Comp. Biochem. Physiol.* **24**, 827–37 (1968).

16. E. L. Sansum and J. R. King. *Physiol. Zool.* **49**, 407–16 (1976).
17. W. M. Hamner. In: M. Menaker (ed.) *Biochronometry*. Washington, DC, National Academy of Sciences, pp. 448–62 (1971).
18. E. Gwinner. *J. Reprod. Fertil.* Suppl. 19, 51–65 (1973).
19. J. T. Rutledge. In: E. T. Pengelley (ed.) *Circannual Clocks*. New York, Academic Press, pp. 297–345 (1974).
20. N. Mrosovsky. In: L. Wang and J. W. Hudson (ed.) *Strategies in the Cold: Natural Torpidity and Thermogenesis*. London, Academic Press, pp. 21–65 (1978).
21. D. S. Farner. *Environ. Res.* **3**, 119–31 (1970).
22. K. Hoffmann. In: J. Aschoff (ed.) *Handbook of Behavioral Neurobiology*, vol. 4: *Biological Rhythms*. New York, Plenum Press. (in the press) (1981).
23. R. G. Schwab. In: M. Menaker (ed.) *Biochronometry*. Washington, DC, National Academy of Sciences, pp. 428–47 (1971).
24. D. S. Farner and B. K. Follett. In: E. J. W. Barrington (ed.) *Hormones and Evolution*. London, Academic Press (1980).
25. P. Berthold, E. Gwinner and H. Klein. *J. Ornithol.* **113**, 170–90 (1972).
26. P. Berthold, E. Gwinner and H. Klein. *J. Ornithol.* **113**, 407–17 (1972).
27. P. Berthold, E. Gwinner and H. Klein. *Vogelwarte* **25**, 297–331 (1970).
28. P. Berthold. *Vogelwarte* **22**, 236–75 (1964).
29. L. Svensson. *Identification Guide to European Passerines*. Stockholm, Naturhistorica Riksmuseet (1970).
30. E. Gwinner. *Z. Tierpsychol.* **38**, 34–43 (1975).
31. E. Witschi and R. A. Miller. *J. Exp. Zool.* **79**, 475–86 (1938).
32. P. Berthold. *Vogelwarte* **30**, 7–10 (1979).
33. D. E. Davis and E. P. Finnie. *J. Mammalogy* **56**, 199–203 (1975).
34. D. E. Davis. *Physiol. Zool.* **40**, 391–402 (1967).
35. E. Gwinner and L. O. Eriksson. *J. Ornithol.* **118**, 60–7 (1977).

DISCUSSION

Elliott: Your experiments on driving the circadian system at $T \neq 24$ h would be more convincing if they were carried out with light sufficient for entrainment, but below the photoperiodic threshold which is known to be higher in at least some birds. For example, the $T > 24$ h cycle may have been read by the circadian system as photostimulatory and this would be expected to accelerate rather than retard at least the growth phase of the gonadal cycle.

Gwinner: The 11 h and 12 h of light used per cycle in our experiments can indeed be expected to accelerate the system during the growth phase of the gonadal cycle. However, it can also be expected that such long photoperiods retard the system during the succeeding phase of photorefractoriness. The net effect on the entire circannual cycle is hard to predict. As emphasized in the paper many of the well-known photoperiodic effects are difficult to reconcile with any frequency demultiplication model.

Nicholls: Do not circannual rhythms result from a succession of photosensitive and photorefractory phases, each of which can be manipulated in duration by changing daylength? This would not imply any analogy to circadian rhythms.

Gwinner: We have experimental evidence indicating that our starlings maintained in a 12 h photoperiod go through a succession of photosensitive and photorefractory phases just like starlings kept under natural photoperiodic variations. Whether the rhythms *result* from this succession we do not know. During a circannual cycle the animals change their responsiveness to long and short photoperiods such that, for instance, long photoperiods accelerate seasonal processes during the photosensitive phase but delay them during the refractory phase.[1] These differential responses can be interpreted in terms of a circannual response curve to zeitgeber stimuli. Provided this interpretation is correct, circannual rhythms would behave in ways almost perfectly analogous to circadian rhythms.

Short: Do you have any evidence of circannual cycles in birds that are independent of gonadal activity? Is the moulting cycle independent of the gonads, and does it persist in gonadectomized birds?

Gwinner: In castrated starlings maintained for 15 months in a constant 12 h photoperiod two successive moults could be observed at an interval of roughly one year. In that respect they behaved like intact controls. The pattern of moult, however, was severely disrupted in the castrates. In garden warblers maintained for 3 years in a constant 10 or 12 h photoperiod one often observes an internal dissociation of the circannual rhythms of moult and testicular size. Moult may become advanced relative to the testicular cycle and eventually coincide with all phases of the testicular cycle.[2] This suggests to us that these two rhythms may be primarily independent rhythmic phenomena.

Bromage: I would like to direct both a comment and a question to Dr Gwinner. Dr Whitehead and myself have results in a teleost fish, the rainbow trout, which are superficially similar to his on birds. Thus, a constant LD 12 : 12 produces an earlier spawning in these fish when compared with fish on a normal 1 year seasonal régime. We have also shown that we can produce more than one spawning a year by placing fish on a number of 6-monthly seasonal cycles. I wonder whether you have tested constant régimes with light periods longer or shorter than 12 hours.

Gwinner: We have kept garden warblers and blackcaps in LD 10 : 14, LD 12 : 12 and LD 16 : 8. Circannual rhythms in moult and migratory restlessness were observed under all three conditions, whereas a circannual rhythm in body weight was expressed only in the garden warblers.[2] In both species and under all three conditions testicular size showed periodic variations but the period length varied greatly from about 10 months to about 4 months.[3] In European starlings, circannual rhythms in testicular size and moult persisted in an LD 12 : 12 and LD 11 : 11 as reported in this paper, but not under photoperiods longer or shorter than 12 hours.[4] The reasons for these differences among species and conditions are not yet understood.

REFERENCES

1. E. Gwinner. *J. Reprod. Fertil.* Suppl. **19**, 51–65 (1973).
2. P. Berthold, E. Gwinner and H. Klein. *J. Ornithol.* **113**, 170–90 (1972).
3. P. Berthold, E. Gwinner and H. Klein. *J. Ornithol.* **113**, 407–17 (1972).
4. R. G. Schwab. In: Menaker M. (ed.) *Biochronometry.* Washington, DC, National Academy of Sciences, pp. 428–47 (1971).

NEURAL AND ENDOCRINE COMPONENTS OF CIRCADIAN CLOCKS IN BIRDS

Michael Menaker, David J. Hudson and Joseph S. Takahashi

Interdisciplinary Program in Neurosciences, Department of Biology,
University of Oregon, Eugene, USA

Summary

Several structures of diencephalic origin play roles in the avian circadian system. The pineal gland acts as a circadian pacemaker in the house sparrow. Its removal abolishes free-running locomotor rhythmicity, which then can be restored by transplantation of a donor's pineal. The phase of the restored rhythm is determined by the donor. The chicken pineal is both light sensitive and rhythmic *in vitro*. The pineal enzyme, serotonin-N-acetyltransferase (NAT) and the putative pineal hormone melatonin, the production of which is regulated by NAT, have circadian rhythms under both static and flow-through organ culture conditions. In sparrows, lesion of the suprachiasmatic nucleus (SCN), a bilateral, anterior hypothalamic structure that receives a direct neural input from the retina, also abolishes free-running locomotor rhythmicity. Hypothalamic lesions in Japanese quail seem to have similar effects. The retinas of chickens show circadian rhythms of NAT activity and melatonin content *in vivo* which persist following pinealectomy.

Generalizations about avian circadian organization are made difficult by the variable effects of pinealectomy on behavioural circadian rhythmicity in different groups of birds. Several species of passerine birds respond as do house sparrows; however, in starlings rhythmicity is not abolished by pinealectomy, although it is severely disrupted. In chickens and Japanese quail, pinealectomy is apparently without effect on behavioural rhythmicity. In spite of these difficulties, it is possible to propose very preliminary models of how the pineal, the SCN and perhaps the retinas might interact to form the core of the circadian system. There is evidence that these structures are also involved in the circadian systems of other vertebrate classes. The need for further comparative work is clearly indicated.

More is known about the physiology of the circadian system of birds than about that of any other vertebrate class. In spite of this, the complexity and variability already evident in the limited information available about the circadian systems of only a few avian species argues that generalizations about vertebrates as a group will not come easily. Among birds, extra-retinal input appears to be general[1] and in several species of passerine birds the pineal gland is the dominant circadian pacemaking oscillator.[2] On the other hand, the role of the pineal is less clear in

Biological Clocks in Seasonal Reproductive Cycles 171–183 (1981) (ed. B. K. and D. E. Follett: Bristol, Wright).

starlings than in the other passerines,[3] and in two gallinaceous species, chickens and quail, the presently available evidence indicates that the pineal does not play a role in circadian control of locomotor rhythmicity in spite of the robust rhythmicity that has been documented within the gland itself.[2,4,5] The suprachiasmatic nuclei (SCN) have been shown to be involved in the circadian systems of both sparrows[6] and perhaps quail[5] but, at least in sparrows, probably play a substantially different role than they do in mammals. Underwood[7] has documented similar variability among the several lizard species that he has studied. In some mammals the SCN appear to be the dominant pacemaker.[8,9] The mammalian pineal's major role in the control of reproduction may be limited to those species that utilize photoperiodic cues.[10] In spite of this variability, or rather because its extent seems clearly delimited, one has the sense that circadian physiology in the vertebrates has by now become firmly grounded in an appropriate anatomical substrate. Most of the phenomena of central interest are apparently occurring in neural and neuro-endocrine processes in structures of diencephalic origin. The details of their interactions must of necessity form the subject matter of this aspect of circadian biology for some time to come.

Birds possess sensitive extra-retinal photoreceptors, as do all non-mammalian vertebrates. In sparrows these photoreceptors have been shown to contribute photic information for entrainment of the circadian system to light cycles and to mediate partially the changes in period that occur with changes in light intensity.[11,12] Although they have not been specifically localized, extra-retinal photoreceptors are known to be in the brain.[13] The pineal gland is a prime candidate as a site of extra-retinal photoreception. It is photoreceptive in lower vertebrates[14] and in several bird species its ultrastructure suggests photoreceptive functions.[15] Although the isolated pineal gland has now been shown to be sensitive to light,[16–19] removal of the pineal gland from blind sparrows (*Passer domesticus*) does not eliminate entrainment to light cycles.[20] Although the pineal almost certainly contributes to extra-retinal photoreception, it is not necessary for entrainment. Pineal photoreceptors are known to be involved in the control of biochemical rhythms in the gland but their broader role in the circadian system has yet to be determined.

Pineal removal, which has only small effects on the entrained locomotor rhythm, nevertheless has profound effects on the circadian system of the sparrow. Both the circadian rhythm of locomotor activity[21] and that of deep body temperature[22] are eliminated by pineal removal. Although pinealectomized sparrows produce arrhythmic locomotor activity in constant conditions, they entrain to light cycles and on short photoperiods their activity onsets anticipate the time of lights-on. This suggests that locomotor activity is not directly driven by the light cycle. When pinealectomized sparrows are transferred from light cycles into constant darkness, the locomotor activity rhythm 'decays' over several days into arrhythmicity. These two lines of evidence suggest that there are oscillators remaining after pinealectomy that can be entrained. The decay and subsequent loss of rhythmicity in constant conditions may indicate that these oscillators are not self-sustaining, and damp out in constant conditions. The complete loss of persistent self-sustained rhythmicity after pinealectomy could reflect the loss of a circadian pacemaker. Recent evidence strongly supports this interpretation.

Neural input to the pineal is from sympathetic fibres arising in the superior cervical ganglia[23] and the only known output pathway is a tract of small unmyelin-ated fibres in the pineal stalk.[24] Zimmerman and Menaker[25] have disrupted both the input and the output pathways without effect on the circadian system, suggesting that the role of the pineal in the production of the locomotor activity rhythm does not require neural connections. Apparently the pineal is hormonally coupled to the rest of the system.

Confirmation of the role of hormonal output from the pineal gland has been obtained by the demonstration that pineal transplantation into the anterior chamber of the eye of an arrhythmic (pinealectomized) host sparrow restores rhythmic locomotor activity[25,26] (figure 1). In some cases rhythmicity is restored in

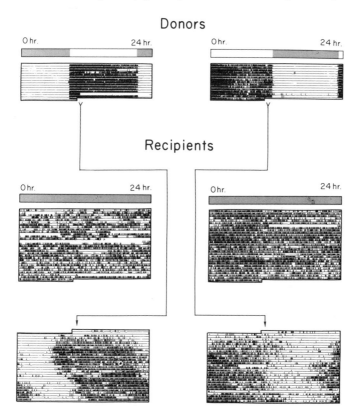

Figure 1. Experimental design for transplantation of pineals from donors on different light schedules. The donors' light cycles and activity records are shown in the top panel. The pinealectomized hosts were held in constant darkness. Their activity records before transplantation are shown in the middle panel, and after transplantation in the bottom panel. (From Zimmerman and Menaker.[27])

as few as one or two days after transplantation. One interpretation of these observations is that the transplanted pineal gland contains a pacemaker that hormonally drives the locomotor rhythm in the host. An alternative is that the pacemaker which drives the rhythm is outside the pineal but expression of the rhythm requires the presence of some pineal product. If the pacemaker is within the pineal, it should be possible by transplanting the pineal to transplant some pacemaker characteristic from the donor to the host.

Phase is an easily measured pacemaker characteristic and one can make the clear prediction that if the pineal contains a pacemaker, the phase of the host after pineal transplantation should correspond to the previous phase of the donor. Zimmerman and Menaker[27] entrained groups of donor sparrows to two light cycles (LD 12 : 12) that were 10 h out of phase. Transplantations were performed during the 2 h period during which both donor groups were exposed to light. The phases of the restored rhythms in the host sparrows were not related to the time of surgery, nor were they distributed randomly; instead the phases of the recipients were closely related to the phases of their respective donors (figure 2). Thus it appears that the

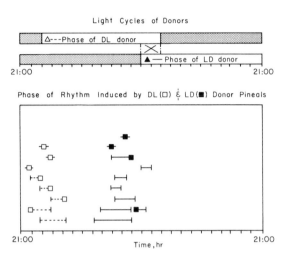

Figure 2. Phases of activity onset of hosts following pineal transplantation from donors on the two different light schedules shown in figure 1. (The time axis has been redrawn to aid visualization of the relationship between activity onset and donor's previous light cycle.) All surgery was performed during the overlapping light period. The symbols indicate the variability in the precision of the estimate of activity onset. When activity onset was clearly defined, a square (■) represents the best estimate of its phase; a horizontal line indicates a conservative estimate of range for those phases that could be determined more precisely. (From Zimmerman and Menaker.[27])

transplanted pineal gland carries temporal information from the donor to the host. The simplest interpretation is that the pineal gland contains a pacemaker capable, through hormonal pathways, of imposing circadian rhythmicity on the locomotor behaviour of sparrows.

The major hormonal product of the pineal is melatonin. Rhythmic production of melatonin by the avian pineal almost certainly results in circadian fluctuations in the levels of melatonin in the serum of sparrows (Norris and Menaker, unpublished results) and of chickens.[28-32] Melatonin that is synthesized in the pineal could therefore act at remote sites and could serve as a coupling agent between the pineal and the rest of the circadian system. Two experiments have clearly demonstrated that exogenously administered melatonin affects the circadian system of birds. When melatonin is administered continuously to sparrows, low doses shorten the free-running period of the activity rhythms, and high doses induce continuous activity.[33] In starlings, a species in which pinealectomy severely disrupts but does not abolish circadian rhythmicity,[3] rhythmic administration of melatonin by daily injections appears to entrain the activity rhythms of pinealectomized birds.[34] Although it remains to be determined whether these effects of exogenously administered melatonin are physiological, they are consistent with the idea that melatonin is the hormone that couples the pineal with the other components of the circadian system.

A good deal is known about the physiology and biochemistry of the avian pineal gland. The synthesis of melatonin in the chicken pineal appears to be regulated by the activity of the enzyme, serotonin N-acetyltransferase.[35] A dramatic circadian rhythm in the activity of this enzyme persists *in vivo* for at least two cycles in constant darkness.[36] This rhythm of enzyme activity is in many ways similar to the rhythm of locomotor activity of sparrows: it entrains to light cycles, it free-runs in constant darkness and it damps out in bright constant light. Pinealectomy, which in chickens reduces serum melatonin to below detectable levels,[30] abolishes free-running locomotor rhythmicity in sparrows. Both circadian entrainment in sparrows and the inhibition of pineal N-acetyltransferase activity in chickens by light are mediated, in part, by extra-retinal photoreceptors.[37,38] Regulation of pineal N-acetyltransferase—which in mammals is under the control of noradrenaline from the sympathetic fibres that innervate the pineal—appears to be free of sympathetic control in chickens.[39] Likewise, destruction of sympathetic inputs to the sparrow pineal does not affect behavioural circadian rhythmicity. [25] Although it is not at all clear how the oscillation in N-acetyltransferase activity in the avian pineal is generated, it is consistent with the data to conclude that the controlling mechanism is within rather than outside the pineal.

Recent experiments have established several important properties of the isolated pineal gland in organ culture. Binkley et al.[40] have shown that chicken pineals explanted into organ culture at night, when N-acetyltransferase activity is high, retain high levels of enzyme activity until the time that activity normally declines *in vivo*. The timing of the decline in N-acetyltransferase activity *in vitro* is related to the light cycle to which the chicks were previously exposed and not to the time the glands were placed into culture. This experiment suggests that chicken pineals contained at least an interval timer that was synchronized to the previous *in vivo* light cycle. Four laboratories have now independently documented a rise and fall in N-acetyltransferase activity in isolated pineals held *in vitro* for at least 24 hours in constant darkness.[16-18,41] Pineals, explanted and cultured late in the daytime when enzyme levels are low, express an endogenous rhythm of N-acetyltransferase activity which peaks at night. Kasal et al.[41] have shown that the rhythm persists for two cycles in constant darkness with a circadian period *in vitro*. However, the difficulty in measuring rhythms in populations of isolated glands (unavoidable when N-acetyltransferase activity is used as an end-point) and the fact that the rhythm appears damped, have precluded a definitive conclusion that the avian pineal is a self-sustained oscillator.

We have developed a flow-through culture system for the isolated pineal from which we can measure the release of melatonin continuously from superfused glands over long periods of time.[19] Chicken pineals release melatonin rhythmically and these rhythms persist *in vitro* with a circadian oscillation. In light cycles the release of melatonin is strongly rhythmic (figure 3). However, in constant conditions the amplitude of the rhythm is lower and appears to be damping (figure 4). It does not seem likely that damping is due to poor culture conditions because in light/dark cycles the rhythm is not damped and persists with high amplitude. Previous experiments using N-acetyltransferase activity as an assay for pineal rhythmicity *in vitro* have yielded equivocal results in constant conditions. While Kasal et al.[41] were able to find two cycles of N-acetyltransferase rhythms in constant darkness, Binkley et al.[16] and Deguchi[17] reported that they were unable to

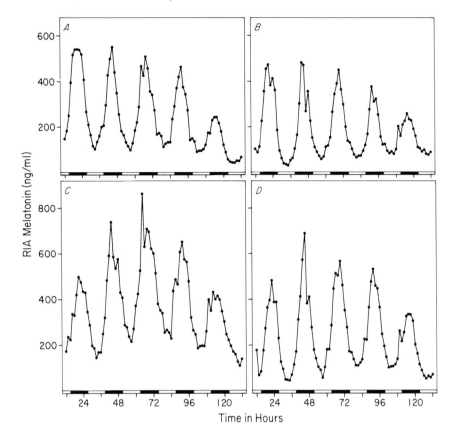

Figure 3. Rhythms of melatonin release from four different individually isolated chicken pineals cultured in a flow-through superfusion system. The *in vitro* LD 12 : 12 (L = 350–500 lux cool white fluorescent: D = 0 lux) light cycle is indicated at the bottom of each panel. Samples of perfusate were collected for 90 min. Concentration of melatonin in the culture medium was determined by radioimmunoasay (RIA). Each point represents a single RIA determination and is plotted at the onset of the collection interval. The flow rate in this experiment was 0·25 ml/h. (From Takahashi et al.[19])

find persistent circadian rhythms. The results from individually isolated pineals suggest that the negative findings using N-acetyltransferase are due in part to the difficulty in measuring low amplitude rhythms in experiments assaying population rhythms. Furthermore, these results clearly show that the damping of the rhythm is a process that occurs within an individual pineal gland.

Whatever mechanism underlies the damping in constant conditions of the chicken pineal melatonin rhythm *in vitro*, it is clear that rhythmic light input maintains the amplitude and synchronizes the rhythm. In addition to the synchronizing effects of light, acute light exposure at night when melatonin levels are high rapidly inhibits melatonin release[19] and N-acetyltransferase activity.[17] In addition, exposure of pineal glands to bright constant light *in vitro* reduces the rise in N-acetyltransferase activity.[16–18]. Thus, light has at least two effects upon the isolated pineal: cyclic light input synchronizes the rhythm and acute light exposure at night rapidly inhibits melatonin release and/or synthesis.

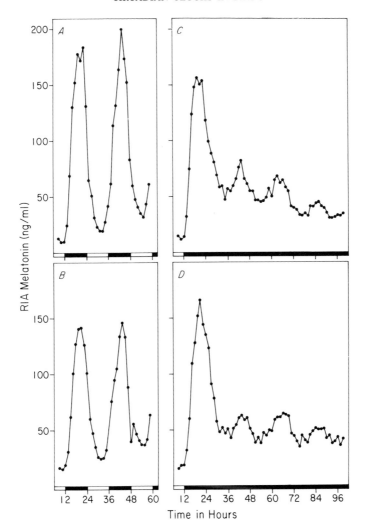

Figure 4. Comparison of the rhythms of melatonin release in a light cycle and in constant conditions. The light treatments are indicated at the bottom of each panel. *A* and *B* are two individual pineals from a 60 h LD 12 : 12 (L = 350–500 lux cool white fluorescent: D = 1–5 lux red light) experiment. *C* and *D* are two individual pineals from a 96 h experiment in constant conditions (dim red illumination, 1–5 lux). Pineals in both experiments came from the same group of chickens. The flow rate was 0·5 ml/h. The flow rate in these experiments was twice as high as that in the experiments shown in figure 3, which accounts for most of the difference in the melatonin concentration in the perfusate. Collection interval and points plotted are as in figure 3. (From Takahashi et al.[19])

The regulation of melatonin synthesis in the chicken pineal is not well understood. Although in the mammalian pineal β-adrenergic receptors appear to regulate the activity of the enzyme N-acetyltransferase (which is the rate-limiting step in the synthesis of melatonin), this is clearly not the case in the chicken pineal. Application of noradrenaline to organ-cultured chicken pineals does not stimulate N-acetyltransferase activity as it does in the rat.[16,42–44] In organ-cultured chicken pineals noradrenaline decreases N-acetyltransferase activity and this inhibitory effect appears to be mediated by an α-adrenergic receptor.[44] Weak evidence suggests that cyclic nucleotides are involved in regulating N-acetyltransferase activity in the chicken pineal. Phosphodiesterase inhibitors (theophylline, 3-isobutyl-1-methyl xanthine, RO 20–1724), cyclic AMP analogues (monobutyryl cAMP, dibutyryl cAMP) and cholera toxin cause increases in N-acetyltransferase activity in organ-cultured pineals.[16,44,45] Wainwright[46] has recently shown that the cGMP content of cultured chicken pineals expresses a daily rhythm which is correlated with the rhythm of N-acetyltransferase activity. He was unable to find a rhythm in cAMP content. The experiments of Deguchi[44] and Wainwright[46] suggest that cyclic nucleotides affect N-acetyltransferase activity; however, the evidence is indirect at the present time.

Taken together, the results described above support the view that the chicken pineal contains a circadian oscillator and a photoreceptor, and has the capacity for melatonin biosynthesis. Deguchi[47] has presented evidence which suggests that the circadian oscillation of N-acetyltransferase activity and the response to light are properties of pineal cells. He has shown that the N-acetyltransferase activity of dispersed cell cultures of chicken pineal glands expresses a circadian rhythm and responds to environmental light. The behaviour of pineal cell cultures is virtually the same as that of pineal organ cultures. Although Deguchi's results are very interesting, they do not yet demonstrate that circadian rhythmicity is a property of individual cells because the culture preparations contain numerous cells of many types. Therefore, many types of intercellular communication among the pineal cells could have occurred. Furthermore, such interactions, if they exist, would be difficult to eliminate experimentally. Nevertheless, the cell culture experiments clearly demonstrate that the structural integrity of the pineal *as an organ* is not necessary for circadian rhythmicity or photoreceptive response.

The structure of the chicken pineal gland is relatively well described.[15,48–50] The gland is encapsulated within a connective tissue sheath. Septa of collagen penetrate inwards from the capsule and divide the pineal into many lobules. Within these lobules are numerous cell clusters or follicles that are supported by connective tissue partitions arising from proliferations of the lobule septa. A network of nerve fibres and an extensive capillary system are interlaced within the connective tissue partitions.

Ultrastructural studies indicate that the follicles contain three identifiable cell types—ependymal, secretory and photoreceptor-like cells.[51] In adult chickens, the ependymal cells have apical cilia with the 9 + 2 arrangement of microtubules, the secretory cells have numerous secretory granules throughout the cytoplasm and the photoreceptor-like cells have synaptic contacts along their basal border and have modified 9 + 0 cilia sometimes connecting the apical portion of the cell to a membranous lamellar whorl located in the lumen of the follicle.[51] The photoreceptor-like cells resemble degenerate versions of the cone-like photoreceptor cells found in lower vertebrate pineals.[14,52] In 2–5-day-old chicks, there appear to be more photoreceptor-like cells (relative to secretory cells) than in the adult.[53] The photoreceptor-like cells of young chicks did not appear to have membranous whorls attached but almost always had 9 + 0 cilia ending in a cytoplasmic bulb. The ultrastructural studies indicate that the pineal of the chicken contains cells with structural features corresponding to both secretory and photoreceptive functions. At the present time no clear correlation at the cellular level can be made between structure and function in the avian pineal.

In spite of the fact that rhythmicity within the chicken pineal is robust and well

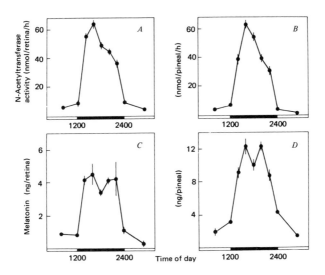

Figure 5. Rhythms of N-acetyltransferase activity and melatonin content of the retina (*A* and *C*) and pineal gland (*B* and *D*) of chicks in LD 12 : 12. The light cycle to which the birds were exposed is shown below each panel (*n* = 5 for each group; symbols represent the mean ± s.e.). (From Hamm[58].)

documented, it is not yet clear what role, if any, the gland plays in the circadian system of gallinaceous birds.[2] Paradoxically, pinealectomy of either chickens or their close relatives, Japanese quail, does not appear to affect their locomotor rhythmicity.[4,5] For this reason it is important to consider whether there might be other sources of rhythmically produced melatonin (or other 'pineal' products) that might be responsible for continued rhythmicity in the absence of the pineal.

Melatonin measured by radioimmunoassay has been found in the retina, brain and Harderian gland of chickens.[54] The enzymes necessary for melatonin biosynthesis, N-acetyltransferase and hydroxyindole-0-methyltransferase are present in retina of chicks.[55,56] It is therefore possible that melatonin is synthesized in tissues other than the pineal.

Hamm and Menaker[57] have found a large amplitude circadian rhythm of indoleamine metabolism in the chicken retina. N-acetyltransferase activity and melatonin content are fifteen times higher at night than during the day in light/dark conditions (figure 5). The properties of retinal N-acetyltransferase are similar to those of pineal N-acetyltransferase: both are rhythmic and are inhibited by light. The retinal N-acetyltransferase rhythm is not dependent upon the presence of the pineal, suggesting that retinal melatonin is synthesized in the retina and is not taken up from the circulation.

The function of the retinal melatonin rhythm is not clear. Retinal melatonin could be released into the circulation and contribute to the circadian rhythm of serum melatonin in chickens. However, there are two reports of failure to find a serum melatonin rhythm after pinealectomy.[28,30] Because melatonin levels in the serum are low, assay methods with the required sensitivity and specificity have been unavailable until recently. Radioimmunoassay methods for melatonin are extremely sensitive; however, they are by no means completely specific. Melatonin, measured by radioimmunoassay, has been reported in the plasma of pinealectomized rats;[59] however, it was not detected by a much more specific assay, gas chromatography-negative ionization mass spectrometry.[60] Clearly, caution is required in drawing conclusions about the sources of circulating melatonin, especially when radioimmunoassay is the only method of measurement employed. Although melatonin can be synthesized in non-pineal tissues of rats,[54] melatonin in the plasma originates solely from the pineal.[60] Whether a similar situation exists in chickens remains to be determined.

If retinal melatonin is not released into the circulation, it could act within the eye. The metabolism of retinal photoreceptors in vertebrates is dramatically rhythmic. In chickens, rod outer segments shed packets of discs during the first few hours of light in a light/dark cycle, and cone outer segments shed during the dark period[61] (figure 6). Because melatonin induces movements of cones similar to those

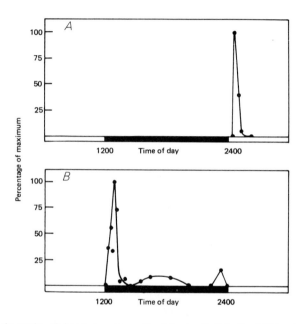

Figure 6. Changes in the number of phagosomes observed in pigment epithelial cells of chick retina at different times of day. A, Number of stage I phagosomes that were derived from rod outer segments. The maximum number of phagosomes from rods was 5·9 phagosomes per 100 visual cells. B, Number of stage I phagosomes that were derived from cone outer segments. The maximum number of phagosomes from cones was 12·8 phagosomes per 100 visual cells. (Redrawn from Young.[61])

caused by light[62] and produces pigment aggregation in the retinal pigment epithelium of fish[63] and mammals,[64] it seems possible that retinal melatonin is involved in the regulation of circadian rhythms of photoreceptor metabolism.[57]

Because the suprachiasmatic nuclei (SCN) appear to function as pacemaking oscillators in rodents their role in avian circadian systems has recently been investigated.[5,6] In house sparrows, SCN lesions abolish the free-running locomotor rhythm but lesioned birds still entrain to light cycles (figure 7). The effects of SCN lesion are thus very similar to those of pinealectomy. In Japanese quail, although pinealectomy is apparently without effect, hypothalamic lesions abolish rhythmicity as they do in sparrows.[5] There are two classes of explanation of these facts: either the circadian systems of the two species are radically different, or there is an underlying commonality that is not yet revealed by the data. While it may be

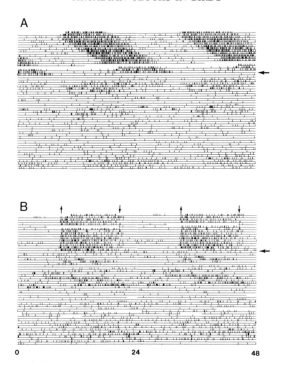

Figure 7. The effect of suprachiasmatic nucleus lesions on the activity of house sparrows. *A*, The effect of a lesion on the circadian activity in constant darkness. The day of the operation is indicated by the arrow on the right. *B*, Entrainment of a lesioned sparrow. The bird was exposed to light cycles for 15 days (LD 12 : 12 indicated by small arrows). The arrow on the right indicates the day on which the bird was transferred to constant darkness.

correct, the first of these possibilities is repugnant. Perhaps the simplest form of the second class of explanation is that the SCN are, in both species, target organs for the hormonal output of a dominant circadian pacemaker—the pineal gland in the case of sparrows, and an as yet unidentified structure in Japanese quail. Particularly in view of the data from organ-cultured chicken pineals, a somewhat more complex hypothesis is more likely, i.e. in both species, both the pineal and the SCN contain oscillators, and in both species the SCN is coupled to locomotor centres; in sparrows the SCN oscillators must be synchronized by rhythmic hormonal output from the pineal, while in the quail's SCN the oscillators are more tightly coupled among themselves and therefore maintain synchrony and thus 'system' rhythmicity even in the absence of pineal input. There are, of course, other models that could account for the current facts. More important than specific modelling at this point would appear to be identification of experimental approaches and biological systems particularly suited for exploration of general features of avian circadian organization. Because its response to pinealectomy is roughly intermediate between that of sparrows and quail, the starling recommends itself as particularly interesting for further work.

The most that can be expected from a comparative approach to any complex physiological system is that it will delineate the range of adaptations that natural selection has been able to devise and thus provide a framework into which new observations can be fitted. Avian circadian systems appear quite diverse and yet are amenable to analysis with presently available techniques; therefore they should repay further comparative study. Perhaps they will provide a model for, or at least indicate the range of variability within, circadian organization among the vertebrates in general.

ACKNOWLEDGEMENTS

This work was supported by National Institutes of Health Grants HD–03803 and HD–07727 to M. Menaker; National Research Service Award 7 F32 NS06096–02 to D. J. Hudson; and a National Science Foundation Graduate Fellowship and National Institute of General Medical Sciences National Research Service Award 5T32 GM 07257 in Systems and Integrative Biology to J. S. Takahashi.

REFERENCES

1. M. Menaker and H. Underwood. *Photochem. Photobiol.* **23**, 299–306 (1976).
2. J. S. Takahashi and M. Menaker. *Fed. Proc.* **38**, 2583–8 (1979).
3. E. Gwinner. *J. Comp. Physiol.* **126**, 123–9 (1978).
4. S. E. MacBride. PhD Thesis, University of Pittsburgh (1973).
5. S. M. Simpson and B. K. Follett. *Proc XVII Int. Ornithol. Congr.* (in the press) (1981).
6. J. S. Takahashi and M. Menaker. In: M. Suda, O. Hayaishi and H. Nakagawa (ed.). *Biological Rhythms and their Central Mechanisms.* Amsterdam, Elsevier/North-Holland, pp. 95–109 (1979).
7. H. Underwood. This volume, pp. 137–52.
8. B. Rusak and I. Zucker. *Physiol. Rev.* **59**, 449–526 (1979).
9. M. Menaker, J. S. Takahashi and A. Eskin. *Rev. Physiol.* **40**, 501–26 (1978).
10. F. W. Turek and C. S. Campbell. *Biol. Reprod.* **20**, 32–50 (1979).
11. J. P. McMillan, J. A. Elliott and M. Menaker. *J. Comp. Physiol.* **102**, 257–62 (1975).
12. J. P. McMillan, H. C. Keatts and M. Menaker. *J. Comp. Physiol.* **102**, 251–6 (1975).
13. M. Menaker. *Proc. Natl Acad. Sci. USA* **59**, 414–21 (1968).
14. E. Dodt. In: H. Autrum, R. Jung, W. R. Lowenstein, D. M. McKay and H. L. Tueber (ed.). *Handbook of Sensory Physiology*, New York, Springer-Verlag, pp. 113–40 (1973).
15. M. Menaker and A. Oksche. In: D. S. Farner and J. R. King (ed.). *Avian Biology*, New York, Academic Press, pp. 79–118 (1974).
16. S. Binkley, J. B. Riebman and K. B. Reilly. *Science* **202**, 1198–201 (1978).
17. T. Deguchi. *Science* **203**, 1245–7 (1979).
18. S. D. Wainwright and L. K. Wainwright. *Can. J. Biochem.* **57**, 700–9 (1979).
19. J. S. Takahashi, H. Hamm and M. Menaker. *Proc. Natl Acad. Sci. USA* **77**, 2319–22 (1980).
20. M. Menaker. *Biol. Reprod.* **4**, 295–308 (1971).
21. S. Gaston and M. Menaker. *Science* **160**, 1125–7 (1968).
22. S. Binkley, E. Kluth and M. Menaker. *Science* **174**, 311–14 (1971).
23. M. Ueck. *Z. Zellforsch.* **105**, 276–302 (1970).
24. M. Ueck and H. Kobayashi. *Z. Zellforsch.* **129**, 140–60 (1972).
25. N. H. Zimmerman and M. Menaker. *Science* **190**, 477–9 (1975).
26. M. Menaker and N. H. Zimmerman. *Am. Zool.* **16**, 45–55 (1976).
27. N. H. Zimmerman and M. Menaker. *Proc. Natl Acad. Sci USA* **76**, 999–1003 (1979).
28. R. W. Pelham, C. L. Ralph and I. M. Campbell. *Biochem. Biophys. Res. Comm.* **46**, 1236–41 (1972).
29. C. L. Ralph, R. W. Pelham, S. E. MacBride and D. P. Reilly. *J. Endocrinol.* **63**, 319–24 (1974).
30. R. W. Pelham. *Endocrinology* **96**, 543–6 (1975).
31. C. L. Ralph. *Am. Zool.* **16**, 35–43 (1976).
32. D. J. Kennaway, R. G. Frith, G. Phillipou, C. D. Matthews and R. F. Seamark. *Endocrinology* **101**, 119–26 (1977).
33. F. W. Turek, J. P. McMillan and M. Menaker. *Science* **191**, 1441–3 (1976).
34. E. Gwinner and I. Benzinger. *J. Comp. Physiol.* **127**, 315–28 (1978).
35. S. Binkley, S. E. MacBride, D. C. Klein and C. L. Ralph. *Science* **181**, 273–5 (1973).
36. S. Binkley and E. B. Geller. *J. Comp. Physiol.* **99**, 67–70 (1975).

37. S. Binkley, S. E. MacBride, D. C. Klein and C. L. Ralph. *Endocrinology*. **96**, 848–53 (1975).
38. C. L. Ralph, S. Binkley, S. E. MacBride and D. C. Klein. *Endocrinology* **97**, 1373–8 (1975).
39. S. Binkley. *Fed. Proc.* **35**, 2347–52 (1976).
40. S. Binkley, J. B. Riebman and K. B. Reilly. *Science* **197**, 1181–3 (1977).
41. C. A. Kasal, M. Menaker and J. R. Perez-Polo. *Science* **203**, 656–8 (1979).
42. J. Axelrod. *Science* **184**, 1341–8 (1974).
43. J. Axelrod and M. Zatz. In: G. Litwack (ed.). *Biochemical Actions of Hormones,* vol. IV. New York, Academic Press, pp. 249–68 (1977).
44. T. Deguchi. *J. Neurochem.* **33**, 45–51 (1979).
45. S. D. Wainwright and L. K. Wainwright. *Can. J. Biochem.* **56**, 685–90 (1978).
46. S. D. Wainwright. *Nature* **285**, 478–80 (1980).
47. T. Deguchi. *Nature* **282**, 94–6 (1979).
48. C. W. Beattie and F. H. Glenny. *Anat. Anz.* **118**, 396–404 (1966).
49. J. Boya and J. Calvo. *Acta Anat.* **101**, 1–9 (1978).
50. P. A. L. Wight and G. M. MacKenzie. *J. Anat.* **108**, 261–73 (1971).
51. M. B. Bischoff. *J. Ultrastruct. Res.* **28**, 16–26 (1969).
52. R. M. Eakin. *The Third Eye.* Berkeley, University of California Press (1973).
53. J. Boya and L. Zamorano. *Acta Anat.* **92**, 202–26 (1975).
54. S. F. Pang, G. M. Brown, L. J. Grota, J. W. Chambers and R. L. Rodman. *Neuroendocrinology* **23**, 1–13 (1977).
55. S. Binkley, M. Hryshchyshyn and K. Reilly. *Nature* **281**, 479–81 (1979).
56. W. B. Quay. *Life Sci.* **4**, 983–91 (1965).
57. H. Hamm and M. Menaker. *Proc. Natl Acad. Sci USA* **77**, 4998–5002 (1980).
58. H. E. Hamm. PhD Thesis, University of Texas at Austin (1980).
59. Y. Ozaki and H. J. Lynch. *Endocrinology* **99**, 641–4 (1976).
60. A. J. Lewy, M. Tetsuo, S. P. Markey, F. K. Goodwin and I. J. Kopin. *J. Clin. Endocrinol. Metabol.* **50**, 204–5 (1980).
61. R. W. Young. *Invest. Ophthalmol. Visual Sci.* **17**, 105–16 (1978).
62. W. B. Quay and R. W. McLeod. *Anat. Rec.* **160**, 491 (1968).
63. C. Cheze and M. Ali. *Can. J. Zool.* **54**, 475–81 (1976).
64. S. F. Pang and D. R. Yew. *Experientia* **35**, 231–3 (1979).

DISCUSSION

Turek: Is there any effect on patterns of circadian rhythm of activity in birds or lizards following blinding in DD?

Underwood: Experiments on the effects of blinding on lizards in DD are currently under investigation.

Kawamura: May I present some data obtained by Dr S. Ebihara in my laboratory showing the effects of SCN lesions and pinealectomy in the Java sparrow (*Padda oryzivora*):

In constant dim light, free-running circadian activity rhythms were abolished after SCN lesions.

In SCN-lesioned birds (pineal organ intact), after exposure to an LD 12 : 12 cycle, no circadian rhythms were observed.

However, in pinealectomized birds with an intact SCN, during exposure to an LD 12 : 12 cycle, some anticipatory activity was observed before the light period began. After exposure to an LD 12 : 12 cycle, circadian rhythms of locomotor activity in constant dim light often showed splitting.

PHOTOPERIODIC TIME MEASUREMENT AND GONADOTROPHIN SECRETION IN QUAIL

B. K. Follett, J. E. Robinson, S. M. Simpson and C. R. Harlow

ARC Research Group on Photoperiodism and Reproduction,
Department of Zoology, University of Bristol

Summary

Birds use their circadian system to measure photoperiodic time, and the first section considers three of the tests capable of showing this: the Nanda–Hamner protocol, the single light pulse technique and the *T* experiment. While the actual circadian rhythm of photoperiodic sensitivity can be visualized with the single pulse design, an understanding of the way in which a circadian rhythm (or rhythms) measures daylength probably necessitates our knowing the physiological changes which occur during photo-induction. Analysis of the temporal pattern of LH secretion in photostimulated Japanese quail has shown that during the first two long days there is a daily rhythm in the secretion of LH which may closely reflect the processes underlying daylength measurement. Induction is caused by the light occurring 12–16 h after dawn and this leads to a release of LH some hours later. These experiments suggest that the circadian period of photoperiodic sensitivity in quail lies in the middle of the subjective night. After the first few days of photostimulation LH (and FSH) secretion is no longer rhythmic but is maximal at all times. It would appear, therefore, that at some stage either the time measuring system or the neural circuits regulating LH-RH secretion become continuously active under long days. Transferring birds from long to short days does not cause LH secretion to cease immediately but rather to decrease slowly over a 10-day period. This 'carry-over' is an important property of the system, and its existence means that the gonads can be grown under schedules involving long days interspersed with short days. Finally, some consideration is given to the fact that the critical daylength in quail alters seasonally from about 12 h in the spring to 14·5 h in later summer. This shift, in what was considered previously to be an unalterable threshold, can be mimicked in the laboratory. It is not known how such a shift occurs, but the favoured hypothesis is that some phase change in the photoperiodic clock is involved.

INTRODUCTION

Seventeen years ago Hamner[1] upset many cherished ideas by showing that a higher vertebrate, the house finch (*Carpodacus mexicanus*), measured photoperiodic time not with an hour-glass but with a system involving circadian rhythms. Using a

Biological Clocks in Seasonal Reproductive Cycles 185–201 (1981) (ed. B. K. and D. E. Follett: Bristol, Wright).

protocol that combines a short photoperiod (6 or 8 h) with long periods of darkness ranging up to 66 h, and which was first developed by Nanda and Hamner[2] for plants, he found testicular growth to occur in LD 6 : 6, 6 : 30 and 6 : 54 but not in LD 6 : 18, 6 : 42 and 6 : 66. This result is understandable only if daylength is being measured with the circadian system: were an hour-glass involved no growth would have occurred in any cycle. Whether daylength is measured in all photoperiodic birds using the circadian system is unknown and the number exposed to the full Nanda–Hamner protocol is so few[1,3-5] that caution is needed before concluding that there is a universal timing mechanism. In addition, unease attaches to many of the actual experiments themselves because induction is often weak and the extent of gonadal growth can be quite slight. Does this raise the possibility of other time-measuring systems existing within birds, or can such results be explained on technical grounds? One unpublished experiment of Barbara Dobson encourages us to take the latter view. She followed the evolution of LH secretion in a range of Nanda–Hamner photoperiods (figure 1) and found that the LD 6 : 30 and 6 : 54

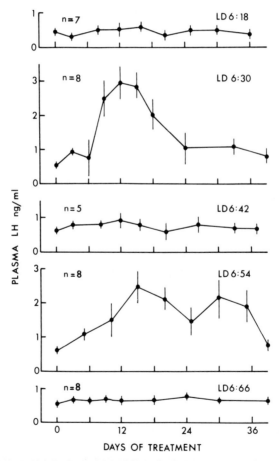

Figure 1. A Nanda–Hamner experiment with Japanese quail. Plasma LH is shown for five schedules; note how induction occurs only in LD 6 : 30 and 6 : 54, but even here it disappears after some cycles. (B. Dobson, unpublished experiment.)

cycles were initially highly inductive but that they later became non-stimulatory and the LH levels fell away. This was presumably due either to the complex entrainment problems that beset such schedules[6] or to a damping in the photoperiodic oscillators. At the end of the experiment the testes were larger than in LD 6 : 18, 6 : 42 and 6 : 66 but all showed histological signs of regression. Inevitably, this result questions whether a negative Nanda–Hamner experiment[7,8] can be used as a powerful argument against circadian involvement in the measurement of daylength. Such a conclusion might have been drawn from the experiment shown in figure 1 had the quail been killed 2 weeks later.

A derivative of the Nanda–Hamner protocol is the 'single-pulse' technique, carried out so far only in the white-crowned sparrow[9] and the short-day plant, *Chenopodium rubrum*.[10] This test relies upon detecting induction after exposure to only one pulse of light and, if this condition can be met, a direct demonstration of the circadian rhythm in photoperiodic sensitivity to light is possible. In the experiment with white-crowned sparrows[9] each animal provided one data point. The bird was released from LD 8 : 16 into a DD free run and some time later given one 8 h pulse of light. This light tested whether or not the bird was photoperiodically sensitive at the time when the pulse was given, any elevation of the level of plasma LH 6 h after the end of the pulse indicating induction. In effect, the light pulse was being used as a probe to assess the status and position of the circadian photoperiodic rhythm. This is shown in figure 2 as a population rhythm which was

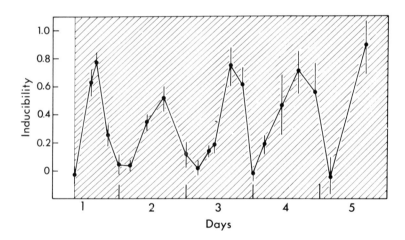

Figure 2. An experiment in white-crowned sparrows that shows the circadian rhythm in photoperiodic sensitivity. The curve was obtained by releasing sparrows ($n = 198$) on day 1 from LD 8 : 16 into darkness. At one of 21 subsequent times the rhythm was assayed by exposing the bird to an 8 h light pulse and measuring the degree of induction (on a scale 0 to 1) by the change in plasma LH. The number of birds at each point ranges from 6 to 16, mean ± s.e. mean. (Redrawn from Follett et al.[9])

determined from the responses of 198 free-running birds each of which had been exposed to light at one of 21 times. There seems to be little doubt that photoperiodic sensitivity is circadian based, the rhythm free-running for five cycles and showing no signs of damping out. The light pulse not only indicated whether the bird was in an inductive phase, but also gave a quantitative measure of 'photoperiodic sensitivity', this term being used because 'photo-inducible rhythm', the more common phrase, has strong overtones of the external coincidence model. Unfortunately, however, the pulse was slightly too long to indicate the precise duration of the photoperiodically sensitive phase or of its free-running period, although some of the responses early on days 3 and 4 suggest that τ may be less than 24 h, and the increasing variance seen as the free run becomes longer may indicate some individual variation in τ. Extension of this model clearly requires a system where induction can be made to occur with a very short light pulse. Japanese quail respond to 1 long day (*see* figures 3–5), 'as well as showing induction to night-interruptions of 15 min if these are given for a week or so,[3,11,12] but marrying up these properties in a thoroughly reliable way has as yet proved impossible. One recent idea is to use schedules combining long and short days (*see* figure 8) in order to produce a highly responsive bird.

The so-called T experiment, originally developed by Pittendrigh and Minis,[13] is another powerful method of showing whether circadian rhythms are involved in measuring daylength. It has proved valuable with the golden hamster[14] and the flesh-fly, *Sarcophaga argyrostoma*,[15] to the extent of allowing the length of the photosensitive phase to be mapped out in these species, but T experiments seem to be less successful in birds. In house finches, a cycle of 26 h (LD 6 : 20) caused significantly greater testicular growth than a shorter 22 h cycle (LD 6 : 16) but the left testes weighed only 3·80 mg after a month of treatment, substantially different from the 50–100 mg obtained with long days (LD 18 : 6). Subsequently, Farner et al.,[16] using 3 h photoperiods in various T cycles, found quite extensive testicular growth (30–120 mg) in house sparrows on long cycles of 25 and 28 h, the rate on LD 3 : 25 being 40 per cent of that during normal long days. In white-crowned sparrows, however, cycles of LD 3 : 23 were not inductive at all. Our own experiences with quail[17] have been generally similar with T cycles, using either a 1 h (T from 19·1 to 25·7 h) or 3 h (T from 21 to 36 h) photoperiod, not causing any testicular growth. As expected, lengthening the T cycle produces a phase advance in the circadian oscillators regulating locomotor activity. The shift is considerable with the longer T cycles and amounts to an advance of nearly 7 h in house sparrows transferred from LD 3 : 21 to 3 : 23,[16] and of 5 h in quail moved from LD 3 : 21 to 3 : 24.[17] Even longer cycles lead to relative coordination or to a breakdown in the rhythm as the limits of entrainment are exceeded. In the case of the circadian rhythm responsible for timing oviposition in the quail an alteration of T from 24 h (LD 13 : 10) to 27 h (LD 13 : 14) advances the mean oviposition time by 7 h.[17] The extent of the phase shifts in these other rhythms and the lack of gonadal growth in quail exposed to very long T cycles (e.g. LD 3 : 33) argues that the phase response curve of the photoperiodic oscillator(s) may be different from those regulating activity and oviposition. This suspicion is reinforced by finding that the critical daylength in T cycles of 30 h is approximately 10·5 h compared with 11·5–12 h under normal 24 h cycles,[17] i.e. a phase shift of about only 1 h. It is not clear whether this means that the photoperiodic oscillator(s) are unrelated to the locomotor clock, or that both are coupled differently to some master oscillator.[18] The lack of a simple phase relationship between the two is important practically because it means that measuring another overt oscillator, such as locomotor activity, whilst giving a qualitative picture of the bird's circadian system can only have limited usefulness in a quantitative sense. This point does not apply to hamsters[14] or to the flesh-fly[15] where the respective activity and eclosion rhythms closely predict the phase of photoperiodic sensitivity.

While many of the experiments above present a strong case for invoking circadian rhythms in avian photoperiodic time-measurement they do not show how the rhythms carry out the process. Two fairly explicit models, however, have been developed and these tend to dominate thinking on this problem. The first model, 'external coincidence', involves a circadian rhythm of photo-inducibility, one phase of which (ϕ_i) is illuminated by long days (causing induction), but not by short days. The essence of the second or 'internal coincidence' model is that, as daylength changes throughout the year, there are alterations in the phase relationships of the many circadian oscillators within the organism. Induction is imagined to occur when two (or more) of these oscillators are placed in a special phase relationship. Virtually all the formal photoperiodic evidence can be fitted to either model but, nevertheless, the view is often expressed that one or other of the models is more or less applicable in a given instance.[15,16] The apparent simplicity of 'external coincidence' is appealing, not least to undergraduates learning about photoperiodism, and this, together with its historical precedence,[19] has tended to make it the most well known and frequently quoted model. A key problem, however, is to know what is meant physiologically by a 'rhythm of photo-inducibility'. While the 'internal coincidence' model is sometimes thought to be conceptually more difficult, its great strength lies in the fact that entrainment theory demands a re-arrangement of the phase relationships between the circadian oscillators as the photoperiod alters, and Pittendrigh, in the opening chapter of this volume, describes how and why this occurs. At the moment, therefore, the general concept of 'internal coincidence' is more attractive, although Pittendrigh stresses that it would be surprising if the detailed mechanisms of photoperiodic time measurement were identical in all species.

Where does this leave the practising physiologist? Perhaps with the feeling that distinguishing between the various models is becoming ever more difficult without an understanding of the concrete events underlying photo-induction. It is for this reason that we have tried to determine how the gonadotrophic hormones are secreted temporally in photostimulated quail. Plasma levels of LH mirror fairly precisely the hypothalamic secretion of LH-RH and seem to be as close a reflection of the photoperiodic clock as can be obtained at the present time. With such an approach it is at least feasible to inquire whether or not LH is secreted rhythmically, perhaps reflecting some kind of daily 'coincidence' within the circadian system, or whether LH levels remain relatively constant—a possibility if coincidence leads to a neural state where a switch is thrown triggering continuous secretion of hormone. As will become clear, however, understanding the results is clouded by difficulty in distinguishing changes in the clock machinery from those in the neural circuits regulating LH-RH secretion. The data offer a little help in the debate about 'internal' versus 'external' coincidence but do not distinguish critically between the alternatives. Approaches such as chronic neurophysiological recording within the brain may soon be feasible but this technique also has drawbacks when used alone. The ideal might be a combination of the two, whereby neurophysiological changes can be correlated directly to alterations in gonadotrophin secretion.

PATTERNS OF GONADOTROPHIN SECRETION DURING PHOTO-INDUCTION

Exposing birds to long days increases the levels of plasma LH and FSH some sevenfold.[20] Secretion is held in check by steroid feedback and the full extent of the photoperiodic response can be appreciated by removing the gonads. In gonadecto-mized quail levels of LH rise from 0·5 ng/ml on short days to over 100 ng/ml on long days: the equivalent changes in FSH secretion being from 20 to about 5000 ng/ml. Returning the quail to short days suppresses gonadotrophin secretion and the levels are basal again within 2 weeks. While this is the overall pattern, a detailed analysis reveals three separate phases.

During the first few days following transfer from LD 8 : 16 to LD 20 : 4[21–23] there is a daily rhythmicity in LH (and FSH) secretion (figure 3). Some variation exists

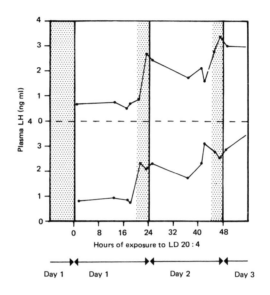

Figure 3. LH concentrations in two male quail during the first 2 days of exposure to LD 20 : 4.[22] Darkness is shown by the shading.

from bird to bird (an ill-understood but potentially important point[21]) but the following picture is generally seen. During much of the first long day LH levels are basal but at about hour 20 they start to rise and continue to do so through the short night and the early part of the next day. During this second day, plasma LH levels either remain constant or fall, but at about hour 18 they begin to rise again. The third day is sometimes like the second but subsequently the rhythmicity breaks down and hormone levels are maximal at all times.

It seems likely that the rhythmic secretion on days 1 and 2 reflects the circadian system underlying daylength measurement fairly directly, and so experiments (figure 4) have been undertaken to investigate the timing of the LH rise under various photoperiods. When birds were transferred from LD 8 : 16 and blood samples taken 12, 19 and 25 h after dawn, the timing of the rise was similar whether the quail were stimulated with 16 h, 20 h or continuous light. With 14·7 h of light there was also a significant rise in LH by hour 25, but this did not occur with photoperiods of 12 or 13·3 h. Two conclusions were drawn: first, that light between 12 and 15 h after dawn is sufficient to trigger a photoperiodic response and, secondly, that the actual rise in LH is not seen until some time after hour 15. The converse experiment was also carried out, with groups of short-day quail exposed

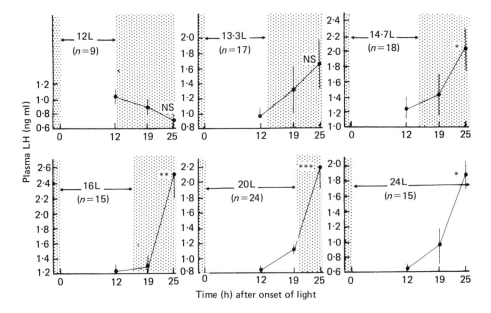

Figure 4. A series of experiments to show the timing of the rise in LH after transfer from LD 8 : 16 to one of various photoperiods. The number of birds is shown in parenthesis together with the duration of the photoperiod. (Means ± s.e. mean.) The notation by the value at 25 h indicates the result of comparing it with the mean level at 12 h: n.s. not significant; *$P < 0.05$; **$P < 0.01$; ***$P < 0.001$. Shaded areas indicate darkness. (From Follett et al.[22])

to 24 h of continuous light or to schedules which included 4 h of darkness 12–16 h after dawn, or 8 h of darkness from hours 12 to 20. The latter treatment totally blocked the LH rise seen in the control birds on continuous light whilst a 4 h block of darkness greatly attenuated the first-day response. This suggests that the period of photoperiodic sensitivity in quail probably extends for 4–6 h beginning 12 h after dawn, agreeing with the earlier estimate based on asymmetric skeleton photoperiods.[3,11,12]

The apparent lag in the LH rise after the first photostimulation is an intriguing feature of the data.[21–23] As figure 5 indicates, there is no trace of an increase in LH secretion until hour 22 and yet the light causing this induction has occurred 4–7 h previously. The most obvious candidate for the cause of the lag is the neuro-endocrine/endocrine machinery but various tests indicate it to be competent all the time and capable of responding with an almost immediate release of LH to either hypothalamic stimulation or to an LH-RH injection. This suggests that the lag is either in the higher neural machinery controlling LH-RH secretion or within the clock. Could it be, for example, that the light from hours 12 to 16 causes various phase changes which lead to coincidence at about hour 20 and an immediate increase in the output of LH-RH? The photoperiodic 'gate' from hours 12 to 16 is a target for experiments designed to elicit the neural events occurring therein; already it is known that sodium pentobarbitone given during this period stops the LH rise in quail.[22]

These physiological events following induction suggest that the photoperiodic system in birds may differ from that in some insects. Saunders[15] has completed experiments in *Sarcophaga argyrostoma* which he interprets with an external

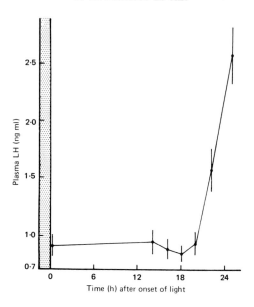

Figure 5. This experiment delineates more precisely the timing of the LH rise during the first long day. Nineteen quail were moved to continuous light at time zero. Note the abrupt rise between hours 20 and 22 ($P < 0.001$), as well as between 22 and 25 ($P < 0.001$). (Mean ± s.e. mean.) (Taken from Follett et al.[22])

coincidence model in which the photo-inducible phase is imagined to lie in the late subjective night, induction under long days or asymmetric skeletons occurring because the photo-inducible phase is delayed so that it is actually illuminated in the hours following dawn. Such a model cannot apply to quail since plasma levels of LH have already increased before the end of the first day, and whatever coincidence device operates it must have been established by the middle of the subjective night. This point is relevant not only to the physiological basis of daylength measurement but also to the interpretation of our formal photoperiodic experiments[11,12,17,22,24] which differs from that offered by the entomologists.[15,25]

The second phase of gonadotrophin secretion develops during the first week of photostimulation. The daily rise late in the photoperiod becomes obscured and secretion of LH and FSH no longer shows a diurnal rhythm but is maximal at all times. This applies to intact or gonadectomized quail of both sexes during a variety of photoperiodic schedules[26] (and unpublished experiments), as well as to photo-stimulated ducks.[27] The lack of rhythmicity argues strongly against the hypothesis that hormone release takes place only when 'coincidence' occurs each day and suggests that there is a more complex neural circuitry between the clock and the LH-RH secreting neurones.

Another complication is that LH-RH is secreted primarily from the hypothalamus of birds and mammals in short pulses every hour or two,[26,28–31] the balance between pulsatile and tonic secretion varying from species to species. In the sheep, for example, pulses of LH-RH are dominant, with the result that plasma LH levels rise abruptly to a peak and then decay virtually to zero until the next LH-RH pulse[30,31] whereas in man, pulsatile release is still dominant but some tonic secretion also occurs.[29] In quail pulses are less evident.[26] All this is relevant to photoperiodism since it has been suggested that in sheep short days stimulate LH and FSH secretion by altering the output from a pulse generator, thereby increasing the frequency of LH-RH pulses.[30,31] This hypothesis has experimental support since the frequency does increase in Soay sheep from summer to winter.[31] There is no marked diurnal rhythm in the pattern of pulses.

It seems likely that changes in pulsatile secretion also occur in photostimulated quail, and, in fact, the cellular machinery involved in pituitary LH release may require intermittent stimulation to function properly. Continuous bathing with LH-RH of a quail's pituitary gland *in vitro* soon leads to a loss in responsiveness and LH secretion falls away despite the fact that LH-RH is still present; this does not occur with intermittent stimulation.[32] It may emerge, therefore, that induction occurs when the photoperiodic clock alters the output from a neural generator regulating LH-RH secretion. In quail, the output may involve enhanced tonic secretion as well as more frequent bursts of activity whilst in sheep the whole effect may be a pulsatile release. This idea is pursued more fully in the article by Goodman and Karsch.[30] One way of testing the idea might be to determine whether other environmental factors can be used to enhance LH output and, if so, whether the detailed pattern of secretion is similar to that during photostimulation.

The third phase of photo-induced gonadotrophin secretion only becomes apparent when birds are transferred from long to short days. Instead of LH and FSH diminishing rapidly the levels only decrease slowly, reaching short-day concentrations after a week or 10 days (figure 6). The basis for this so-called 'carry-over' lies

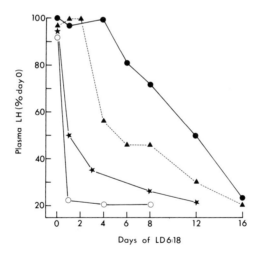

Figure 6. The 'carry-over' of LH secretion in quail. Data have been taken from various published[20,26] and unpublished experiments. They have been normalized so that the LH level at the time of transfer to LD 8 : 16 is 100 per cent. In practice the absolute concentrations at the end of the pre-treatments ranged from 1·5 to 2·5 ng/ml. Note the difference between the 'carry-over' in intact mature birds (●) with the changes following hypothalamic deafferentation (○). Photoperiodic stimulation with 4 days of LD 20 : 4 (▲) or with an asymmetric skeleton for 2 weeks (★) results in a much shorter 'carry-over'.

within the brain rather than in the pituitary gland since hypothalamic deafferentation causes a rapid fall in plasma LH. To some extent the duration of 'carry-over' depends upon the photostimulatory conditions imposed upon the birds since it is shorter in quail that have been exposed to LD 20 : 4 for 2 or 4 days rather than for 8 days, and is minimal in birds where an asymmetric skeleton photoperiod has been used to induce gonadal growth (figure 6).

This finding, as well as those obtained during the first week of stimulation, suggests that photoperiodic induction becomes progressively more intense with time.[22] The first long-day stimulus elicits a burst of LH secretion lasting for some hours, the second stimulus increases the length of the burst, and, by day 4 or 5, levels of LH are raised at all times. Once well established, LH-RH secretion can continue for some days regardless of the photoperiodic conditions. One prediction from the 'carry-over' hypothesis is that gonadal growth should be possible with schedules where long days and short days are mixed; a treatment tested first in white-crowned sparrows.[33] Recently, we have reinvestigated the phenomenon in quail, measuring LH daily in groups of birds exposed to LD 16 : 8 every day, or to schedules where they received one long day every second, fourth or sixth day. As figure 7 indicates, growth of the ovaries and oviducts is only slightly diminished in

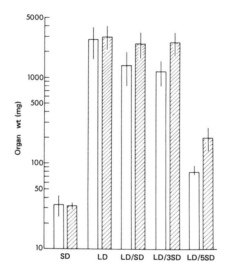

Figure 7. Ovarian (plain bars) and oviducal (hatched bars) weights in quail exposed to schedules that intersperse short days (SD) with long days (LD). All birds were killed after 22 days of treatment; $n = 7$ or 8 in each case. (Mean ± s.e. mean.)

cycles of 1 long and 3 short days compared with continuous long days even though the latter birds received 4 times as many stimulatory photoperiods. With only 1 long day every 6 days growth was much less and only marginally greater than growth on short days. The LH measurements in the birds on a schedule of 1 long and 3 short days are shown in figure 8 for 3 successive cycles. It is clear that in the first 5 cycles the long day caused an increase in LH secretion ($P < 0.05$) and this was usually followed by a decrease during the 3 short days. The exact timing of the rise was checked in the third cycle and found to be similar to that under continuous long days (c.f. figure 8 and figures 3–5). In the latter cycles the pattern of hormone release changed and LH levels were high at all times. Just as previously, therefore, induction occurred progressively until the processes underlying 'carry-over' were fully developed.

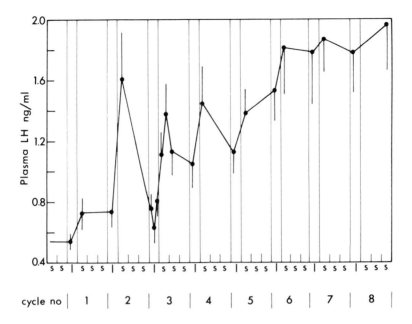

Figure 8. Plasma LH concentrations in a group of quail ($n = 8$) exposed to 1 LD/3 SD. Note the rises in plasma LH following the LD stimulus in the first five cycles. After this rise there was usually a decrease in LH in the following short days. In cycle 6 the LD (LD 20 : 4) was replaced by the schedule 8 L : 5·5 D : 0·5 L : 10 D; this did not cause a significant rise in LH. During the third cycle samples were taken more frequently to compare the timing of the rise with that found under normal long days (*see* figures 3–5). (Mean ± s.e. mean.)

'Carry-over' is certainly not confined to quail and white-crowned sparrows and exists in a stronger form in species such as the house sparrow. Here, transfer of mature males to darkness has no effect on testicular size even after a month of treatment.[16] Even more dramatic examples occur, of course, in those plants that flower some weeks after exposure to only 1 long day.[34] Once again the question arises as to whether 'carry-over' reflects the neuroendocrine circuitry or is a property of a fully saturated photoperiodic clock. Some experiments with juncos[35] and white-crowned sparrows[36] point towards the latter possibility. 'Carry-over' is significantly less under short days than in continuous darkness, a result that may be interpreted in terms of the faster uncoupling of oscillators under cycles of light and darkness. To be set against these data are those from quail[26] and house sparrows[16] where the length of 'carry-over' is independent of whether or not the birds are transferred to darkness or short days.

SEASONAL SHIFT IN THE CRITICAL DAYLENGTH

The critical daylength for gonadal growth in quail living on LD 8 : 16 is just under 12 h and, until quite recently, we assumed that this was a largely unalterable threshold characteristic of this species. However, it was then observed[20,37] that under natural daylengths the gonads of quail regress not when the photoperiod has fallen below 12 h but when it is about 14·5 h; this phenomenon serving as an adaptation allowing the quail to terminate reproduction in the late summer rather than in the autumn. Since Japanese quail remain reproductively active for years when kept on long days and are not, therefore, photorefractory in the classic sense, why do they undergo regression in August when the daylengths are apparently still maximally stimulatory? Various experiments suggest that there is a seasonal shift in the critical daylength from 12 h in February/March to 14·5 h in August/September, and back again to 12 h by the following spring.

In a first experiment[20] quail were exposed to simulated annual photocycles using Sangamo Weston 'solar dial' time clocks (52 °N). In birds on increasing daylengths a significant rise in plasma levels of LH occurred when the photoperiod reached 12·7 h but in the sexually mature quail exposed to shortening days it decreased rapidly once the photoperiod fell below LD 14 : 10 (results similar to those found in the natural situation). It could then be shown that exposing quail to long days of a fixed duration also shifted the critical daylength.[20] For quail on LD 8 : 16 a 13 h photoperiod (LD 13 : 11) is highly stimulatory, but for birds on LD 16 : 8 the photoperiod is read as 'short' and there is testicular regression. However, photoperiods such as LD 13 : 11 are not read as 'short' immediately after the quail are exposed to LD 16 : 8 but only after about 2 months. The extent of the apparent shift in threshold is proportional to the length of the long day[24] and treatment with LD 20 : 4 for 10 weeks leads to a situation where photoperiods of LD 13 : 11 or less are read as 'short', LD 14 : 10 causes a major decrease in testicular size, LD 15 : 9 only a minor depression in LH secretion while LD 17 : 7 and LD 20 : 4 continue to be read as 'long' days. One other feature of refractoriness in the quail is that whilst LD 13 : 11 is read as 'short' for about 6 weeks, the birds eventually come to regard it as 'long', presumably because the critical daylength has shortened again, and the gonads regrow.

Castrated quail under both natural and fixed photoperiods show a shift in critical daylength identical to that in intact birds. The shift also occurs in quail transferred to LD 16 : 8, even though gonadotrophin secretion has been maintained at a minimal level by an implant of sex steroid.[24] These experiments argue that the shift has little to do with the activity of the hypothalamo-hypophysial system and that its basis must either be in the neural circuits regulating LH-RH output or in the photoperiodic clock mechanism. The latter is a particularly attractive hypothesis and critical tests must be devised to determine whether the circadian rhythm of photoperiodic sensitivity does change its characteristics with time and photoperiod. There seem to be few, if any, precedents in circadian studies[38] for a situation where an entrained rhythm changes its phase progressively over many weeks, but there remains the attractive conjecture that refractoriness in quail may have some affinity with the mechanisms underlying after-effects.

REFERENCES

1. W. M. Hamner. *Science* **142,** 1294–5 (1963).
2. K. K. Nanda and K. C. Hamner. *Bot. Gaz. (Chicago)* **120,** 14–25 (1958).
3. B. K. Follett and P. J. Sharp. *Nature (Lond.)* **223,** 968–71 (1969).
4. F. W. Turek. *J. Comp. Physiol.* **92,** 59–64 (1974).
5. E. Gwinner and L.-O. Eriksson. *J. Ornithol.* **118,** 60–7 (1977).
6. W. M. Hamner and J. T. Enright. *J. Exp. Biol.* **46,** 43–61 (1967).
7. S. D. Skopik and M. F. Bowen. *J. Comp. Physiol.* **111,** 248–59 (1976).
8. H. Underwood. This volume, pp. 137–52.
9. B. K. Follett, P. W. Mattocks and D. S. Farner. *Proc. Natl Acad. Sci. USA* **71,** 1666–9 (1974).
10. J. M. Kinet. *Nature* **236,** 406–7 (1972).

11. B. K. Follett. *J. Reprod. Fertil.* Suppl. **19**, 5–18 (1973).
12. B. K. Follett. *Gen. Comp. Endocrinol.* (in the press) (1981).
13. C. S. Pittendrigh and D. H. Minis. *Am. Nat.* **98**, 261–94 (1964).
14. J. A. Elliott. This volume, pp. 203–17.
15. D. S. Saunders. *J. Comp. Physiol.* **132**, 179–89 (1979).
16. D. S. Farner, R. S. Donham, R. A. Lewis, P. W. Mattocks, T. R. Darden and J. P. Smith. *Physiol. Zool.* **50**, 247–68 (1977).
17. S. M. Simpson and B. K. Follett. (In preparation.)
18. C. S. Pittendrigh. This volume, pp. 1–35.
19. E. Bünning. *Cold Spring Harbor Symp. Quant. Biol.* **25**, 249–56 (1960).
20. B. K. Follett and J. E. Robinson. *Prog. Reprod. Biol.* **5**, 39–61 (1980).
21. T. J. Nicholls and B. K. Follett. *J. Comp. Physiol.* **93**, 301–13 (1974).
22. B. K. Follett, D. T. Davies and B. Gledhill. *J. Endocrinol.* **74**, 449–60 (1977).
23. M. Wada. *Gen. Comp. Endocrinol.* **39**, 141–9 (1979).
24. J. E. Robinson. PhD Thesis, University of Wales (1980).
25. B. Dumortier and J. Brunnarius. This volume, pp. 83–99.
26. B. Gledhill and B. K. Follett. *J. Endocrinol.* **71**, 245–57 (1976).
27. J. Balthazart, J.-C. Hendrick and P. Deviche. *Gen. Comp. Endocrinol.* **32**, 376–89 (1977).
28. D. J. Dierschke, A. N. Bhattacharya, L. E. Atkinson and E. Knobil. *Endocrinology* **91**, 793–800 (1972).
29. S. C. C. Yen, C. C. Tsai, F. Naftolin, G. Vandenberg and L. Ajabor. *J. Clin. Endocrinol. Metabol.* **34**, 671–5 (1972).
30. R. L. Goodman and F. J. Karsch. This volume, pp. 223–36.
31. G. A. Lincoln and R. V. Short. *Rec. Prog. Horm. Res.* **36**, 1–52 (1980).
32. P. J. Bentley and B. K. Follett. (In preparation.)
33. B. K. Follett, D. S. Farner and M. L. Morton. *Biol. Bull.* **133**, 330–42 (1967).
34. D. Vince-Prue. *Photoperiodism in Plants.* New York, McGraw-Hill (1975).
35. A. Wolfson. *Rec. Prog. Horm. Res.* **12**, 177–244 (1966).
36. F. W. Turek. In: I. Assenmacher and D. S. Farner (ed.) *Environmental Endocrinology.* Berlin, Springer-Verlag, pp. 144–52 (1978).
37. B. K. Follett and S. L. Maung. *J. Endocrinol.* **78**, 267–80 (1978).
38. J. Aschoff. In: D. T. Krieger (ed.) *Endocrine Rhythms.* New York, Raven Press, pp. 1–61 (1979).

DISCUSSION

Short: Two rather separate questions. First, is the annual change in photoperiodic threshold confined to the male bird? The reason I ask is that in red deer and sheep, which are autumn breeders, the testes of the male start to decline in size and endocrine activity in November, whereas the females of those species can continue to show normal ovulatory cycles until the spring. This suggests a sex difference in the timing of the cycle. Secondly, one could also attempt to explain the gonadal switch-off in endocrine terms, either by hypothalamic down-regulation of pituitary activity or pituitary down-regulation of gonadal activity following a period of maximal stimulation, thus resulting in a period of desensitization.

Robinson: We have concentrated the majority of our work on the male quail. However, in laboratory experiments with the female we find that they respond to a decreased photoperiod in a manner similar to that of the male and thus we have detected no differences in the sexes in the initiation of timing of regression.

Follett: I would strongly concur with your feeling that we should look at the possibility that the refractoriness could occur through endocrine changes, not through effects on the clock. So far we have looked at castrates and steroid-implanted quail. To be brief, neither of these treatments, despite having radically different effects on gonadotrophin secretion, changes refractoriness from that seen in intact quail.

Turek: We have also observed testicular regression when white-throated sparrows are transferred from LD 20 : 4 to LD 13 : 11, whereas testicular growth is initiated when sparrows are transferred from LD 8 : 16 to LD 13 : 11. An old hypothesis of Hamner is that there is a seasonal change in the photo-inducible phase. Have you done any night interruption experiments which might indicate that the photo-inducible phase can be altered by previous lighting history?

Follett: We have done one such experiment so far but it was not satisfactory and gave very variable results.

Engelmann: In trying to test whether a change in the strength of the coupling of different oscillators might be the reason for your findings, wouldn't it be worth while to drive the system at the border of entrainment, or has this been done already?

Follett: Thank you for your suggestion. We were considering this experiment but it has not yet been undertaken.

Bittman: Using a night interruption protocol, Lofts and Murton[1] reported different photo-inducible phases for LH and FSH in the greenfinch. Is there any point of the quail's annual cycle in which you see different critical daylengths for induction of release of different pituitary hormones? Do you know of any instances of staggered gonadotrophin rises in the quail's seasonal cycles and/or their consequences for the gonads? Do these data compel the conclusion that there exists either even more coinciding oscillators or more photosensitive phases than we have envisaged?

Follett: We have looked carefully at these points and can find no situation, as yet, where the secretion of LH and FSH can be separated photoperiodically in a consistent fashion, although there are hints that on marginally stimulatory long days FSH secretion is less elevated than that of LH.

Davies: You have demonstrated that it takes several weeks for the system to develop the capacity to respond to a reduced photoperiod when the initial schedule is LD 16 : 8. Will this capacity to respond develop more quickly if the animals are maintained on longer photoperiods of say LD 18 : 6 or 20 : 4?

Follett: Yes, the speed of changing the threshold as well as the amount of change is faster the longer the photoperiod.

Glover and Heaf (comment): I should like to add some comments on other differences in the plasma biochemical composition of Japanese quail exposed to long days compared with that for birds on short days which probably affect the response of the birds on transfer to an intermediate but potentially stimulatory photoperiod (e.g. LD 13 : 11). These differences relate to the retinol and thyroxine transport proteins secreted specifically by the liver. The rate of their secretion varies in response to photoperiod as indicated below.

Retinol-binding protein (RBP) and thyroxine-binding prealbumin (TBPA) which form a 1 : 1 complex in plasma[2] are both concerned with the transport of vitamin A, essential for the proper development and maintenance of reproduction in birds.[3] It has been observed that the concentration of RBP increased approximately twofold[4,5] whereas that of TBPA decreased by approximately 50 per cent[6] in the sexually active compared to the quiescent phase of the annual breeding cycle of both male and female Japanese quail (*see* figure 1). The changes influence considerably the supply of these key ligands to the respective target tissues. The increase in retinol supply in the plasma occurs around the same time as that of the gonadotrophins, but there is no direct correlation between them. Furthermore, 17β-oestradiol and progesterone have no effect on the synthesis of RBP by Japanese quail liver *in vivo*[7] (also Heaf, unpublished observations). These observations, together with previous work in sheep,[8] indicated that the stimulus for increased synthesis of RBP by the liver is most likely controlled by an unidentified factor related more directly to the hypothalamo-pituitary hormone axis in parallel with the gonadotrophin changes rather than secondary to gonadal development.

This seems to be even more true of the factor responsible for the smooth variations in TBPA concentration, which shows a very high inverse correlation with length of photoperiod, e.g. $r = -0.9$ for seasonal changes[6] and -0.99 for shortened cycles[9] in Japanese quail. Here the decrease in plasma concentration with increasing photoperiod, and *vice versa*, occurred gradually throughout the annual cycle except for the mid-summer period when a minimal threshold value appeared to be reached. Sudden changes in the length of photoperiod from LD 7 : 17 to LD 20 : 4 or in the opposite direction have also been shown to cause an immediate change in plasma TBPA concentration of groups of birds detectable within 3 days.[9] Again, when the photoperiod was kept constant at LD 7 : 17 but the lighting intensity reduced about eightfold from 73 μW/cm^2 to 9 μW/cm^2, then the mean concentration of TBPA in a group of 15 immature male birds increased approximately 25 per cent over 1 week and was significant ($P < 0.001$) after 4 days (Carroll, Heaf and Glover, unpublished observations). The extreme sensitivity of the thyroxine-binding prealbumin synthesis in the liver to changes in the lighting environment, particularly when the photoperiod is less than 14 h, suggests that it too is under direct control of the hypothalamo-pituitary hormone axis. Work is in progress in an attempt to identify the mediating factors.

Thyroxine-binding prealbumin is a specific and major carrier of thyroxine (T4) in plasma of birds.[10] It has a much greater affinity for the hormone than albumin, the other major carrier,[11] consequently changes in its concentration will affect the supply of free thyroxine in plasma for target tissues. We have observed in a preliminary study a lowering of total T4 concentration in Japanese quail when TBPA concentration declines, but the effect on the concentration of the free form of thyroxine needs to be determined. These observations are in keeping with the reported seasonal changes in thyroxine level of the woodchuck.[12]

Thus changing photoperiod controls the peripheral distribution of retinol and free thyroxine through their carrier proteins synthesized in the liver. This is separate from but additional to the control system mediated by thyrotrophin which governs the production of thyroxine in the thyroid gland. The distribution control system helps to maintain the right balance of these key ligands to provide the appropriate biochemical environment for the gonads to develop under the stimulus of the gonadotrophins.

The question is now raised as to the specific roles and importance of these two proteins in relation to reproduction in seasonal breeding animals. Some of these may be as follows:

1. The cyclic increase in supply of retinol via RBP appears to be necessary for the proper differentiation, rapid growth and maintenance of gonadal tissues. Regression of the tissues is in turn associated with a fall in plasma retinol supply back to normal tissue maintenance levels.

2. The accessibility of retinol to gonadal tissues is probably also influenced by TBPA. When the latter is in excess RBP is fully complexed with it, but in the reverse situation excess RBP does not bind so readily to the remaining three potential sites as to the first on the prealbumin molecule. Hence a reduction in the concentration of TBPA with lengthening photoperiod relative to that of RBP will enhance the concentration of free RBP available to bind to tissue receptors for the transfer of retinol. This type of change is particularly striking in female birds where

a substantial amount of free retinol has been shown to be transferred to the eggs.[13] It is also noteworthy that the fall in the TBPA to RBP ratio below unity occurs at the time of the development of the gonads indicated by the horizontal bars in figure 1 (which show egg-laying period) and rises above 1 again later when regression sets

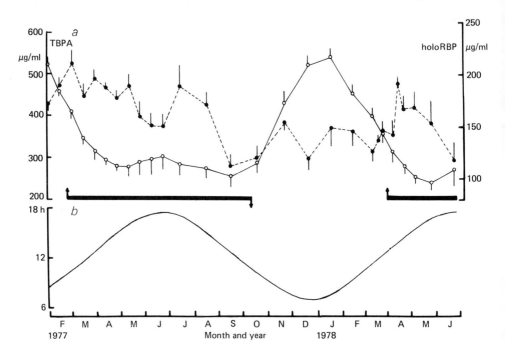

Figure 1. The seasonal cyclic changes in mean ± s.e. mean concentrations of thyroxine-binding prealbumin and retinol-binding protein in plasma of 6 Japanese quail (3 males and 3 females) (*a*), compared with the annual changes in photoperiod (*b*). The central bars indicate the egg-laying period of females. The RBP values are plotted on a scale 2·5 × that of TBPA to indicate directly the periods when the molar concentration of RBP exceeds that of TBPA.

in. Thus TBPA may have a role in helping to control the peripheral distribution of retinol in seasonal breeding animals.

3. The reduction in TBPA with lengthening photoperiod in Japanese quail will result in some lowering of the plasma-free thyroxine (T4) level and possibly also triiodothyronine (T3), so affecting indirectly the general level of metabolism, and possibly the sensitivity of some steroid hormone receptors on target cells (e.g. a high level of thyroxine has been shown to shield the prostrate from inhibitory effects of excessive levels of circulating oestrogen above that of androgen.[14]

This retinal and thyroxine transport system seems to form a useful parameter for assessing the capacity of Japanese quail and possibly other animals to become sexually active or quiescent at specific times in the annual cycle.

Brown-Grant: Are there any experiments which demonstrate a role for the suprachiasmatic nucleus in birds in regulating a circadian function other than activity?

Davies: Unlike the situation in rodents, the suprachiasmatic nucleus of most birds is not a discrete structure but a scatter of cells within the supra-optic region immediately anterior to the optic chiasma. This imposes technical difficulties for the experimenter to ensure complete destruction of the cells following electrolytic lesioning. In female quail, destruction of the supra-optic region (SOR) has no effect on tonic gonadotrophin secretion regulating photo-induced ovarian development, but such lesioned birds fail to ovulate owing to a block of the cyclic LH ovulatory surge. A polyfollicular syndrome is observed where up to 10 large yolky follicles can be found. Such SOR lesions do not disrupt the photoperiodic response, however, since when these birds are transferred to a short photoperiod of LD 8 : 16 tonic gonadotrophin secretion is blocked and the ovary undergoes complete regression.

REFERENCES

1. R. K. Murton, B. Lofts and N. J. Westwood. *J. Zool. (Lond.)* **161**, 125 (1970).
2. M. Kanai, A. Raz and D. S. Goodman. *J. Clin. Invest.* **47**, 2025–44 (1968).
3. J. N. Thompson, J. McC. Howell and G. A. J. Pitt. *Proc. R. Soc. B.* **159**, 510 (1964).
4. D. J. Heaf and J. Glover. *J. Endocrinol.* **83**, 323–30 (1979).
5. J. Glover, D. J. Heaf and S. Large. *Br. J. Nutr.* **43**, 357–67 (1980).
6. M. El-Sayed, D. J. Heaf and J. Glover. *Gen. Comp. Endocrinol.* **41**, 539–45 (1980).
7. D. J. Heaf and J. Glover. *J. Endocrinol.* (in the press) (1980).
8. J. Glover, C. Jay, R. C. Kershaw and P. E. B. Reilly. *Br. J. Nutr.* **36**, 137 (1976).
9. D. J. Heaf, M. M. El-Sayed, B. Phythian and J. Glover. *J. Endocrinol.* (submitted) (1981).
10. L. S. Farer, J. Robbins, R. S. Blumberg and J. E. Rall. *Endocrinology* **70**, 686–96 (1962).
11. J. Robbins, S.-Y. Cheng, M. C. Gershengorn, D. Clinoer, H. J. Cahnmann and H. Edelnoch. *Rec. Prog. Horm. Res.* **34**, 477–520 (1978).
12. R. A. Young, E. Darnforth jun., P. P. Krupp, R. Frink and E. A. H. Sims. *Endocrinology* **104**, 996–9 (1979).
13. D. J. Heaf, B. Phythian, M. M. El-Sayed and J. Glover. *Int. J. Biochem.* **12**, 439–43 (1980).
14. A. B. Kar. *Ind. J. Exptl Biol.* **4**, 7.

CIRCADIAN RHYTHMS, ENTRAINMENT AND PHOTOPERIODISM IN THE SYRIAN HAMSTER

Jeffrey A. Elliott

Hopkins Marine Station of Stanford University, Pacific Grove, California, USA

Summary

The Syrian hamster (*Mesocricetus auratus*) is a vernal breeder which requires an alternation of long and short daily photoperiods for expression of its annual reproductive cycle. Induction of testicular function by long photoperiods depends on a remarkably precise clock mechanism for the measurement of daylength. The photoperiodic clock is based on a circadian rhythm of responsiveness to light (CRPP) which is entrained to the light/dark (LD) cycle and persists in continuous darkness as a free-running rhythm. The photo-inducible phase (ϕ_i) of the CRPP has been mapped with respect to the circadian activity cycle and this information used to construct an unusually substantive and specific model for the mechanism of daylength measurement by a circadian system. The implications and limitations of this model are explored. Part of this analysis also shows that the hamster's photoperiodic clock is extremely responsive to light, a 15 min light pulse given once every 10 days in darkness being sufficient for photoperiodic induction of gonadal activity if, and only if, it is timed to coincide with ϕ_i.

INTRODUCTION

One of the central problems in photoperiodism concerns the mechanism of photoperiodic time measurements (PTM). How do organisms measure daylength? Two conceptually distinct mechanisms have been proposed for the photoperiodic clock and each has been found to occur in some organisms. In some insects[1,2] time measurement is effected by a passive hour-glass timer that requires daily resetting by light and measures the duration of darkness. Evidence in favour of a non-oscillatory timer has also been found in a vertebrate. In the lizard, *Anolis carolinensis*, photoperiodic control of gonadal activity appears to depend on an hour-glass-like process that measures the duration of light.[3] However, in the Syrian hamster, as in many species of plants, insects and birds, daylength is measured by a clock that is based on an endogenous, self-sustained circadian oscillator.[4,5]

The Syrian hamster (*Mesocricetus auratus*) is a vernal breeder which requires alternation of long and short daily photoperiods for expression of its annual reproductive cycle. In male hamsters, long photoperiods ($\geq 12 \cdot 5$ h light/day) maintain reproductive competence indefinitely, while transfer to short photoperiods provokes regression of the testes within 10 weeks followed by a recovery of

Biological Clocks in Seasonal Reproductive Cycles 203-17 (1981) (ed. B. K. and D. E. Follett: Bristol, Wright).

gonadal function by 25 weeks. The regrowth of the testes can also be induced photoperiodically. This is observed when hamsters with regressed testes are returned to long photoperiods. Even though a return to long days is not required for the recovery of the testes, it is essential for the occurrence of the annual cycle because a hamster with newly recrudesced gonads is refractory and must experience 10–20 weeks of long photoperiods before short photoperiods will induce a second regression.[6–8] The adult hamster is therefore responsive to photoperiod throughout the year, although the form of the response changes with the phase of the annual cycle. All these effects of photperiod involve a clock for daylength measurement which is a function of the hamster's circadian system.[4,5,9,10]

This paper is concerned with an analysis of the photoperiodic clock in the hamster and its relationship to the entrainment of the circadian activity rhythm by light. The analysis shows that daylength is measured with reference to a circadian rhythm of photoperiodic photoresponsivity (CRPP) that bears an essentially constant phase relationship to the activity cycle. Photoperiodic induction of testicular function depends on the coincidence of light with the photo-inducible phase (ϕ_i) of the rhythm. Recent experiments show that the photoperiodic clock is remarkably sensitive to light. The circadian time of a light stimulus is of much greater importance than its duration or frequency.

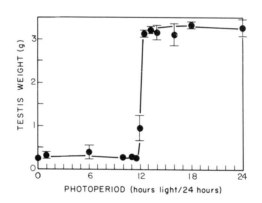

Figure 1. Photoperiodic response curve. Each point represents the mean ($\pm$ s.e.) paired testis weight of a group of hamsters exposed to the indicated photoperiod for the previous 12–14 weeks.[5]

CIRCADIAN BASIS OF THE PHOTOPERIODIC CLOCK

Figure 1 illustrates the precision of the time measurement in hamster photoperiodism. When male hamsters are subjected to photoperiods ranging in short steps from 0 to 24 h light per day there is a sharp 'break' in the response between photoperiods of 12·0 and 12·5 h—the minimum photoperiod required for maintenance of gonadal activity (figure 1) or for the photoperiodic induction of gonadal recrudescence is 12·5 h.[5,11] The steepness of the curve in the region of the critical daylength and the flatness of the response to shorter and longer photoperiods are both attractive features. The first indicates considerable accuracy and homogeneity in the clock mechanism. The second suggests that the hamster's photoperiodic clock may be comparatively simple in its basic design.

The circadian basis of hamster photoperiodic time measurement was first demonstrated using the so-called Nanda–Hamner resonance protocol.[4] Male hamsters were exposed to LD cycles consisting of a 6 h photoperiod coupled to dark periods of 18, 30, 42 or 54 h. The testes regressed when the period (T) of the LD cycle was 24 h (LD 6 : 18) or 48 h (LD 6 : 42) but the gonads of males subjected to T cycles of 36 h (LD 6 : 30) and 60 h (LD 6 : 54) remained large and functional. Since photoperiodic induction of gonadal activity rises and falls as a periodic (modulo 24 h) function of T,[12] these results indicate that hamster PTM is based on a circadian rhythm of photoperiodic photoresponsivity (CRPP). The data rule out an hour-glass mechanism by demonstrating that hamsters do not measure the absolute duration of the light or dark phases of the LD cycle or the L/D ratio.[4] Similar results were obtained in studies of the photoperiodic induction of gonadal growth in hamsters and voles and for the photoperiodic termination of photorefrac-toriness in hamsters.[9,10,13]

When a circadian clock is involved in photoperiodic time measurement, as is clearly the case in hamsters, a deeper understanding of the mechanism involved is only possible if a clear picture can be obtained both of the entrainment of the circadian clock by light cycles and of photoperiodic induction of the reproductive system.[14-16] Since the behaviour of the CRPP cannot be monitored independently of the reproductive response, the hamster's locomotor activity rhythm (wheel-running) is used to study entrainment of the clock and to provide a continuous assay of the phase of the CRPP. This requires the assumption that the CRPP and the activity rhythm retain a relatively fixed phase relationship under entrainment to different LD cycles. This assumption is supported by all the available evidence.

CHARACTERISTICS OF THE CIRCADIAN ACTIVITY RHYTHM

The formal properties of the circadian clock system (pacemaker) regulating hamster wheel-running activity have been the subject of extensive studies.[5,17-22] Here our concern is only with those general characteristics which have played a role in the design and execution of experiments undertaken to study the mechanism of hamster photoperiodic time measurement.

The circadian activity rhythm of the hamster (figure 2) is well known for its precision and stability under constant conditions.[18] Intra-individual variation in the free-running period (τ) in continuous darkness (DD) is also remarkably low.[17,18] When this was studied in 69 hamsters released into DD from prolonged entrainment to cycles of 24 h (LD 14 : 10 or LD 6 : 18) τ values ranged from 23·90 h to 24·37 h with a population mean of 24·12 h ($\pm$ 0·04 h s.e.mean).[17] In the following discussion the value $\bar{\tau} = 24\cdot1$ h is taken as the species average and unless stated otherwise this is the assumed value of τ in DD. In entrainment the period of the free-running rhythm (τ_{DD}) comes to match the period (T) of the light cycle and a stable phase relationship (ψ_{RL}) is established between the rhythm and the light cycle.[20]

To simplify discussion of phase relationships we will adopt a circadian time (CT) scale in which one full cycle (360°) of the activity rhythm with period τ is divided into 24 circadian hours each lasting $\tau/24$ hours of real time. The phase reference point for the rhythm is the daily onset of nocturnal wheel-running activity which is

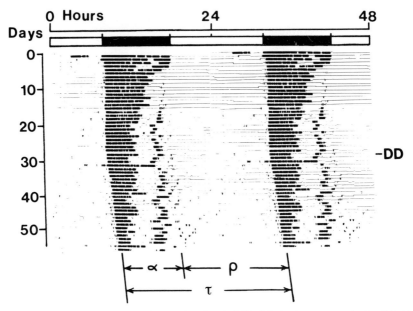

Figure 2. Circadian rhythm of wheel-running activity of an individual Syrian hamster entrained to LD 14 : 10 and free-running in DD. Each line on the left is one day's record (0–24 h) cut from noon to noon and pasted beneath the record of the preceding day. For illustration, the entire record is shown again on the right (24–48 h). The heavy bands indicate nearly continuous running. Days 0–27 show 4 weeks of entrainment to the LD 14 : 10 light cycle diagrammed above. Days 28–56 show the rhythm free-running in DD. The activity cycle is divided into activity time (α) and rest time (ϱ). In this example the period of the free run (τ_{DD}) is 24·07 h and α is 8·5 h.[22]

designated CT 12 and marks the beginning of the hamster's subjective night (CT 12 to CT 24). The half-cycle immediately preceding activity onset is designated the subjective day (CT 0 to CT 12) and corresponds closely to that portion of the cycle that occurs in the light in entrainment to LD 12 : 12.[5]

The phase of the free-running rhythm can be reset by a single light pulse (figure 3). The magnitude and direction of the phase shift ($\Delta\phi$) response is characteristic of the circadian phase (CT) of the pulse and this dependence is summarized in a phase response curve (PRC).[19,23] Phase delays ($-\Delta\phi$) are elicited by light in the late subjective day and early subjective night (CT 10–CT 16). Phase advances ($+\Delta\phi$) peak in the middle of the subjective night (CT 18) and continue with diminishing magnitude into the early subjective day (CT 2). In the remainder of the subjective day (CT 2–CT 10) light pulses have little or no effect on the phase of the rhythm. The hamster PRC for 1 h pulses (figure 3, *inset*) is not substantially different in shape or amplitude from the PRCs measured for 10 min and 15 min light pulses.[19,23]

The PRC describes a circadian rhythm of response to light which mediates the entrainment of a circadian rhythm to a light cycle.[14] In each cycle of entrainment the phase of the rhythm is reset by an amount ($\Delta\phi$) equal to the difference between τ and T ($\tau - T = \Delta\phi$). For a given value of τ and T and a standard light pulse of fixed duration and intensity there is a unique circadian phase (CT) at which light must fall to produce stable entrainment. Thus the circadian time of a standard light pulse varies systematically as a function of T and if τ and T are known it can be predicted from the PRC. This relationship can be exploited to study the characteristics of the CRPP.[5,14,16]

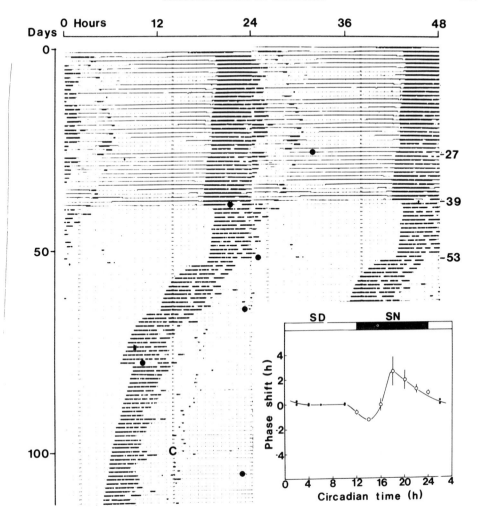

Figure 3. Effect of single light pulses on the phase of the free-running rhythm. The solid circles (●) indicate the time (midpoint) of single light pulses given on days 27, 39, 53, 65, 78 and 106 of DD. The pulse on day 78 was 2 h long; all other pulses were 1 h. Phase advances ($+\Delta\phi$) were elicited by pulses 1, 3 and 4; phase delays ($-\Delta\phi$) by pulses 2 and 5. The last pulse and a control disturbance (C) on day 100 (cage moved with the aid of a dim red light) failed to cause detectable phase shifts. The inset summarizes the data from 18 animals and many ($n = 47$) estimates of the $\Delta\phi$ responses to 1 h light pulses (50–100 lux) given at different phases of the cycle. Open circles (○) plot the mean ± s.e. for $\Delta\phi$ responses averaged over 2-hourly bins of circadian time. Solid circles (●) plot individual responses. An eye-fit line drawn through the points gives a phase response curve.[22]

MAPPING THE PHOTO-INDUCIBLE PHASE

Entrainment and photoperiodic induction when *T* is close to τ

When PTM is based on a circadian rhythm of photo-inducibility it should be possible to describe the rhythm in some detail by examining the photoperiodic effect of brief stimuli applied at numerous selected phase points in the circadian cycle. To do this, hamsters were entrained to 1 h light pulses given in LD cycles with different T values. Systematically varying T within the range of entrainment (i.e. between 23·2 and 24·8 h) produced a systematic change in the circadian time

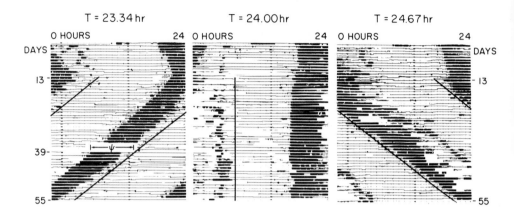

Figure 4. Entrainment of the activity rhythm to light cycles with different periods (T). Activity recording began on the hamster's last day in LD 6 : 18 (lights-on hours 8–14). Days 2–12 of each record show the rhythm free-running in DD. T cycles were initiated on day 13 and are represented by the heavy lines which connect the midpoints of 1 h light pulses. Note the dependence of ψ on T.[5]

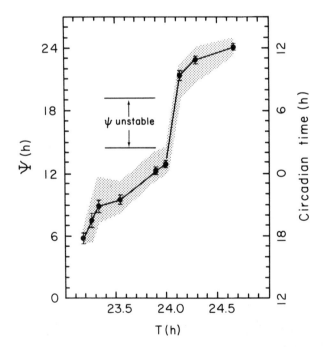

Figure 5. Dependence of ψ on T. Each point represents the mean and 95 per cent confidence interval (vertical bars) of ψ for a particular value of T. The shaded area represents the range of all observations. Rotating the figure 90 ° clockwise shows how the circadian time of the light pulse (right ordinate) varies as a function of T.[17]

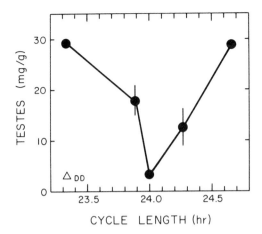

Figure 6. Testicular response as a function of T (1 h light/cycle). Each point represents the mean testis weight of a group of 10–12 hamsters exposed to one of five different Ts and sacrificed after 64 days. The testes regressed fully in DD (Δ) and in LD 1 : 23 (T 24), but regression was partially or completely blocked in all non-24 h T cycles.[5]

at which the pulse occurred (figure 4). In this way the circadian time of light can be purposefully controlled to examine the photoperiodic effect of light at nearly all phases of the circadian cycle (figure 5).

As expected the photoperiodic testicular response also depends markedly on T (figure 6). When T is precisely 24 h (LD 1 : 23) or different from 24 h but still very close to $\bar{\tau}$ (e.g. $T = 23\cdot93$ h, $T = 24\cdot14$ h) the testes regress with the same time course as in DD, indicating that these light cycles are 'read' as short days. In T cycles farther from $\bar{\tau}$ regression is blocked. For example, it is completely blocked in $T = 23\cdot34$ and $T = 24\cdot66$ and partially blocked in $T = 23\cdot89$ and $T = 24\cdot27$. The same pattern of response was found for the photoperiodic induction of testicular recrudescence by T cycles: A *single* hour of light per *circadian* cycle can induce rapid regrowth of the testes but it does so only when T differs from $\bar{\tau}$ by $\geqslant 0\cdot5$ h.[11,17]

The photoperiodic response to different T cycles is also strongly correlated with the circadian time of light. Figure 7 summarizes the results of a series of experiments examining the effectiveness of various T cycles for the photoperiodic induction of testicular growth. Plotting the photoperiodic response as a function of the mean circadian time of the light pulse yields a *circadian* photoperiodic response curve which may be viewed as an empirical representation of the circadian rhythm of photo-inducibility (CRPP) at the basis of hamster PTM.

The photo-inducible phase (ϕ_i) of the CRPP corresponds closely to the subjective night phase of the activity rhythm and also to the responsive region of the PRC. In figure 7B ϕ_i can be defined operationally as that portion of the cycle where a 1 h light pulse induces recrudescence in $\geqslant 50$ per cent of the animals. According to this criterion, ϕ_i commences shortly before activity onset and lasts for approximately 10·5 h (CT 11·5 to CT 22). When the same criterion ($\geqslant 50$ per cent induction) is applied to data on the effectiveness of various T cycles for the photoperiodic maintenance of testicular function ϕ_i is found to begin a little earlier and to last longer (~CT 11 to CT 23) encompassing nearly a full 180° of the circadian cycle.[5,17] These results leave no doubt that a circadian rhythm of

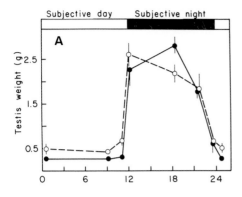

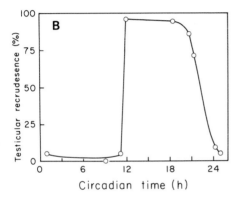

Circadian time (h)

Figure 7. Circadian photoperiodic response curve for testicular recrudescence. **A,** Hamsters with regressed testes were transferred to one of seven different T cycles and sacrificed after 40 days of exposure. Each point represents the mean testis weight of a group of 9–10 hamsters subjected to a particular T cycle plotted as a function of the mean circadian time of the light pulse. ○, Hamsters housed in individual activity cages; ●, hamsters housed in groups in cages without wheels. **B,** Summary curve. The graph gives the percentage of animals showing testicular growth for groups of 17–40 animals subjected to a particular T cycle plotted as a function of the mean circadian time of the light pulse.[17]

responsiveness to light participates in hamster PTM and that the phase and duration of ϕ_i can be defined with respect to the circadian activity cycle.

The response to single pulses

There are two portions of the circadian cycle not examined in T experiments. The 'dead zone' of the hamster PRC lies between CT 2 and CT 10 (*see* figure 3): a 1 h light pulse falling in this region fails to shift the phase of the free-running rhythm, cannot entrain the rhythm and therefore cannot be tested for photoperiodic effect using the T cycle paradigm. However, this portion of the circadian cycle is known to be non-responsive to photoperiodic stimuli because it is illuminated when the circadian system is entrained by standard non-stimulatory photoperiods such as LD 6 : 18 or LD 10 : 14.[5] Figure 7 also lacks points in the early subjective night between CT 12 and CT 18. This 'gap' in the curve corresponds to the transition between delays and advances in the PRC. Stable entrainment is not possible with 1 h light pulses falling in this portion of the PRC because the slope is too steep and in the wrong direction for a systematic approach to entrainment equilibrium.[20] Thus although there is a $\Delta\phi$ response to light in this region, its photoperiodic photosensitivity could not be tested in T experiments.

To test this portion of the cycle for photo-inducibility, groups of hamsters were transferred from LD 14 : 10 to DD and then pulsed with light (15 min) at different circadian times approximately once every 10 days during the first month of DD.

The animals remained in uninterrupted darkness until 6 weeks after their initial transfer to DD when they were moved to a room maintained on constant dim red light of an intensity (≈ 0.5 lux) previously found to be below photoperiodic threshold (Elliott, Stetson and Menaker, unpublished observations).

Light pulses which induce large phase shifts in the free-running rhythm also significantly delay the time course of gonadal regression in DD (figure 8). Ten

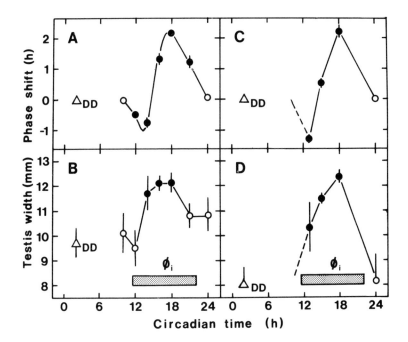

Figure 8. Results of two experiments showing that light pulses that cause large phase shifts in the free-running rhythm also inhibit testicular regression. **A,** Phase response curve for 15 min light pulses. Eight groups of animals ($n = 42$) were released into DD after entrainment to LD 14 : 10. One group (DD) received no light pulses. The remaining groups each received pulses at a different circadian time (CT) with each animal in a group receiving a light pulse at the same CT on days 5, 15 and 28 of DD. Each point plots the group mean ± s.e. of the average of the three $\Delta\phi$ values for all animals pulsed at the indicated CT. Closed circles indicate values significantly different from the DD control group. **B,** Mean testis width at week 10 of DD plotted as a function of the circadian time of the light pulses. The shaded bar represents the location of ϕ_i on the circadian time scale as estimated from figure 7B. Panels **C** and **D** give comparable data for pulses given on days 10, 20 and 30 of DD. (*See* text.)

weeks after transfer to DD hamsters pulsed with light at CT times 13, 14, 15, 16 or 18 had significantly larger testes than controls receiving no pulses ($P < 0.05$ to $P < 0.001$). Pulses at CT 10, 12, 21 or 24 failed to alter the time course of gonadal regression in DD. These data emphasize the overriding importance of the circadian time of the light stimulus as opposed to its duration or frequency: a 15 min pulse once every ten circadian cycles is sufficient for induction provided it coincides with ϕ_i. In addition, the data indicate that the region of the subjective night which could not be tested in T experiments is nevertheless highly responsive to photoperiodic stimulation, suggesting that ϕ_i is continuous through the early subjective night (CT 13 – CT 18). Thus they also confirm the results of T experiments (figure 7) by indicating that ϕ_i encompasses a major fraction of the subjective night.[5,17] In reporting these observations, however, it should be noted that failure to obtain photoperiodic induction with pulses at CT 12 and CT 21 is not consistent with the data from entrainment to different T cycles (figure 7). The physiological basis of this discrepancy is not known, but the shorter duration and infrequent presentation of the stimulus in the single pulse experiments would appear to be the most likely explanation. In any case, the phase and duration of ϕ_i in the hamster contrasts with the situation in the flesh-fly *Sarcophaga* where T cycle data indicate that ϕ_i is restricted to a brief portion ($\sim$ CT 20–CT 21) of the late subjective night.[24] To my knowledge comparable data pinpointing the location of ϕ_i in the circadian cycle are not available for other species although it should be noted that T cycle experiments with the house sparrow, *Passer domesticus*, are consistent with the assumption that ϕ_i lies in the early subjective night.[25]

A MODEL FOR THE PHOTOPERIODIC CLOCK

The foregoing analysis (delineation of ϕ_i) leads to an unusually specific and substantive formal model for the mechanism of daylength measurement by a circadian system. This model assumes: (*a*) that the phase and duration of ϕ_i with respect to activity are defined by the circadian photoperiodic response curves measured in T cycle experiments (e.g. figure 7B, 50 per cent induction) and (*b*) that photoperiodic induction is contingent on the coincidence of light and ϕ_i. This model, based on the response to T cycles close to τ, is also entirely consistent with the photoperiodic response and the entrainment of the activity rhythm to other kinds of LD cycles (e.g. resonance light cycles, symmetric skeleton photoperiods) regarded as critical tests for the participation of circadian rhythms in PTM.[4,5,17,22] The model also accounts nicely for the response to 24 h LD cycles with different photoperiods and is consistent with the critical daylength defined in these experiments.[5] While a review of these tests of the model is beyond the scope of this paper, several implications and limitations of the model warrant further discussion.

Except for the long duration of ϕ_i, the model resembles the external coincidence model of Pittendrigh and Minis[14,15,16] since, in the most simple interpretation, PTM is mediated by a single oscillation responding to the external light cycle. In this case ϕ_i may be a property of the pacemaker regulating the activity rhythm or it may be a portion of a separate rhythm in the circadian system which is driven by the same pacemaker. On the other hand, none of the available evidence excludes the possibility that hamster PTM involves two or more circadian oscillations whose mutual phase relationships change with photoperiod.[26] In this interpretation ϕ_i represents that portion of the activity cycle during which a brief exposure to light suffices to shift the multiple internal oscillations into an inductive configuration (internal coincidence).

The phase and duration of ϕ_i have been defined operationally using data from populations of animals entrained to different T cycles. The extent to which this information is influenced by stimulus parameters (e.g., duration, frequency) and other particular conditions of the experiments is not known. The extent of inter-individual variation in the phase and duration of ϕ_i is also unknown. The experiments detected a longer ϕ_i for maintenance then for recrudescence of the testes. This correlates with the observation that a longer photoperiod ($T = 24$ h) is required to stimulate regrowth than to maintain testes that are already large and metabolically active: i.e., a 12·5 h photoperiod is fully effective for maintenance but induces a sub-maximal rate of growth.[11] This difference may reflect a change in the time-measuring machinery itself or alternatively it may indicate that the response of the neuroendocrine-gonadal system to input from the clock may differ depending on the phase of the annual cycle.[7,9]

There is a close temporal correspondence between ϕ_i (figure 7) and the responsive portion of the PRC (figures 3, 8). Both responses to light are maximal in the subjective night and minimal in the subjective day. Since the $\Delta\phi$ required for entrainment to different Ts can be calculated from the relationship $\tau - T = \Delta\phi$ it follows that the dependence of the photoperiodic response on T involves a correlation between induction and the magnitude of the $\Delta\phi$ required for entrainment. The strength of this correlation has been tested and found to be highly significant in each of nine separate T experiments but its functional significance is not known.[17] It may be that ϕ_i is a property of the pacemaker regulating the activity rhythm which suggests a direct functional relationship between entrainment of the pacemaker (its $\Delta\phi$ response) and photoperiodic regulation of reproduction. Alternatively, ϕ_i may be a property of a separate rhythm driven by the activity pacemaker. In this case it is easy to imagine that natural selection has acted independently on this rhythm to favour a response to light that is maximal in the subjective night. One advantage of ϕ_i being a property of a driven rhythm would be that natural selection could adjust the photoperiodic response without directly affecting the activity pacemaker.[12] Since, in rodents, a single pacemaker (SCN) appears to regulate a diversity of physiological and behavioural rhythms in addition to activity and photoperiodic photo-inducibility,[27] the opportunity for independent evolutionary change in the photoperiodic mechanism may be essential. If selection on the photoperiodic response were to change pacemaker period (τ), the timing of a large array of functions unrelated to photoperiodism could be adversely affected.[12]

The sensitivity of hamster PTM *per se* to change in pacemaker τ can be predicted from the dependence of the photoperiodic response on T. A 2 per cent ($0\cdot5$ h) change in T (figures 6, 7) is sufficient to disrupt PTM so that a short photoperiod becomes inductive.[5,17] Thus, Eskes and Zucker[28] predicted and found that a 2 per cent change in τ induced by ingestion of D_2O also caused a short photoperiod (LD 10 : 14) to become photostimulatory when T is 24 h. From this one can speculate that natural selection could achieve a substantial change in the timing of the breeding season by producing a small change in τ; a larger $\Delta\tau$ (≈2 per cent) would abolish seasonality completely. However, the same results could be achieved with small ($\leq0\cdot5$ h) adjustments in the duration of ϕ_i or its phase with respect to the pacemaker and this would not have to involve a change in pacemaker τ. The demonstrated sensitivity of hamster PTM to change in τ (and T) may provide a partial explanation for the low inter-individual variability of τ in hamsters as compared to, for example, house mice[19] which are non-photoperiodic.

There are two characteristics of the hamster's photoperiodic clock which have potential adaptive significance and which could not have been predicted from the photoperiodic response curve (figure 1) or any existing theoretical model for the circadian basis of PTM.[16,26] It is tempting to speculate that the first of these, the long duration of ϕ_i, may derive from the close proximity of the species $\bar{\tau}$ to 24 h,[17,18]—i.e., its function may be to permit animals with τs on either side of 24 h to measure photoperiod with comparable accuracy. Alternatively, both (a) the length of ϕ_i and (b) the hamster's responsiveness to very brief and infrequent exposure to light may be more directly related to restrictions imposed by the hamster's nocturnal and fossorial habits. Could it be that these characteristics help ensure adaptive timing of the breeding season in a species which may not see daylight on a daily basis?[22] It will be of interest to see what generalizations emerge as data are accumulated on other species, thereby permitting some comparison of nocturnal and diurnal forms.

PTM AND PACEMAKER COMPLEXITY

Several characteristics of the circadian activity rhythms of hamsters and other vertebrates (e.g. splitting, history dependence of τ and α) are accommodated by a model in which the pacemaker comprises two mutually coupled oscillators which are differentially entrained by light.[21] One of these is coupled to external light in the evening (E), the other (M) to light in the morning. When mutually coupled E and M share a system period (τ_{EM}) that is intermediate between the values (τ_E and τ_M) each would express separately as free-running oscillators. In this model changes in the free-running period of the activity rhythm (τ_{EM}) and changes in the duration of activity time (α) reflect changes in the phase relationship (ψ_{EM}) between E and M. This accounts for the history dependence of τ and α since ψ_{EM} necessarily changes as photoperiod changes. This model can also accommodate the role of the circadian system in PTM by incorporating the hypothesis that photoperiodic induction is dependent on ψ_{EM} values characteristic of entrainment to long photoperiods.[21]

Figure 9 shows that light pulses at CT 16 and CT 18 which delay gonadal

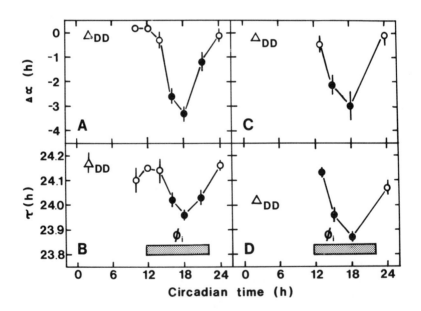

Figure 9. After effects on τ and α associated with light pulses at different circadian times. Panels **A** and **B** show data for pulses at CTs 10, 12, 14, 15, 18, 21 and 24 (c.f. figure 8**A**, **B**). Panels **C** and **D** give data for pulses at CTs 13, 15, 18 and 24 (c.f. figure 8**C**, **D**). α was measured from eye-fitted lines marking onsets and cut-offs of activity[18] on the 4–5 days immediately preceeding (α_1) and following (α_2) a light pulse. $\Delta\alpha$ is a measure of the transient change (compression) of α following a light pulse (i.e. $\Delta\alpha = \alpha_2 - \alpha_1$). τ was estimated from 7 to 10 days of the 'steady-state' free run beginning several days after the final light pulse. (For other details *see* text and figure 8.)

regression in DD (*see* figure 8) also compress α and shorten τ. This correlation also holds for pulses at CT 13, 15 and 18 except that pulses at CT 13 which cause phase delays lengthen rather than shorten τ and do not compress α (Elliott, unpublished observations). On the other hand, pulses at CT 10, 12 and 24 which fail to maintain the testes also do not change τ and α. However, the correlation between after-effects on τ and α and photoperiodic induction is not a perfect one. Pulses at CT 21 also changed τ and α but failed to delay gonadal regression while pulses at CT 14 delayed regression but failed to change τ and α. These results suggest that if changes in ψ_{EM} are critically involved in photoperiodic induction either (*a*) they are not accurately reflected in the after-effects on τ and α, or (*b*) the sensitivity with which these parameters can be measured is insufficient. Compression of α also occurred along with photoperiodic induction in Ts far from τ but in Ts closer to τ induction occurred in the absence of a detectable change in α.[11,17] In general, about all one can conclude from these comparisons is that the circadian time of the light pulse is a more reliable predictor of the gonadal response than are the after-effects on τ and α. It is not necessary to invoke pacemaker complexity to account for hamster PTM.

OUTLOOK

The photoperiodic clock of the Syrian hamster can be described in terms of a circadian rhythm of responsiveness to light which is entrained to the daily photoperiod. On long photoperiods the photo-inducible phase (ϕ_i) of this rhythm is illuminated and the gonads are maintained; on short photoperiods ϕ_i lies in the dark and the gonads regress. While this model accounts for the circadian basis and precision of daylength measurement in the hamster, it says nothing about the concrete physiological mechanisms involved. Thus the major challenge ahead is to understand the neural and endocrine machinery for the time measurement itself and for the transfer of photoperiodic information from the circadian system to the reproductive system.

As reviewed elsewhere in this volume[27,29-31] considerable progress has already been made in this direction. The pineal gland and its hormone melatonin clearly are essential for the photoperiodic control of reproduction in hamsters and ferrets and so are the suprachiasmatic nuclei (SCN) of the anterior hypothalamus. The SCN are also essential for the normal LD entrainment of circadian rhythms and for their persistence as free-running oscillations in constant conditions.[32] Indeed, recent evidence leaves no doubt that these nuclei function as a central pacemaker or as part of the pacemaking system for a diversity of circadian rhythms in mammals,[27] including the rhythm of pineal melatonin synthesis. Thus, it seems certain that the SCN are directly involved in PTM. In cooperation with the pineal gland they must also be involved in the pathway (network?) by which the circadian system regulates the activity of the reproductive system.[29] It is more difficult to characterize the role of the pineal gland. Exogenous melatonin can induce gonadal regression and the circadian pattern of pineal melatonin synthesis is regulated by photoperiod but it is not clear whether photoperiodically induced changes in the pattern are of functional significance in regulating the activity of the neuroendocrine-gonadal system.[29] For example, the nocturnal peak in pineal melatonin content lasts about as long and occurs at about the same circadian time (relative to activity) in LD 14 : 10 as it does in LD 10 : 14 or in DD, so that the only major change is in the phase of the melatonin peak and of locomotor activity with respect to the photoperiod[33,34] (Elliot and Tamarkin, unpublished observations). However, it remains possible that the action of the pineal depends on the phase relationship of the melatonin rhythm to some other rhythm whose phase relationship to activity changes as a function of photoperiod (*see* figure 20, in Pittendrigh p. 27). Clearly much more work remains to be done, but the time may yet arrive when we shall be able to understand the physiological basis of photoperiodic time measurement in terms of the action of light on the circadian system which includes the SCN and the pineal gland.

ACKNOWLEDGEMENTS

This work was supported by a series of research grants from the National Institutes of Health and by a grant from the Whitehall Foundation. I thank K. Franklin, D. Rath, K. Pekarek and A. M. Coletti for technical assistance.

REFERENCES

1. A. D. Lees. *Nature (Lond.)* **210**, 986–989 (1966).
2. S. D. Skopik and M. F. Bowen. *J. Comp. Physiol.* **111**, 249–59 (1976).
3. H. Underwood. This volume, pp. 137–52.
4. J. A. Elliott, M. H. Stetson and M. Menaker. *Science* **178**, 771–3 (1972).
5. J. A. Elliott. *Fed. Proc.* **35**, 2339–46 (1976).
6. R. J. Reiter. *Anat. Rec.* **173**, 365–71 (1972).
7. F. W. Turek, J. A. Elliott, J. D. Alvis and M. Menaker. *Biol. Reprod.* **13**, 475–81 (1975).
8. M. H. Stetson, M. Watson-Whitmyre and K. S. Matt. *J. Exp. Zool.* **202**, 81–8 (1977).
9. M. H. Stetson, J. A. Elliott and M. Menaker. *Biol. Reprod.* **13**, 324–39 (1975).
10. M. H. Stetson, K. S. Matt and M. Watson-Whitmyre. *Biol. Reprod.* **14**, 531–7 (1976).
11. J. A. Elliot. PhD Thesis, University of Texas at Austin (1974).
12. C. S. Pittendrigh. This volume, pp. 1–35.
13. C. A. Grocock and J. R. Clarke. *J. Reprod. Fertil.* **43**, 461–70 (1974).
14. C. S. Pittendrigh and D. H. Minis. *Am. Nat.* **98**, 261–94 (1964).
15. C. S. Pittendrigh. *Z. Pflangenphysiol.* **54**, 275–307 (1966).
16. C. S. Pittendrigh and D. H. Minis. In: M. Menaker (ed.) *Biochronometry.* Washington, DC, National Academy of Sciences, pp. 212–50 (1971).
17. J. A. Elliott and M. Menaker. Submitted to *Biol. Reprod.* (1981).
18. C. S. Pittendrigh and S. Daan. *J. Comp. Physiol.* **106**, 223–52 (1976).
19. S. Daan and C. S. Pittendrigh. *J. Comp. Physiol.* **106**, 253–6 (1976).
20. C. S. Pittendrigh and S. Daan. *J. Comp. Physiol.* **106**, 291–331 (1976).
21. C. S. Pittendrigh and S. Daan. *J. Comp. Physiol.* **106**, 333–55 (1976).
22. J. A. Elliott and B. D. Goldman. In: N. T. Adler (ed.) *Neuroendocrinology of Reproduction: Physiology and Behaviour.* Plenum Press, New York (in the press) (1981).
23. P. J. DeCoursey. *J. Cell. Comp. Physiol.* **63**, 189–96 (1964).
24. D. S. Saunders. *J. Comp. Physiol.* **132**, 179–89 (1979).
25. D. S. Farner, R. S. Lewis, P. W. Mattocks, T. R. Darden and J. P. Smith. *Physiol. Zool.* **50**, 247–68 (1977).
26. C. S. Pittendrigh. *Proc. Natl Acad. Sci. USA* **69**, 2734–7 (1972).
27. S. T. Inouyé and H. Kawamura. *Proc. Natl Acad. Sci. USA* **76**, 999–1003 (1979).
28. G. A. Eskes and I. Zucker. *Proc. Natl Acad. Sci. USA* **75**, 1034–8 (1978).
29. K. Hoffmann. This volume, pp. 237–50.
30. F. W. Turek and G. B. Ellis. This volume, pp. 251–60.
31. J. Herbert. This voume, pp. 261–76.
32. B. Rusak and I. Zucker. *Physiol. Rev.* **59**, 449–526 (1979).
33. L. Tamarkin, S. M. Rappert and D. C. Klein. *Endocrinology* **104**, 385–9 (1979).
34. L. Tamarkin, S. M. Rapper and D. C. Klein et al. *Endocrinology* **107**, 1525–9 (1980).

DISCUSSION

Turek: Why does the coincidence of light at circadian time (CT) 21 not maintain testicular growth in your pulse experiments, but it does in your *T* experiments?

Elliott: In *T* experiments the hamsters saw a 1 h light pulse once in every circadian cycle of entrainment whereas in the pulse experiments we are concerned with a 15 min pulse that is seen only once every 10–12 circadian cycles and only three times during a 10 week exposure to continuous darkness. These differences in protocol may explain the failure of gonadal maintenance observed with pulses at CT 21. This implies a reduced responsiveness at CT 21 as compared to CT 18 which, although suggested, was not as clearly indicated in the *T* experiments.

Short: How do hamsters perceive light in the wild? As burrowing, nocturnally active animals, are they sensitive to the low levels of light that must enter their burrows? If they can respond to very low light intensities, why don't they respond to moonlight?

Elliott: One has always assumed that these animals can entrain to the day/night cycle by periodically sampling the light intensity at or near the burrow entrance around the times of sunset and dawn. This supposition has recently gained some support from several unpublished studies (Pittendrigh, Goldman and others) in which hamsters were housed in simulated burrow systems. In Menaker's laboratory in Texas we examined the influence of light intensity on the photoperiodic testicular response to long days (LD 14 : 10) and constant light. The threshold for testicular maintenance is approximately 1·0 lux in LL and 0·1 lux in LD. I would guess that if you were to test the threshold under a cycle containing 10 h of bright light (100 lux) and 14 h of dim light and in which the intensity of the dim light was varied you would find a photoperiodic threshold of 1·0 lux. However, I have yet to do the experiment so your question about the photoperiodic effect of moonlight is an interesting one.

Gwinner: What is the minimum duration of light, given during the photosensitive phase, to induce or maintain large testes in hamsters? Considering the fact that hamsters in nature see only very little light one might expect that very brief light pulses are effective.

Elliott: In my work with hamsters free-running in continuous darkness and with entrainment to symmetric skeletons I have never used pulses shorter than 15 min. However, unpublished work from Goldman's laboratory indicates that a 30 s pulse interrupting the night of an LD 10 : 14 photoperiod is effective for both entrainment and photoperiodic induction. DeCoursey demonstrated entrainment to a 1 s flash in the flying squirrel, *Glaucomys volons*, but it remains to be seen whether this brief exposure to light can be effective for photoperiodic induction in the Syrian hamster.

GENERAL DISCUSSION—MAMMALS

Led by B. Rusak

Department of Psychology, Dalhousie University, Halifax, Nova Scotia

Since circadian organization is probably involved in the photoperiodic responses of many rodents,[1] an adequate description of the physiological basis of photoperiodism will require an understanding of the neural mechanisms regulating circadian rhythmicity. Among rodents the suprachiasmatic nuclei (SCN) have been identified as central to circadian organization. This conclusion is based on the observation that SCN-lesioned animals do not generate either behavioural or physiological circadian rhythms[2,3] (*see* Rusak[4,5] for reviews).

Although the lesion-based data are quite convincing on this point, it is necessary to stress the inherent ambiguity of such data; the loss of a particular function after ablation of a tissue is subject to multiple interpretations.[5,6] Choosing among these interpretations requires both careful observation of lesioned animals in a variety of environmental conditions and evidence derived from studies that do not involve destruction of the structure whose function is at issue. Dr Kawamura's neurophysiological studies[7] of the rat SCN are an important contribution to this process of refining our hypotheses about SCN function.

It is necessary to be explicit about any hypothesis of SCN function that is being considered. One hypothesis is that the SCN are the generators of rodent circadian rhythms, and that these rhythms are entrained to illumination cycles via the direct visual projection to the SCN, the retino-hypothalamic tract (RHT). This hypothesis has had great heuristic value and is very attractive because of its simplicity, but it is not adequate to account for the data now available on SCN-lesioned rodents. Rather, these data indicate that the SCN are part of a complex of circadian oscillators in the nervous system and that this oscillator-complex can be synchronized via several afferent pathways. A full description of these data cannot be repeated here, but the main findings and their implications for our understanding of SCN function can be summarized briefly.[5,8–10]

Photic synchronization

There is evidence that at least some rhythms can be synchronized to daily light/dark (LD) cycles after SCN (and presumably RHT) ablation; synchronization of hamster activity rhythms may be quite regular or may be transient and unstable.[4,8] These observations alone do not imply the existence of a circadian mechanism outside the SCN; synchronization could be explained as the passive response of a non-rhythmic system to an imposed illumination cycle. However, several features of these data are not compatible with this explanation and indicate instead that the system being synchronized is endogenously rhythmic. The activity records of SCN-lesioned hamsters include phase leads of activity onset relative to the beginning of the D phase, transient cycles during resynchronization to a shifted LD cycle and the breakdown of the rhythm over several cycles after release into constant illumination. These features may result from the presence of either damped or self-sustained circadian oscillators that are synchronized by the LD cycle.

The occurrence of photic synchronization in SCN-lesioned animals implies the existence of visual projections outside the RHT–SCN system that are capable of mediating light effects on rhythms. The indication of a role in photic synchronization for projections outside the RHT is reinforced by independent evidence that interruption of portions of the classic visual system can significantly alter aspects of entrainment and phase shifting of hamster activity rhythms.[9,10]

Non-photic synchronization

The activity of SCN-lesioned rodents may be synchronized by either photic or non-photic cycles. Restriction of food availability to a brief daily period can synchronize several rhythms in both intact and SCN-lesioned rats.[11–13] Daily activity bouts occur in anticipation of the restriction phase and only restriction schedules with circadian periods are effective in synchronizing behaviour. These results indicate the persistence of an endogenous timing mechanism with a circadian period in SCN-lesioned rats. In a few animals the activity rhythm free-runs for several cycles with a circadian period after the restriction schedule is discontinued,[11] just as photically synchronized rhythms in SCN-lesioned hamsters may persist for several cycles in constant conditions.[8] The eventual loss of the rhythms may be attributed either to the gradual loss of synchrony among multiple self-sustained oscillators or to the fact that the driving oscillators are strongly damped. Additional evidence described below suggests that these oscillators outside the SCN are in fact self-sustained.

Multiple circadian oscillators

The activity of SCN-lesioned hamsters released from an LD cycle into constant conditions sometimes breaks up into recognizable components that free-run with independent circadian periods. These diverging components may establish various complex patterns, including ultradian rhythms and apparent arrhythmicity. Within these patterns one can often identify circadian components that maintain relatively stable periods over many days. Similar activity components sometimes break away from synchronization to an LD cycle and free-run with a circadian period even in the presence of the cycle.[4,8]

These complex patterns are difficult to explain except as the result of the interactions of a population of circadian oscillators outside the SCN. These oscillators may be capable of coupling into relatively unstable subpopulations, but SCN-lesioned animals appear to lack an internal mechanism that can maintain stable phase relations among the entire population. The resultant dissociation among component oscillators may ultimately produce apparent behavioural arrhythmicity.

It is important to note that the absence of normal circadian rhythms need not reflect the absence of endogenous circadian oscillators. No overt circadian rhythm in rodents has been reported to survive SCN ablation under constant conditions, yet such lesioned animals still possess circadian oscillators. The loss of overt circadian rhythms may reflect the inability to integrate a population of oscillators into a single, coherent, circadian framework.

This analysis suggests the hypothesis that the SCN normally provide this internal integration by regulating coupling among elements of the oscillator population.[8] After SCN ablation, both photic and non-photic circadian environmental cycles can synchronize at least a portion of this population; the resulting integration may persist for several cycles in constant conditions. However, in the absence of either an external or an internal source of integration, the component oscillators dissociate and overt circadian rhythms are lost.

Conclusion.

The findings summarized here indicate that the rodent circadian system includes multiple pathways that mediate photic and non-photic synchronization of a population of coupled circadian oscillators. The SCN appear to function both as primary targets of photic input and as integrative mechanisms that are responsible for maintaining normal coupling among other oscillators. Analysis of the neuro-physiological basis for rodent circadian organization will require identification of these oscillators and of their interactions with the SCN in this complex multi-oscillator system. Addressing the full complexity of circadian organization itself is a necessary step in describing the physiological basis for the role of circadian rhythms in rodent photoperiodism.

REFERENCES

1. J. A. Elliott. *Fed. Proc.* **35,** 2339–46 (1976).
2. R. Y. Moore and V. B. Eichler. *Brain Res.* **42,** 201–6 (1972).
3. F. K. Stephan and I. Zucker. *Proc. Natl Acad. Sci. USA* **69,** 1583–6 (1972).
4. B. Rusak. *Fed. Proc.* **38,** 2589–95 (1979).
5. B. Rusak and I. Zucker. *Physiol. Rev.* **59,** 449–526 (1979).
6. T. A. Schoenfeld and L. W. Hamilton. *Physiol. Behav.* **18,** 951–67 (1977).
7. S. T. Inouyé and H. Kawamura. *Proc. Natl Acad. Sci. USA* **76,** 5962–6 (1980).
8. B. Rusak. *J. Comp. Physiol.* **118,** 145–64 (1977).
9. B. Rusak. *J. Comp. Physiol.* **118,** 165–72 (1977).
10. I. Zucker, B. Rusak and R. G. King jun. In: A. H. Riesen and R. F. Thompson (eds.). *Advances in Psychobiology,* vol 3. New York, Wiley, pp. 35–74 (1976).
11. Z. Boulos, A. M. Rosenwasser and M. Terman. *Behav. Brain Res.* **1,** 39–65 (1980).
12. D. T. Krieger, H. Hauser and L. C. Krey. *Science* **197,** 398–9 (1977).
13. F. K. Stephan, J. M. Swann and C. L. Sisk. *Behav. Neural Biol.* **25,** 545–54 (1979).

DISCUSSION

Brown-Grant: Lithium, used clinically in the control of recurrent mania, may cause hypothyroidism with compensatory thyroid gland enlargement. Maybe you have made your hamsters hypothyroid and this might affect the 'coupling' of various oscillators.

Brady: Dr Rusak, you showed a model with an effector being driven by an SCN-oscillator system, and jokingly identified the effector as a leg. I wonder whether you, Dr Menaker or Dr Kawamura would like to suggest what you think the nature of the downstream control is between the SCN and the eventual 'effector' (the leg), bearing in mind that rhythms in many behaviours are stopped by SCN ablations.

Rusak: The SCN may couple in different ways to different effector systems. SCN efferents may continuously modulate neural activity in their targets and thereby gradually change the probabilities of various behaviours and the levels of hormones released throughout the day. Other effectors may be activated by a discrete signal from the SCN. Since we are completely ignorant of the neurophysiological basis for coupling between the SCN and any effector system, we don't know whether these are really different mechanisms or just modified versions of a single mechanism. For example, if one adds a trigger mechanism between the continuously modulated output of the SCN and an effector, one can achieve discrete timing or gating of an event (such as the luteinizing hormone surge in female rodents) without requiring a discrete signal from the SCN.

Bittman (Comment following an interchange by Pittendrigh/Rusak/Elliott, the question being as to whether any effects of melatonin on the circadian system are not gonadally mediated): We are liable to remain handicapped by our inability to resolve which (and how many) oscillators subserve photoperiodic time measurement (PTM). If a rhythm, e.g. locomotor activity, is affected by the pineal independently of gonadal effects we cannot assume that oscillations more central to PTM behave similarly. It remains possible that the mammalian pineal retains the capacity to address a subset of oscillations whose functions concern photoperiodism and reproductive function but whose operation we cannot directly observe.

Turek: Until we know site of action of melatonin, it is difficult to talk about physiological and pharmacological doses of melatonin. Indeed, melatonin may reach its site of action in an untreated animal via axonal transport, the CSF or the plasma. In addition, we should bear in mind that changing levels of melatonin in the pineal or the plasma may not be the important variable. We know that the photoperiod can alter sensitivity to the feedback effects of steroid hormones, and it may well be that the photoperiod can alter the responsiveness of neural tissue to melatonin.

THE HYPOTHALAMIC PULSE GENERATOR: A KEY DETERMINANT OF REPRODUCTIVE CYCLES IN SHEEP

Robert L. Goodman and Fred J. Karsch

Reproductive Endocrinology Program, Departments of Pathology and Physiology, The University of Michigan, Ann Arbor, USA

Summary

In this paper we develop the thesis that changes in a hypothalamic pulse generator driving episodic LH secretion play a crucial role in the control of reproductive cycles in the ewe. It is postulated that LH-pulse frequency is a critical determinant of follicular development with high-frequency pulses required for initiating the events which lead to ovulation. Both external (photoperiod) and internal (steroids) signals can govern the timing of ovulation by controlling the frequency of the hypothalamic pulse generator. In the breeding season, progesterone decreases LH-pulse frequency whereas oestradiol cannot; oestradiol acting, at least in part, on the pituitary gland reduces pulse amplitude. Therefore, LH pulses in the luteal phase of the oestrous cycle occur too infrequently for initiation of the pre-ovulatory events. In the follicular phase when progesterone has declined, the pulse generator reverts to its endogenous periodicity, high-frequency LH pulses result and ovulation occurs. Thus progesterone determines the bi-weekly rhythm in ovulation responsible for oestrous cyclicity. During anoestrus the inhibitory photoperiod has two effects: (a) it acts independently of ovarian steroids to reduce the frequency of the hypothalamic pulse generator and (b) it allows oestradiol to gain access to this neural oscillator. Consequently, the inhibitory effects of oestradiol increase in anoestrus because it can now reduce both LH-pulse frequency and amplitude. This interaction of oestradiol and photoperiod results in slow LH pulses in intact ewes and thereby prevents ovulation in anoestrus.

INTRODUCTION

While the potential significance of circadian and circannual clocks in seasonal cycles has been recognized for some time, the importance of another 'biological' clock has only recently been realized. This neural oscillator, which has an endogenous period of 30–90 min, is a critical component of the neuroendocrine system governing tonic LH secretion.* Since changes in tonic gonadotrophin secretion determine seasonal variations in gonadal function,[1-4] it is likely that

---

* The term 'tonic' is used to describe LH secretion that is controlled by the inhibitory effects of gonadal steroids. It includes all types of secretion except the pre-ovulatory surge of gonadotrophins.

Biological Clocks in Seasonal Reproductive Cycles 223–236 (1981) (ed. B. K. and D. E. Follett: Bristol, Wright).

alterations in the activity of this neural oscillator underlie annual reproductive rhythms. In this paper we shall briefly consider data leading to the discovery of the neural oscillator and shall then present recent evidence which demonstrates that changes in its frequency control both the timing of ovulation during the oestrous cycle and the annual reproductive cycle of the ewe.

THE HYPOTHALAMIC PULSE GENERATOR

The existence of a neural oscillator (or pulse generator) was first inferred from the striking pulsatile pattern in serum LH concentrations observed in samples collected at frequent intervals;[5] LH levels increase abruptly within a few minutes and then decay exponentially until the beginning of the next pulse (figure 1, left panel). This

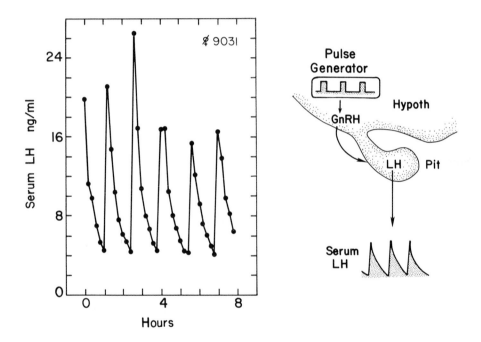

Figure 1. *Left*: Example of pulsatile pattern in LH secretion. LH concentrations in peripheral serum were determined in samples obtained every 12 min from a female sheep 10 days after ovariectomy in the mid-anoestrous season. *Right*: Postulated mechanism underlying pulsatile LH secretion. A pulse generator in the hypothalamus (Hypoth) produces a rhythmic discharge of gonadotrophin-releasing hormone (GnRH) which stimulates episodic release of LH from the anterior pituitary gland (Pit).

pattern is thought to reflect a discrete burst of secretory activity followed by a relatively long period of little or no secretion during which LH concentrations decline as the hormone is metabolized.[5,6] Such episodic secretion of LH now appears to be the rule. It has been described in a wide variety of species including monkeys,[5] man,[7] sheep,[8,9] deer,[10] cows,[11] rabbits,[12] rats[13] and quail,[14] and in most of these it is evident in both intact and gonadectomized males and females.

Attempts to elucidate the mechanisms underlying pulsatile LH secretion quickly led to the concept of a pulse generator producing episodic discharges of gonado-trophin-releasing hormone (GnRH) into the hypophysial portal circulation (figure 1, right panel). The postulation of episodic GnRH secretion was initially based on the ability of centrally active drugs to block LH pulses[15,16] and has since been confirmed by the demonstration of GnRH pulses in hypophysial portal blood.[17] Moreover, intermittent, but not continuous, administration of GnRH is able to maintain LH secretion following destruction of the hypothalamus.[18] Thus, the episodic nature of GnRH secretion, and the consequent pulsatile secretion of LH, may be obligatory to the normal functioning of the hypothalamo-hypophysial axis.

That these GnRH pulses reflect the operation of an endogenous pulse generator now also seems incontrovertible. The most plausible alternative hypothesis is that the LH pulses are the consequence of a 'short loop' feedback whereby a given pulse is initiated when serum LH concentrations from the previous pulse fall below a critical level. This hypothesis, however, is untenable because neither exogenous LH administration,[19] nor increased endogenous LH levels,[20] abolishes LH pulses. The demonstration of GnRH pulses in portal blood[17] also points to an endogenous pulse generator since stalk blood collection, of necessity, disrupts the secretion of LH and thus any feedback loop that could potentially result in discontinuous GnRH secretion.

Although the concept of an endogenous neural oscillator driving episodic LH secretion is now well established, the neural mechanisms underlying this pulse generator remain a mystery. Surgical deafferentation studies have clearly demon-strated that it is resident within the medial basal hypothalamus.[21-23] Furthermore, the arcuate nucleus appears to be of critical importance since lesions of this area abolish LH pulses.[22-24] At this time, these rudimentary anatomical considerations are all that is known about the neurophysiology of this system.

While this hypothalamic pulse generator can operate autonomously, its function is normally modulated by hormonal as well as extra-hypothalamic neural inputs. In fact, such modulation appears to play a crucial role in the normal regulation of seasonal breeding. We shall now consider how external (photoperiod) and internal (steroid) signals alter the *frequency* of this oscillator and, thereby, control ovarian cyclicity in the ewe. It should be emphasized that much of the data to be presented is preliminary; the resulting concepts are thus proposed as working hypotheses, not well-documented theories. At the outset, we should also acknowledge several other investigators who have independently developed similar theories for the control of reproductive function. In particular, Lincoln and Short[4], Knobil[25] and Ryan and Foster[26] have all postulated that changes in the frequency of GnRH (or LH) discharges play a critical role in the regulation of gonadal function. In addition, the work of Baird and McNeilly[27] has contributed greatly to our understanding of the ovarian component of this system.

STEROID INHIBITION OF LH PULSES DURING THE BREEDING SEASON: CONTROL OF OESTROUS CYCLICITY

To put the effects of steroids on pulsatile LH secretion into a physiological perspective, we must digress briefly and consider the current model for steroidal control of the ovine oestrous cycle.

Because progesterone is the primary negative feedback hormone during the breeding season,[28,29] the fall in serum progesterone at luteolysis initiates a sequence

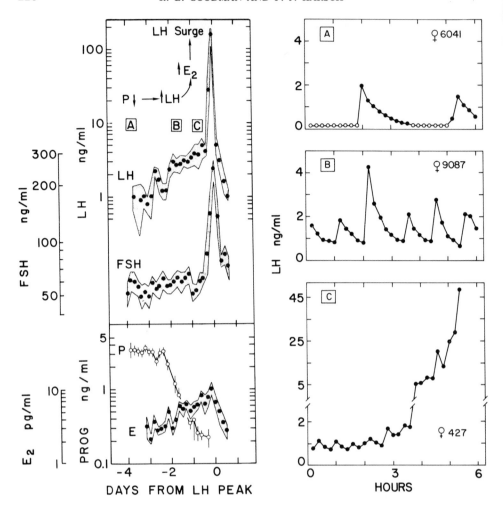

Figure 2. Left: Sequence of endocrine events leading to ovulation in the ewe (*top*). Mean ± s.e. mean serum concentrations of LH, FSH, oestradiol (E_2) and progesterone (P) in samples obtained from six ewes every 4 h. Data are plotted on a logarithmic scale and normalized to pre-ovulatory LH peak. A, B and C denote times that LH pulses were examined. *Right*: Pattern of LH pulses at various times during the oestrous cycle: luteal phase (A); 1–2 days before LH peak (B); onset of LH peak (C) (note break in scale for LH concentration). Data are for representative individual sheep from which samples were obtained at 12 min intervals. ○, Undetectable serum LH concentrations (<0.15 ng/ml). (From Karsch et al.[30] together with some unpublished observations.)

of events culminating in the LH surge that produces ovulation (figure 2, left panel). The sequence begins with a sustained increase in tonic LH secretion resulting from progesterone withdrawal.[28–30] This LH increase stimulates ovarian oestradiol secretion which elicits oestrous behaviour and the pre-ovulatory LH surge.[27,31,32] Thus, progesterone can be viewed as controlling the timing of ovulation during the breeding season. When progesterone is elevated it holds tonic LH secretion in check and thereby prevents ovulation; when it falls, LH rises and ovulation follows within 3–4 days. We will now describe evidence suggesting that progesterone exerts this effect via the hypothalamic pulse generator.

The first piece of the puzzle fell into place with the description of the pattern of pulsatile LH secretion during the oestrous cycle.[8,33,34] As illustrated in figure 2 (right panel), the frequency of LH pulses is inversely correlated with serum progesterone concentrations, whereas pulse amplitude is inversely related to oestradiol. Thus, when progesterone secretion is maximal in the luteal phase, LH pulses occur infrequently. In contrast, during the LH rise that follows luteolysis,

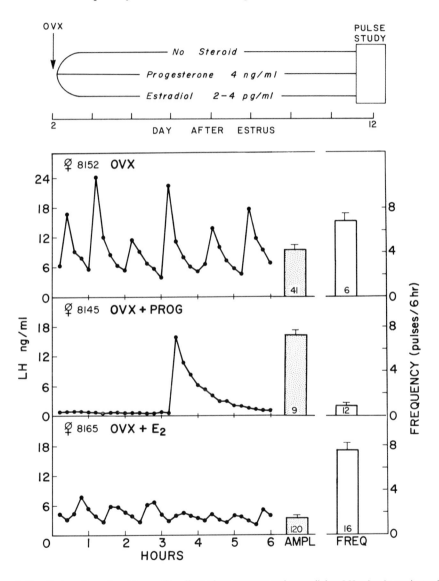

Figure 3. *Top*: Design of experiment to examine effects of progesterone and oestradiol on LH pulses in ovariectomized sheep (OVX), steroids administered via s.c. implants. *Bottom*: LH pulse pattern in representative female treated with no steroid (OVX), progesterone (OVX + PROG), or oestradiol (OVX + E₂). Samples were obtained every 12 min. Bars at right depict mean ± s.e. mean frequency (FREQ) and amplitude (AMPL, peak minus preceding nadir) of LH pulses. Numbers within bars indicate number of pulses (amplitude) or number of 6 h observation periods (frequency). (From Goodman and Karsch.[35])

the frequency of LH pulses increases several fold and the amplitude decreases markedly just prior to the onset of the LH surge.

This characteristic variation in LH pulse pattern raises the possibility that progesterone and oestradiol selectively modulate different components of this system, namely that progesterone decreases LH-pulse frequency while oestradiol reduces pulse amplitude. This possibility was tested in the experiment illustrated in figure 3. Ewes were ovariectomized 2 days after oestrus during the mid-breeding season and immediately treated with steroid-containing Silastic implants which maintained luteal phase levels of either progesterone (4 ng/ml) or oestradiol (2–4 pg/ml). Controls received no steroid treatment. Analysis of LH pulses on days 12 and 13 after oestrus revealed that the two steroids inhibited different aspects of pulsatile LH release. Progesterone selectively reduced frequency whereas oestradiol decreased only amplitude. Additional studies suggest that progesterone acts on the brain to reduce the frequency of episodic GnRH release whereas oestradiol acts, at least in part, upon the anterior pituitary gland to decrease its response to GnRH and thus limit LH pulse amplitude.[35]

These observations can be used to construct a working hypothesis for the control of ovulation during the breeding season (figure 4). An important underlying assumption of this hypothesis is that an LH pulse frequency approximating once an hour (or less) is required for the final stages of follicular development and the

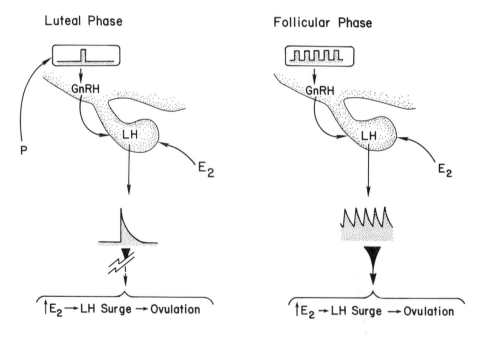

Figure 4. Model for control of ovulation during the oestrous cycle. Left: In the luteal phase, progesterone (P) acts on the hypothalamic pulse generator to reduce frequency of episodic GnRH and LH release. Oestradiol (E₂) acts on the pituitary gland to decrease response to GnRH thus decreasing LH pulse amplitude. Infrequent LH pulses are insufficient for ovulation. Right: In the follicular phase, frequency of the pulse generator increases due to P withdrawal, whereas E₂ still limits pulse amplitude by an action on the pituitary gland. High frequency LH pulses initiate the endocrine sequence which causes ovulation. Note, induction of the LH surge by E₂ is not necessarily due to an action on the pituitary gland.

pre-ovulatory oestradiol rise.* We suggest that throughout much of the cycle luteal progesterone decreases the frequency of the hypothalamic pulse generator and the resulting occasional LH pulse is not sufficient to stimulate sustained oestradiol secretion; hence ovulation cannot occur (figure 4, left panel). When progesterone falls, however, the pulse generator, which is *not slowed* by oestradiol, reverts to its endogenous periodicity of 30–60 min (figure 4, right panel). This high frequency then initiates the sequence of hormonal events, previously described, that ends in ovulation.

MODULATION OF LH PULSES BY PHOTOPERIOD AND STEROIDS: CONTROL OF SEASONAL BREEDING

As mentioned earlier, seasonal reproductive patterns in a number of species are due to photoperiodically controlled changes in tonic gonadotrophin secretion.[1-4] In the ewe, this is manifested largely as a seasonal variation in the ability of oestradiol

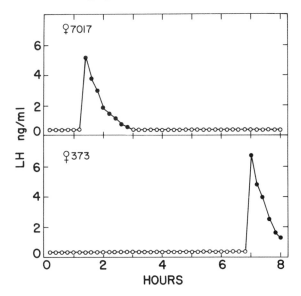

Figure 5. LH pulse pattern in representative intact ewes during mid-anoestrous season. Samples were obtained every 12 min. ○, Undetectable serum LH concentrations (<0·4 ng/ml).

to suppress LH with oestradiol being much more potent in anoestrus.[36] Thus, in intact anoestrous ewes the secretion rate of LH is not sufficient to produce the changes in follicular development and oestradiol secretion seen during the follicular phase of the cycle.[3,31,32] The model just developed raises the intriguing possibility that the deficit in LH secretion in anoestrus, as in the luteal phase, reflects a reduction in the frequency of pulsatile LH secretion. As has been reported previously[34,37] and illustrated in figure 5, LH pulses are indeed infrequent in anoestrus, occurring approximately every 6–12 h in intact ewes. Thus, it appears that, once again, a decrease in the frequency of the pulse generator controlling

---

* This assumption seems reasonable in the light of evidence that hourly injections of LH mimicking the follicular-phase LH rise induce ovulation in prepubertal lambs[26] and anoestrous ewes.[27]

GnRH secretion may be responsible for the absence of ovulation. It is interesting that an analogous situation may exist in the ram. In an elegant series of experiments, Lincoln has provided evidence that changes in LH pulse frequency, emanating from photoperiodically driven shifts in the pulsatile release of GnRH, cause the transitions between the breeding and non-breeding season in the Soay ram.[4]

The slow LH pulses in intact anoestrous ewes raise an important question. What holds LH pulse frequency in check during anoestrus? Progesterone can probably be excluded since circulating levels of this steroid are extremely low at this time of year, but two other possibilities remain. First, inhibitory photoperiods may decrease pulse frequency by mechanisms not involving steroid feedback—a 'direct photoperiodic drive' which it has been suggested occurs in the ram.[4] Alternatively, an ovarian hormone other than progesterone may gain the capacity to inhibit LH pulse frequency in anoestrus.

Effects of photoperiod on pulsatile LH secretion

In a variety of species, serum gonadotrophin concentrations change with season in the apparent absence of gonadal hormones, with *mean* levels decreasing in the non-breeding season to varying degrees depending on the species.[1,2,38] These observations have led to the hypothesis that seasonal breeding results from photoperiodically driven changes in GnRH secretion that are independent of steroid feedback. Because our earlier studies failed to detect seasonal changes in mean LH levels in ovariectomized sheep, we initially concluded that this hypothesis was untenable in the ewe.[32,36] However, the realization that changes in frequency of the hypothalamic pulse generator may be the critical determinant of seasonal breeding raised the possibility that there are reciprocal and off-setting seasonal change in the frequency and amplitude of LH pulses without corresponding differences in mean levels. We have, therefore, investigated the effects of season on the pattern of pulsatile LH secretion in ovariectomized ewes.

In the first experiment, LH pulses were analysed 10 days after castration of breeding season and anoestrous ewes. Pulses of LH increased to approximately one per hour in both seasons and, while the frequency tended to be slightly less in anoestrus, this trend was not statistically significant.[32] We are now re-examining this question using long term (>2 months) gonadectomized ewes in order to eliminate possible residual effects of ovarian steroids.[38] Using this model, a distinct seasonal difference in LH pulses is beginning to emerge (figure 6). Large well-organized pulses occur hourly in anoestrus; more frequent, small amplitude pulses prevail in the breeding season. In fact, the frequency was so high during the breeding season, that our normal 12 min bleeding interval was barely sufficient to demonstrate regular pulses (figure 6, top panel). When the sampling régime was increased to once every 4 min, however, discrete LH pulses became more evident (figure 6, inset).

The occurrence of seasonal changes in gonadotrophin profiles in long term gonadectomized animals must be interpreted with care for a number of reasons discussed elsewhere.[38] Not the least of these is the possibility that minute quantities of extra-gonadal steroids may become capable of acting upon a hypothalamo-hypophysial axis that is exquisitely sensitive to their inhibitory effects in anoestrus (*see below*). Such reservations notwithstanding, the results of this experiment are consistent with the conclusion that environmental (presumably photoperiodic)

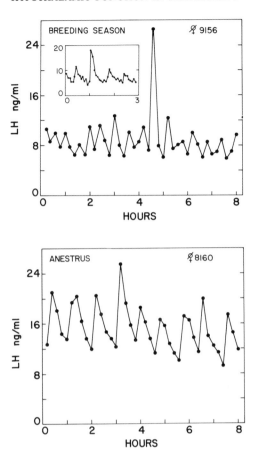

Figure 6. LH pulse pattern in representative long term ovariectomized ewes during (*top*) mid-breeding season and (*bottom*) mid-anoestrus. Samples were obtained at 12 min intervals for 8 h, and in the breeding season, samples were also obtained from the same ewe on another day at 4 min intervals for 3 h (*inset*).

signals can impinge on the hypothalamic pulse generator to modulate its frequency independent of steroid feedback.

Steroid control of LH pulses in anoestrus

While the foregoing studies have suggested that a steroid-independent effect of photoperiod does occur, they have also demonstrated that this effect is insufficient, by itself, to account for the low frequency of LH pulses in intact anoestrous ewes (figure 5). We are, thus, still faced with the question of which ovarian steroid is responsible for suppressing the hypothalamic pulse generator in anoestrus.

Our first candidate to fill this role is oestradiol. Although this steroid may seem an unlikely choice in light of the evidence that it does not reduce LH pulse frequency (figure 3), those studies were performed in the breeding season when oestradiol is a weak inhibitory steroid. Since oestradiol becomes an extremely potent negative feedback hormone in anoestrus,[36] it may gain access to the pulse generator and inhibit the frequency of LH pulses at this time of year. In support of this argument, ovarian oestradiol secretion appears to be sufficient, by itself, to account for the low mean LH levels seen in intact anoestrous ewes.[31,32] We have, therefore, begun to test the hypothesis that the seasonal change in oestradiol negative feedback reflects a change in its ability to decrease the frequency of the hypothalamic pulse generator.

Our first experiment closely resembled that described in figure 3, but only oestradiol treatments were employed. Two different size implants were used. One produced serum oestradiol concentrations (2–4 pg/ml) slightly greater than those in intact anoestrous ewes (1–2 pg/ml); the other provided concentrations less than these (<1 pg/ml). Regardless of dose, oestradiol suppressed LH below the level of detection during the pulse analyses on days 4 and 10 after ovariectomy. While this emphasizes the extreme sensitivity of the anoestrous ewe to the negative feedback action of oestradiol, it also precludes an evaluation of the effects of oestradiol on LH pulse frequency.

We thus conducted a second experiment in which we examined the acute effects of oestradiol in suppressing the elevated levels of LH present in long term gonadectomized ewes. LH pulses were analysed for 4 h prior to insertion of an oestradiol implant, and again 3 days later, during both the breeding and anoestrous seasons. As illustrated in figure 7, oestradiol had no obvious effect on LH pulse frequency during the breeding season, whereas in anoestrus the steroid was remarkably effective in this regard. This finding suggests that the effects of oestradiol on pulsatile LH secretion differ markedly with season, thus providing support for the hypothesis that a seasonal change in the ability of oestradiol to suppress the hypothalamic pulse generator underlies the seasonal variation in the negative feedback action of this steroid.

MODEL FOR THE INTERACTION OF PHOTOPERIOD AND STEROIDS IN THE CONTROL OF PULSATILE LH SECRETION

The foregoing considerations may be used to construct a speculative model to explain how external (photoperiod) and internal (steroids) signals interact to control pulsatile LH secretion and, thereby, determine the reproductive status of the ewe (figure 8). The centrepiece of this model is a hypothalamic pulse generator which produces the episodic discharges of GnRH that drive pulsatile LH secretion. The activity of the pulse generator governs the frequency of LH discharge. LH pulse amplitude may reflect a number of variables which include the amount of GnRH released in each secretory episode, the response of the pituitary gland to GnRH, and perhaps the frequency of impulses arising from the pulse generator itself.

At the ovarian level, LH pulse *frequency* is postulated to be the limiting variable, with an hourly frequency required for the final stages of follicular development and the sustained oestradiol rise which elicits the pre-ovulatory LH surge. Why an hourly frequency is essential is unclear. Possibly it simply reflects the ability of this frequency to produce a sustained elevation in mean LH concentrations.[31,32]

It is proposed that in all seasons the frequency of pulsatile LH secretion reflects a balance between the regulatory actions of both steroidal and photoperiodic signals, the latter of which are transmitted to the pulse generator via a number of intermediary steps. The photoperiodic input is stimulatory (or non-inhibitory) in

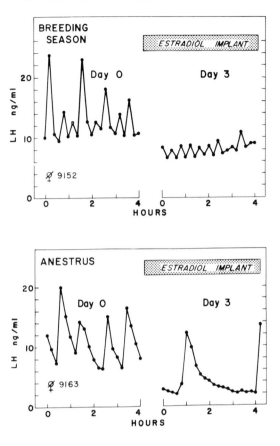

Figure 7. Seasonal difference in effect of oestradiol on LH pulses in representative long term ovariectomized ewes. Samples were obtained every 12 min for 8 h immediately before, and again 3 days after, insertion of an oestradiol implant in the (*top*) mid-breeding season and (*bottom*) mid-anoestrus. The subcutaneous oestradiol implant, which remained in place throughout the study, produced physiological serum oestradiol concentrations of 2–4 pg/ml. Note seasonal difference in LH pulses both before and during oestradiol treatment.

the breeding season and inhibitory (or non-stimulatory) during anoestrus. This input serves two functions; it modulates LH pulse frequency directly and it determines the extent to which certain ovarian steroids can act in this regard.

Negative feedback steroids regulate LH pulse frequency throughout the year, but the specific ones which serve this function change with season. In the breeding season (figure 8, top) when the photoperiodic input is stimulatory, progesterone holds LH pulse frequency in check; oestradiol cannot. The inhibitory effects of oestradiol control LH pulse amplitude and are expressed, at least in part, on the pituitary gland. Thus, at this time of year, progesterone determines the bi-weekly rhythm in ovulation responsible for oestrous cyclicity. When progesterone is high, ovulation does not occur; when it falls, ovulation follows in 3–4 days.

In the anoestrous season (figure 8, bottom), it is suggested that in addition to limiting pulse amplitude, oestradiol acquires the capacity to suppress pulse frequency and that this may account for the photoperiodically driven change in oestradiol negative feedback. Thus, an interaction of photoperiod and oestradiol

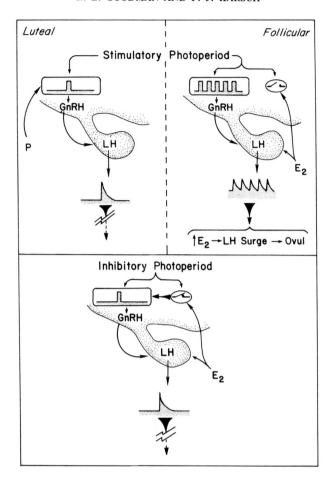

Figure 8. Hypothetical model for interaction of photoperiod and steroids in the control of ovulation in female sheep. *Top*: Under stimulatory photoperiods of the breeding season, there is a dual facilitation of LH pulse frequency: (1) a steroid-independent stimulation of the pulse generator and (2) a photoperiodically imposed blockade of an inhibitory effect of oestradiol on the pulse generator. Thus, LH pulse frequency, which is reduced by progesterone (P) in the luteal phase (*left*) is uninhibited in the follicular phase (*right*) when oestradiol (E_2) predominates. As a result, ovulation can occur when P is absent. *Bottom*: Under inhibitory photoperiods of anoestrus, there is a dual suppression of LH pulse frequency: (1) a direct photic suppression of the pulse generator and (2) an acquired capacity of oestradiol to inhibit the pulse generator. As a result of both factors, LH pulses occur too infrequently for ovulation, despite low levels of P.

on the hypothalamic pulse generator determines the seasonal rhythm in ovulation. When photoperiod is inhibitory (anoestrus), oestradiol gains access to the pulse generator thus holding pulsatile LH secretion in check and preventing ovulation. When photoperiod becomes stimulatory (transition to breeding season), this effect of oestradiol is lost and ovulation occurs. Consequently, a new corpus luteum is formed, serum progesterone levels rise and this steroid supplants oestradiol as the timer of LH pulses and hence ovulation.

This model is undoubtedly greatly oversimplified. For example, we have omitted seasonal changes in other hormones, such as FSH and prolactin, and have ignored potentially important changes in follicular responsiveness to LH. This simplification was to some extent necessary in order to focus attention on what we feel are the rate-limiting steps controlling ovarian function in the ewe. It is also interesting to note that other anovulatory conditions (sexual immaturity,[8] lactation[39]) are characterized by infrequent LH pulses. Thus, changes in the output of the hypothalamic pulse generator may be a fairly widespread mechanism for controlling ovarian cyclicity.

ACKNOWLEDGEMENTS

We are indebted to our colleagues Drs Douglas L. Foster, Eric L. Bittman, Kathleen D. Ryan and Sandra J. Legan whose critical advice and assistance made these studies possible. We should also like to thank Mr Douglas Doop, Ms Marjorie Hepburn and Ms Regina Jakacki for their technical assistance. This work was supported by grants from the NIH (HD–08333, HD–07689 and HD–05615).

REFERENCES

1. B. K. Follett. In: D. B. Crighton, N. B. Haynes, G. R. Foxcroft et al. (ed.) *Control of Ovulation*. London, Butterworths, pp. 267–94 (1978).
2. F. W. Turek and C. S. Campbell. *Biol. Reprod.* **20**, 32–50 (1979).
3. S. J. Legan and F. J. Karsch. *Biol. Reprod.* **20**, 74–85 (1979).
4. G. A. Lincoln and R. V. Short. *Recent Prog. Horm. Res.* **36**, 1–52 (1980).
5. D. J. Dierschke, A. N. Bhattacharya, L. E. Atkinson et al. *Endocrinology* **87**, 850–3 (1970).
6. W. R. Butler, P. V. Malven, L. B. Willett et al. *Endocrinology* **91**, 793–801 (1972).
7. S. C. C. Yen, C. C. Tsai, F. Naftolin et al. *J. Clin. Endocrinol. Metabol.* **34**, 671–5 (1972).
8. D. L. Foster, J. A. Lemons, R. B. Jaffe et al. *Endocrinology* **60**, 101–6 (1974).
9. C. B. Katongole, F. Naftolin and R. V. Short. *J. Endocrinol.* **60**, 101–6 (1974).
10. G. A. Lincoln and R. N. B. Kay. *J. Reprod. Fertil.* **55**, 75–80 (1979).
11. C. H. Rahe, R. E. Owens, H. J. Newton et al. *Endocrinology* **107**, 498–503 (1980).
12. P. H. Rowe, C. R. N. Hopkinson, J. C. Shenton et al. *Steroids* **25**, 313–21 (1975).
13. V. L. Gay and N. A. Sheth. *Endocrinology* **90**, 158–62 (1972).
14. B. Gledhill and B. K. Follett. *J. Endocrinol.* **71**, 245–57 (1976).
15. A. N. Bhattacharya, D. J. Dierschke, T. Yamaji et al. *Endocrinology* **90**, 778–86 (1972).
16. G. A. Schuiling and H. P. Gnodde. *J. Endocrinol.* **70**, 97–104 (1976).
17. P. W. Carmel, S. Araki and M. Ferin. *Endocrinology* **99**, 243–8 (1976).
18. P. E. Belchetz, T. M. Plant, Y. Nakai et al. *Science* **202**, 631–3 (1978).
19. E. Knobil. *Recent Prog. Horm. Res.* **30**, 1–36 (1974).
20. E. L. Piper, J. L. Perkins, D. R. Tugwell et al. *Proc. Soc. Exp. Biol. Med.* **148**, 880–2 (1975).
21. L. C. Krey, W. R. Butler and E. Knobil. *Endocrinology* **96**, 1073–87 (1975).
22. G. L. Jackson, D. Kuehl, K. McDowell et al. *Biol. Reprod.* **17**, 808–19 (1978).
23. B. D. Soper and R. F. Weick. *Endocrinology* **106**, 348–55 (1980).
24. T. M. Plant, L. C. Krey, J. Moossy et al. *Endocrinology* **102**, 52–62 (1978).
25. E. Knobil. *Recent Prog. Horm. Res.* **36**, 53–88 (1980).
26. K. D. Ryan and D. L. Foster. *Fed. Proc.* **39**, 2372–7 (1980).
27. D. T. Baird and A. S. McNeilly. *J. Reprod. Fertil.* (in the press) (1981).
28. D. T. Baird and R. J. Scaramuzzi. *J. Endocrinol.* **70**, 237–45 (1976).
29. R. L. Hauger, F. J. Karsch and D. L. Foster. *Endocrinology* **101**, 807–17 (1977).
30. F. J. Karsch, D. L. Foster, S. J. Legan et al. *Endocrinology* **105**, 421–6 (1979).
31. R. L. Goodman and F. J. Karsch. In: R. J. Reiter and B. K. Follett (ed.) *Seasonal Breeding in Higher Vertebrates*. Basle, Karger, pp. 134–54 (1980).
32. F. J. Karsch, R. L. Goodman and S. J. Legan. *J. Reprod. Fertil.* **58**, 521–35 (1980).
33. D. T. Baird. *Biol. Reprod.* **18**, 359–64 (1978).
34. P. Yuthasastrakosol, W. M. Palmer and B. E. Howland. *J. Reprod. Fertil.* **50**, 319–21 (1977).

35. R. L. Goodman and F. J. Karsch. *Endocrinology* **107,** 1286–90 (1980).
36. S. J. Legan, F. J. Karsch and D. L. Foster. *Endocrinology* **101,** 818–24 (1977).
37. R. J. Scaramuzzi and D. T. Baird. *Endocrinology* **101,** 1801–6 (1977).
38. R. L. Goodman and F. J. Karsch. *J. Reprod. Fertil.* (in the press) (1981).
39. P. J. Wright, P. E. Geytenberk, I. J. Clark et al. Sixth International Congress of Endocrinology, Melbourne. Abstr. 884 (1980).

THE ROLE OF THE PINEAL GLAND IN THE PHOTOPERIODIC CONTROL OF SEASONAL CYCLES IN HAMSTERS

Klaus Hoffmann

Max-Planck-Institut für Verhaltensphysiologie, Andechs,
Federal Republic of Germany

Summary

Pineal involvement in transferring the effects of photoperiodic stimuli to the neuroendocrine axis in hamsters is discussed. Experiments with pinealectomized animals demonstrate unequivocally that the pineal is involved in conducting photoperiodic stimuli, which finally act upon the gonads but also modify other gonad-independent functions. At least in *Phodopus* there is evidence that the pineal conveys not only the inhibitory effects of short photoperiods but also the stimulatory effects of long photoperiods. The latter is an active process rather than arising from the elimination of an inhibition. Taken together these findings suggest that the pineal gland ought not to be considered to have an antigonadotrophic action, but is an important link in the transduction of the photoperiodic message. There is also some evidence for photoperiodic effects which bypass the pineal.

A considerable number of experiments in which melatonin has been used suggest that this pineal indole is involved in transferring the photoperiodic message, and that not only is the amount of melatonin available important but also its temporal pattern of release. While in the golden hamster measurements of pineal melatonin have given equivocal results, in the Djungarian hamster determinations of pineal N-acetyltransferase (NAT) activity in different photoperiods strongly support this concept.

INTRODUCTION

In contrast to birds in which the daily cycle of melatonin synthesis does not necessarily depend upon the innervation of the pineal gland, and can be influenced directly by light, the nocturnal rise in pineal and plasma melatonin content in mammals as well as the activity of pineal N-acetyltransferase, which is considered the rate limiting enzyme in melatonin formation, does depend upon an intact sympathetic innervation.[1-3] Moreover, while in birds the participation of the pineal gland in the photoperiodic mechanism is doubtful, and at best marginal,[4] in mammals the pineal gland is an integral part of the photoperiodic mechanism.[5] In at least four mammalian species (voles, ferrets and two hamsters) there is good evidence that the pineal is involved in the transduction of photoperiodic signals to

Biological Clocks in Seasonal Reproductive Cycles 237–250 (1981) (ed. B. K. and D. E. Follett: Bristol, Wright).

the neuroendocrine axis.[5-11] This involvement has been best studied in hamsters, and an attempt will be made here to summarize the present situation in this group.

While in the majority of studies on pineal participation in photoperiodic effects the golden or Syrian hamster *Mesocricetus auratus* has been used, we have concentrated on the Djungarian hamster *Phodopus sungorus*. *Phodopus* shows, under conditions of natural illumination, a marked annual cycle, not only in gonadal activity and in gonad-dependent structures, but also in pelage colour and body weight (figure 1). All these functions can be influenced by photoperiod.

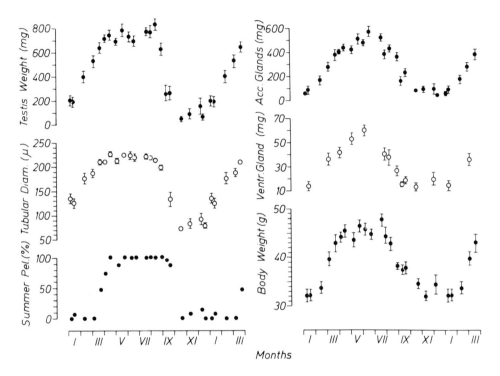

Figure 1. Annual cycles in testicular weight (both testes), seminiferous tubular diameter, pelage colour (percentage of animals in full summer pelt) and weight of the accessory sex glands, ventral marking gland and the body in adult male Djungarian hamsters. Mean ± s.e. mean of 15 animals for each point. The hamsters were maintained at a constant temperature (20 ± 3 °C), but exposed to natural illumination. (From Hoffmann.[11])

There is a further difference between the two species since juvenile development in the golden hamster is not influenced by photoperiod, photoperiodic sensitivity beginning only after puberty, whilst in *Phodopus* sexual maturation is highly dependent upon photoperiod at all ages.[12] Both species show a well-defined critical photoperiod, which is slightly longer in *Phodopus*, reflecting perhaps the higher latitude which this species inhabits (figure 2).

THE EFFECTS OF PINEALECTOMY

Since the pioneering studies by Czyba et al.[13] and by Hoffman and Reiter,[14] it has

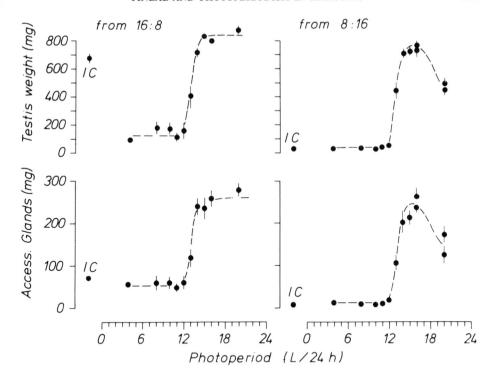

Figure 2. Critical photoperiod in male Djungarian hamsters. Mean paired testicular weight and weight of the accessory glands in hamsters that had lived from birth to 35 days in long (*left*) or short (*right*) photoperiods, and were then exposed for 45 days to the photoperiod indicated. Mean ± s.e. mean for 12–19 hamsters in each group. IC = initial control, state at 35 days. IC value for accessory glands is low in LD 16 : 8 since development of accessory glands is delayed relative to growth of testes.[13] (Based on unpublished data.)

been shown repeatedly in the golden hamster that pinealectomy prevents the gonadal involution that is normally induced by exposure to short photoperiods or by blinding.[5,8,9] Pinealectomy induces premature recrudescence in hamsters maintained in short photoperiods.[15,16] The changes in gonadotrophins and in prolactin, which accompany or anticipate the gonadal responses to short days, have likewise been shown to be abolished or at least attenuated after pinealectomy.[17,18] The increase in negative feedback sensitivity to gonadal steroids which is found in short daylengths is blocked or greatly reduced in pinealectomized hamsters.[19] In fact, more than 50 papers have demonstrated that in male, as well as in female golden hamsters, pinealectomy blocks the effects of short photoperiods, or of blinding, upon gonadal size and activity.

Not only extirpation of the pineal gland, but also interference with its sympathetic innervation, can suppress the effects of short photoperiods. Surgical interruption of the nervi conarii which innervate the pineal, removal of the superior cervical ganglia from which the postganglionic fibres derive, or decentralization of these ganglia, all block the effect of short photoperiods upon the gonads, as does transection of the descending pathways destined for the pineal gland.[9,20] Destruction of the suprachiasmatic nuclei also prevents short photoperiods from causing gonadal regression.[21,22] In general, it can be stated that any procedure which interrupts the nervous pathway from the suprachiasmatic nuclei to the pineal has essentially similar effects to those of pinealectomy.

From such findings it was concluded that the pineal has exclusively 'antigonado-trophic' effects, and that these effects are enhanced by short photoperiods or by darkness (as after blinding) and are abolished or greatly diminished by light. Reiter[9] in particular has expressed this view and has suggested that in long photoperiods the animals are 'physiologically pinealectomized', and hence unable to exhibit the antigonadal potency of the pineal gland. Interruption of the nervous pathways from the suprachiasmatic nuclei is similarly considered to render the gland functionless.

The concept of an exclusively antigonadal action of the pineal gland is at least partially challenged by some observations in *Phodopus*. In this species pinealec-tomy not only prevents involution of gonads and gonad-dependent structures in short photoperiods,[10,11,23] but also inhibits the changes in other functions that are normally induced by short days. This holds for moult into the whitish winter pelage (figure 3) or for the drastic decrease in body weight.[23-25] Both these functions are at

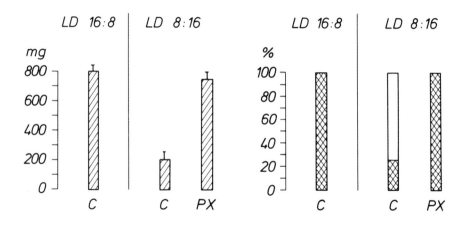

Figure 3. Effect of pinealectomy in Djungarian hamsters. Testicular regression (*left*) and moult into winter pelage (*right*) in short photoperiods are prevented by pinealectomy. (After Hoffmann.[10,23])

least partially independent from gonadal factors for they can be manipulated by photoperiod in gonadectomized animals.[25,26] Such results suggest that the pineal acts not exclusively on the neuroendocrine–gonadal axis, but also on several other functions influenced by photoperiod.

It may be that a common site of action of the pineal underlies such changes. Similar ideas have recently been mentioned in a paper on the effect of blinding and pinealectomy on gonadal as well as on thyroidal activity in the golden hamster.[27]

The concept that the pineal has exclusively inhibitory effects upon the neuro-endocrine axis also needs qualification. While virtually all experiments in *Meso-cricetus* accord with this claim, findings in *Phodopus* suggest that the pineal is at least partially involved in the transduction of the stimulatory effects of long photoperiods. In five independent experiments with males it was found that after the testes and accessory glands had regressed under short photoperiods, subse-quent exposure to long days resulted in a slight but significant delay in recrudescence in those animals that had been pinealectomized prior to exposure to the long days.[11,23,24,28,29] The results of three such experiments are given in figure 4.

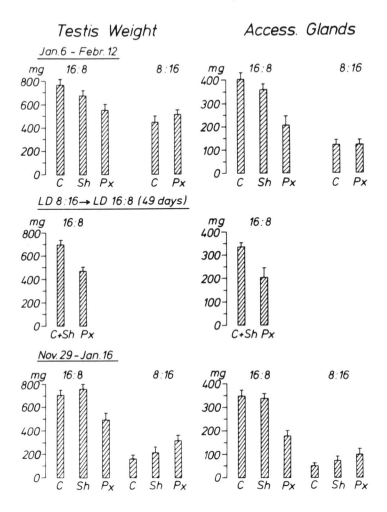

Figure 4. Weight of testes and accessory glands in Djungarian hamsters after exposure to long or short photoperiods. Experiments were started with animals whose testes and accessory glands had regressed due to exposure to short natural (*above* and *below*) or artificial (*middle*) photoperiods. Hamsters were pinealectomized (Px), sham-operated (Sh) or untreated (C) before the beginning of the experiment. Development of testes ($P < 0.05 - < 0.002$) and accessory glands ($P < 0.01 - < 0.002$) in LD 16 : 8 was significantly retarded in pinealectomized animals in all three experiments. *Lower diagram*: In pinealectomized hamsters testicular ($P < 0.05$) and accessory gland ($P < 0.02$) development was significantly retarded in LD 8 : 16 versus LD 16 : 8. (From Hoffmann.[24])

Similarly, in juveniles raised from birth in long photoperiods, testicular development was significantly delayed in pinealectomized animals (figure 5). Such findings are not confined to the Djungarian hamster. In female ferrets, Herbert[31] has shown that the acceleration of the onset of oestrus which can be induced by exposing the animals to long photoperiods in winter, is prevented in pinealectomized animals.

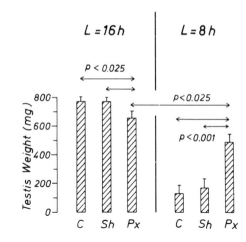

Figure 5. Testicular weight in Djungarian hamsters at 45 days of age that had lived from birth in LD 16 : 8 or LD 18 : 6. At the age of 2 days the animals were pinealectomized (Px), sham-operated (Sh) or left intact (C). (After Brackmann and Hoffmann.[30])

The observations suggest that the pineal not only has inhibitory effects on sexual and other functions but that it is also partly involved in the stimulatory action of long photoperiods. The latter is an active process rather than simply occurring as a result of eliminating an inhibition. Such assumptions are also supported by recent work in the golden hamster.[32]

Figure 5 also shows that pinealectomy of juvenile *Phodopus* partially prevents the inhibitory effects of short photoperiods. However, there is still significantly less testicular development in the pinealectomized hamster in short than in long days. A similar trend can be seen in figure 4, especially in the lower diagram which demonstrates that in pinealectomized animals recrudescence of testes and accessory glands is significantly less in short than in long photoperiods. If all photoperiodic influences were routed via the pineal gland, no such differences should exist. The results indicate that there are photoperiodic effects independent from the pineal gland. Such a conclusion is supported by several reports in the golden hamster which also indicate that pinealectomy does not completely abolish the effects of short photoperiods or of blinding.[17-19,33]

THE EFFECTS OF MELATONIN TREATMENT

It is commonly assumed that in the mammalian pineal gland neuronal information, modified by the condition of illumination, is transduced into chemical messages that are released from the gland. These act at distant target sites, there modifying the functioning of the neuroendocrine axis.[1,5,34,35] The chemical nature of the substance or substances involved in conveying the photoperiodic message have not yet been defined unequivocally but two groups of compounds are under discussion—polypeptides and indoleamines. Of the latter, melatonin in particular has been considered at length, and to date there are more than fifty papers dealing with its effects in hamsters. In most of these papers the effect upon gonadal activity has been measured.

Since injections of melatonin to hamsters had no effects in the early experiments,[36] a great number have implanted the material subcutaneously, either in beeswax or in Silastic tubing, thus making it continuously available. In several such experiments drastic changes in gonadal state have then been observed. Figure 6 shows the effects of melatonin implantation on sexual development in *Phodopus*.

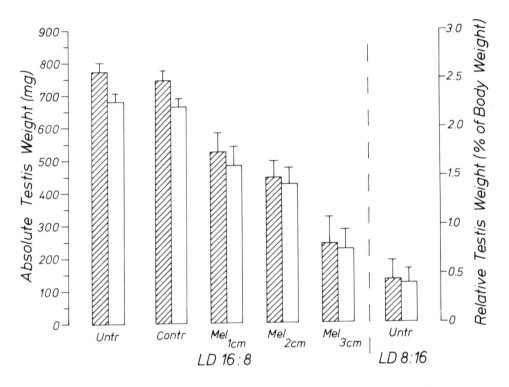

Figure 6. Testis weight in Djungarian hamsters maintained in LD 16 : 8 or LD 8 : 16 from birth and sacrificed at 45 days of age. At the age of 6 days hamsters were implanted with empty (Untr) or melatonin-filled (Mel) Silastic capsules of different lengths or were left untreated (Contr). Open bars, absolute weight; hatched bars, relative weight. (From Brackmann.[37])

There is a dose-dependent inhibition of testicular development in long photoperiods and with the highest dose of melatonin testicular size was not significantly different from that observed in the short-day controls.[37] Such results might suggest that melatonin is antigonadal and that the high levels of melatonin produced in short photoperiods could be responsible for inhibiting sexual development or for causing gonadal involution. However, inhibitory effects were not always observed in *Phodopus* after the implantation of melatonin. Experiments with adults at different seasons of the year gave different results, although the amount and methods of melatonin application were similar. Prior to experimental treatment all animals had been kept under natural illumination, and their physiological state varied accordingly (c.f. figure 1). In winter melatonin prevented or greatly diminished the stimulatory effects of long photoperiod upon gonadal recrudescence as well as upon the increase in body weight and moult into summer pelage.[38] Likewise, towards autumn when the first indications of gonadal involution were noticeable, melatonin implants induced rapid regression of the testes and their accessory glands in hamsters kept in long days.[11,23,29] Such results correspond to those seen in figure 6. However, in the summer when the animals had fully active testes, melatonin implantation not only failed to induce regression in long photoperiods, but even prevented gonadal involution and the change into winter pelage in short photoperiods.[10,23,39] Thus, instead of mimicking short photoperiods, melatonin implantation mimicked pinealectomy. It should be mentioned that all these experiments were repeated and, at the appropriate seasons, gave similar results.[25,29]

In golden hamsters drastic effects of implanting melatonin have also been reported. In particular, the finding that melatonin implants can suppress the effect of short photoperiods has been amply confirmed in *Mesocricetus*, in males as well as in females.[18,33,40,41] It has also been reported that melatonin may hasten recrudescence in short photoperiods.[42,43] On the other hand, in several cases regression of the gonads, similar to that obtained with short days, has been observed in golden hamsters after melatonin implantation.[44–46] In general, it seems that in this species implanted melatonin tends to counteract the photoperiod, i.e. in long photoperiods it induces regression, though this can be somewhat inconsistent, while in short photoperiods it prevents involution or hastens recrudescence.

Two points deserve mentioning. In a series of parallel experiments melatonin induced testicular regression in photoperiodic but not in non-photoperiodic rodents.[47] Secondly, even in photoperiodic species an inhibition of gonadal development, or induction of regression, could be caused by melatonin implants only at times when the animals were sensitive to photoperiod. Thus melatonin inhibited gonadal growth in juvenile *Phodopus* (see figure 6) in which photoperiod is highly effective, but not in immature golden hamsters in which juvenile development is independent of photoperiod.[48] Spontaneous gonadal recrudescence, which in both species cannot be prevented by short photoperiods, also cannot be blocked by melatonin.[38,43] While all these findings indicate that melatonin may somehow be involved in the normal photoperiodic mechanism, they certainly do not support the concept that melatonin is specifically an 'antigonadotrophic' hormone that, in higher doses, leads to gonadal regression or prevents its development. The results reported so far suggest rather that continuously available melatonin interferes with the effects of daylength. This idea is strengthened by reports that implanted melatonin not only prevents gonadal regression in short photoperiods, but may also stop the initiation as well as the termination of photorefractoriness.[49,50]

The continuous release of melatonin from implants differs from the physiological situation, and this may account for some of the perplexing inconsistencies outlined above. More recent observations suggest that not only the amount but also the temporal pattern of melatonin availability are important. As Tamarkin et al.[51–53] first showed, injections of melatonin can be highly effective in golden hamsters if repeated daily, and such injections can mimic the effects of short photoperiods. The most remarkable result was that the outcome depended on the time of melatonin injection. During the forenoon injections were ineffective, but in the afternoon the same procedure induced gonadal regression in animals kept in long days. Injections were also effective if given at the beginning or towards the end of the dark phase of the illumination schedule.

Since injection of melatonin once daily did not induce regression in hamsters that were pinealectomized, or in which the sympathetic nervous pathway to the pineal had been interrupted, it has been argued that the physiological site of action of melatonin could be the pineal itself where it might regulate synthesis and/or release of the 'authentic antigonadotrophic factors'.[54] However, experiments in *Mesocricetus* and *Phodopus* with implanted melatonin have already shown that regression can be induced in pinealectomized hamsters.[29,46] Injections of melatonin could also induce gonadal regression in pinealectomized or ganglionectomized animals when they were given thrice-daily at 3 h intervals.[32,53, 55–57] This refutes the hypothesis that the major site of melatonin action is the pineal, and suggests instead that in intact animals the action of injected melatonin is brought about by the combined effects of endogenous and exogenous melatonin.

It should be mentioned that in several experiments with injected melatonin inconsistencies have been observed. Thus, once-daily injections of melatonin could finally initiate gonadal regression in pinealectomized hamsters if repeated daily for several months.[56] Effects of the same treatment sometimes differed between males and females, or varied from experiment to experiment.[53,55,56] Nevertheless, the generalization that the action of injected melatonin depends on when it is given in the light/dark cycle in intact animals as well as the statement that injections repeated three times daily can cause regression in pinealectomized hamsters, seem well supported. This argues that the temporal pattern of melatonin release from the pineal is an essential part of the photoperiodic message. On the other hand, it should not be concluded from these findings that melatonin is an antigonadotrophic hormone. As with implanted melatonin, properly timed injections were only effective in animals that were sensitive to short photoperiods. Gonadal recrudescence could not be prevented by melatonin injections, and photorefractory hamsters do not respond.[56,58]

Several hypotheses might explain the temporal variation in the effectiveness of injected melatonin. There could be a daily rhythm of sensitivity at the target sites, and whether melatonin causes regression or not depends upon whether it hits the phase of this cycle. This hypothesis, however, has to cope with reports that thrice-daily injections were equally effective during either the day or night in pinealectomized hamsters.[53,55] Moreover, it has been shown that such injections have the same result in hamsters in which the suprachiasmatic nuclei (SCN) have been destroyed.[57] Since these nuclei seem to be essential for the proper expression of many circadian rhythms, as well as their entrainment by light,[59] the latter finding is also at variance with the assumption of a rhythm of melatonin sensitivity at the target site. Another possibility is that the length of time during which melatonin levels are elevated above the baseline concentration is important in conveying the photoperiodic message.

MEASUREMENTS OF PINEAL MELATONIN AND NAT ACTIVITY

The question arises as to whether all these hypotheses accord with the normal physiological rhythm of melatonin and its alterations by photoperiod. No data on plasma melatonin have come to my notice but pineal rhythms in melatonin content in the golden hamster have been determined by radioimmunoassay in two laboratories.[60–63] Melatonin content was low during the daytime and also at the beginning of the dark period, but later reached values about ten times the base level. Surprisingly, no daily rhythm in pineal N-acetyltransferase (NAT) activity was found by one group,[63] and only about a threefold increase coinciding with the melatonin peak by the other.[60] This contrasts markedly with reports in the rat in which peak values twenty or more times higher than base level were found.[1,2] While it has been speculated that melatonin synthesis in the hamster might be regulated by another mechanism, it is also possible that the assay procedure used for determining NAT activity did not reveal the true changes occurring in this species.[63] As in the rat, continuous light completely suppressed the nocturnal rise in pineal melatonin, and light during times of elevated levels induced a rapid decline.[62,63] Superior cervical ganglionectomy also abolished the rhythm of pineal melatonin concentration, values being basal throughout.[61]

When daily patterns of pineal melatonin were compared between short photoperiods that induce testicular regression and long photoperiods that maintain testicular size, no obvious difference was observed.[63] Neither the amplitude of the nocturnal increase nor the times at which the levels were elevated differed markedly; although the authors concede that due to the 2 h sampling interval some differences might have been missed. There was a difference in the phase of the elevated values relative to the light/dark cycle, which might indicate that the position of the melatonin peak is important in the photoperiodic mechanism. Such an assumption, however, has to grapple with the findings mentioned above. Calculations of the total melatonin produced daily within the pineal in short photoperiods, gave values that were less than 1/50 of the minimal dose required to induce regression by daily injections in long photoperiods. However, melatonin produced in the pineal and melatonin injected need not give the same concentrations at the target sites, and temporal patterns of plasma melatonin in short photoperiods and after injection in long photoperiods should be compared. In general, however, the reports on pineal melatonin secretion in golden hamsters do not support the hypothesis very strongly and are equivocal.

A much better correspondence between pineal rhythms and photoperiod has been found in the Djungarian hamster.[64,65] Here pineal NAT activity was determined in different photoperiods (figure 7). Unlike the golden hamster, in *Phodopus* a massive elevation of NAT activity during darkness was observed, similar to that reported in the rat. A comparison of the pattern in long and short photoperiods revealed marked differences (figures 7a, b). While maximal values were higher in long photoperiods the time of elevated NAT activity was distinctly longer in short photoperiods. Moreover, when in short photoperiods the dark time was interrupted daily by 1 min of light at midnight, the pattern of NAT activity was indistinguishable from that in long photoperiods (figures 7a and c). Exposure to the latter schedule of illumination has been shown to mimic fully the effects of complete long photoperiods on gonads, pelage colour and body weight.[66] Figure 7d gives the data for testis size. This means that in *Phodopus*, full photoperiods and skeleton photoperiods not only have the same effect on three functions governed by photoperiod, but also upon the pattern of pineal NAT activity and thus most likely on the pattern of melatonin production and release. This parallelism strongly supports the hypothesis that the temporal pattern of melatonin availability is important in conveying the photoperiodic message.

CONCLUSIONS

The experiments in hamsters unequivocally demonstrate that the pineal is essential in conveying photoperiodic effects. The results in *Phodopus* suggest that the pineal

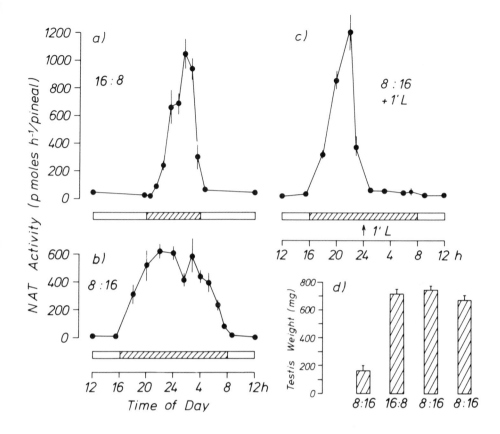

Figure 7. Pineal N-acetyltransferase activity in pineals of Djungarian hamsters which had been maintained for 3–4 weeks in long (*a*) or short (*b*) photoperiods, or in short photoperiods in which the dark time was interrupted by 1 min of light each night at midnight (*c*), and effects of such photoperiods on testicular recrudescence in winter animals (*d*). Horizontal bars represent schedule of illumination. (*a*, *b* and *c* from Hoffmann, Illnerova and Vaněček,[65] *d* from Hoffmann.[66])

does not have an exclusively antigonadotrophic action, but is involved in conferring inhibitory as well as stimulatory effects of the photoperiod. This concept is supported by Herbert's[67] work in ferrets. Pineal action is not exclusively upon the neuroendocrine-gonadal axis, but also on other functions that are influenced by photoperiod.

Though the picture is far from clear, a considerable number of findings suggest that melatonin is involved in conveying the photoperiodic message, and that the temporal pattern of melatonin release is important. It should be stressed that this statement is different from the assumption that melatonin is an antigonadal hormone. In all cases in which melatonin application had effects in *Phodopus*, these concerned not only the gonads and gonad-dependent functions, but also pelage colour and body weight. This suggests that melatonin acts at or above the sites that produce the releasing hormones. Moreover, it should be remembered that effects of melatonin application are only found when the hamsters are sensitive to photoperiod.

ACKNOWLEDGEMENT

The work on *Phodopus* has been supported in part by the Deutsch Forschungs-gemeinschaft, SPP 'Mechanismen Biologischer Uhren'.

REFERENCES

1. D. C. Klein. In: S. M. Reichlin, R. J. Baldessari and J. B. Martin (eds.) *The Hypothalamus*. New York, Raven Press, pp. 303–27 (1978).
2. J. L. Stephens and S. Binkley. *Experientia* **34**, 1523–4 (1978).
3. M. Menaker. This volume, pp. 171–83.
4. W. Gwinner and J. Dittami. *J. Comp. Physiol.* **136**, 345–8 (1980).
5. K. Hoffmann. In: J. Aschoff (ed.) *Handbook of Behavioral Neurobiology*, vol. 5. New York, Plenum Press, pp. 449–73.
6. G. M. Farrar and J. R. Clarke. *Neuroendocrinology* **22**, 134–43 (1976).
7. P. Thorpe and J. Herbert. *J. Endocrinol.* **70**, 255–62 (1962).
8. R. J. Reiter. *Chronobiologia* **1**, 365–95 (1974).
9. R. J. Reiter. In: R. J. Reiter (ed.) *Progress in Reproductive Biology*, vol. 5, *Biological Rhythms*. Basle, Karger, pp. 169–90 (1978).
10. K. Hoffmann. *Naturwissenschaften* **61**, 364–5 (1974).
11. K. Hoffmann. *Progr. Brain Res.* **52**, 397–415 (1978).
12. K. Hoffmann. *J. Reprod. Fertil.* **54**, 29–35 (1978).
13. J. C. Czyba, C. Girod and N. Durand. *CR Séanc. Soc. Biol.* **158**, 742–5 (1964).
14. R. A. Hoffman and R. J. Reiter. *Science* **148**, 1609–11 (1965).
15. R. J. Reiter. *Anat. Rec.* **160**, 13–24 (1968).
16. F. W. Turek and G. B. Ellis. In: A. Steinberger and E. Steinberger (eds.) *Testicular Development, Structure and Function*. New York, Raven Press, pp. 389–93 (1980).
17. R. J. Reiter and L. Y. Johnson. *Hormone Res.* **5**, 311–20 (1974).
18. R. J. Reiter. In: A. Kumar (ed.) *Neuroendocrine Regulation of Fertility*. Basle, Karger, pp. 169–90 (1976).
19. F. W. Turek. *Endocrinology* **104**, 636–40 (1979).
20. R. J. Reiter. *Ann. Endocrinol.* **33**, 571–81 (1972).
21. B. Rusak and L. P. Morin. *Biol. Reprod.* **15**, 366–74 (1976).
22. M. H. Stetson and M. Watson-Whitmyre. *Science* **191**, 197–9 (1976).
23. K. Hoffmann. *Nova Acta Leopold.* **46**, 217–29 (1977).
24. K. Hoffmann. In: I. Assenmacher and D. S. Farner (eds.) *Environmental Endocrinology*. Berlin, Springer-Verlag, pp. 94–102 (1978).
25. K. Hoffmann. Unpublished experiments.
26. K. Hoffmann. *Naturwissenschaften* **65**, 494 (1978).
27. J. Vriend, R. J. Reiter and G. R. Anderson. *Gen. Comp. Endocrinol.* **38**, 189–95 (1979).
28. K. Hoffmann and I. Küderling. *Experientia* **31**, 122–3 (1975).
29. K. Hoffmann and I. Küderling. *Naturwissenschaften* **64**, 339–40 (1977).
30. M. Brackmann and K. Hoffmann. *Naturwissenschaften* **64**, 341–2 (1977).
31. J. Herbert. In: G. E. W. Wolstenholme and J. Knight (eds.) *The Pineal Gland*. Edinburgh, Churchill Livingstone, pp. 303–27 (1971).
32. E. L. Bittman and I. Zucker. *Science* (in the press) (1981).
33. R. J. Reiter, M. K. Vaughan, D. E. Blask and I. Y. Johnson. *Science* **185**, 1169–71 (1974).
34. R. J. Reiter. In: A. Steinberger and E. Steinberger (eds.) *Testicular Development, Structure and Function*. New York, Raven Press, pp. 395–400 (1980).
35. J. Axelrod. *Science* **184**, 1341–8 (1974).
36. R. J. Reiter. In: E. Knobil and W. H. Sawyer (eds.) *Handbook of Physiology: Endocrinology IV*, Part 2. Washington, DC, American Physiological Society, pp. 365–95 (1974).
37. M. Brackmann. *Naturwissenschaften* **64**, 624–5 (1957).
38. K. Hoffmann. *J. Comp. Physiol.* **85**, 267–82 (1973).
39. K. Hoffmann. *Int. J. Chronobiol.* **1**, 333 (1973).
40. R. J. Reiter, A. J. Lukaszyk, M. K. Vaughan and D. E. Blask. *Am. Zool.* **16**, 93–101 (1976).
41. R. J. Reiter, M. D. Rollag, E. S. Panke and A. F. Banks. *J. Neural Transm.* Suppl. 13, 209–23 (1978).
42. R. J. Reiter, M. K. Vaughan, P. K. Rudeen and R. C. Philo. *Am. J. Anat.* **147**, 235–42 (1976).

43. F. W. Turek and S. H. Losee. *Biol. Reprod.* **18**, 299–305 (1978).
44. F. W. Turek, C. Desjardins and M. Menaker. *Science* **190**, 280–2 (1975).
45. F. W. Turek, C. Desjardins and M. Menaker. *Proc. Soc. Biol. Med.* **151**, 502–6 (1976).
46. F. W. Turek. *Proc. Soc. Biol. Med.* **155**, 31–4 (1977).
47. F. W. Turek, C. Desjardins and M. Menaker. *Biol. Reprod.* **15**, 94–7 (1976).
48. F. W. Turek. *Biol. Reprod.* **20**, 1119–22 (1979).
49. F. W. Turek and S. H. Losee. *Biol. Reprod.* **20**, 611–16 (1979).
50. S. H. Losee and F. W. Turek. *Acta Endocrinol.* (in the press) (1981).
51. L. Tamarkin, W. K. Estrom, A. I. Hamill and B. D. Goldman. *Endocrinology* **99**, 1534–41 (1976).
52. L. Tamarkin, N. G. Lefebvre, C. W. Hollister and B. D. Goldman. *Endocrinology* **101**, 631–4 (1977).
53. L. Tamarkin, C. W. Hollister, N. G. Lefebvre and B. D. Goldman. *Science* **198**, 935–55 (1977).
54. R. J. Reiter, D. E. Blask, L. Y. Johnson, P. K. Rudeen, M. K. Vaughan and P. J. Waring. *Neuroendocrinology* **22**, 107–116 (1976).
55. B. Goldman, V. Hall, C. Hollister, P. Roychoudhury, L. Tamarkin and W. Westrom. *Endocrinology* **104**, 82–8 (1979).
56. E. L. Bittman. *Science* **202**, 648–50 (1978).
57. E. L. Bittman, B. D. Goldman and I. Zucker. *Biol. Reprod.* **21**, 647–56 (1979).
58. R. J. Reiter, L. J. Petterborg and R. C. Philo. *Life Sci.* **25**, 1571–6 (1979).
59. B. Rusak and I. Zucker. *Physiol. Rev.* **59**, 449–526 (1979).
60. E. S. Panke, R. J. Reiter, M. D. Rollag and T. W. Panke. *Endocrinol. Res. Comm.* **5**, 311–24 (1978).
61. E. S. Panke, M. D. Rollag and R. J. Reiter. *Endocrinology* **104**, 194–7 (1979).
62. M. D. Rollag, E. S. Panke, W. Trakulrungsi, C. Trakulrungsi and R. J. Reiter. *Endocrinology* **106**, 231–6 (1980).
63. L. Tamarkin, S. P. Reppert and D. C. Klein. *Endocrinology* **104**, 385–9 (1979).
64. K. Hoffmann, H. Illnerová and J. Vaněček. *Naturwissenschaften* **67**, 408–9 (1980).
65. K. Hoffmann, H. Illnerová and J. Vaněček. *Biol. Reprod.* (in the press) (1981).
66. K. Hoffmann. *Experientia* **35**, 1529–30 (1979).
67. J. Herbert. This volume, pp. 261–76.

DISCUSSION

Bittman: Our recent findings in *Mesocricetus auratus* (Bittman and Zucker, to be published) lead me to concur with your conclusion that the role of the pineal gland can more reasonably be stated to be mediation of photoperiodic phasing of the reproductive cycle than production of an antigonadotrophin in short days. In previous work we employed a melatonin injection technique to show that spontaneous recrudescence and refractoriness result from loss of sensitivity of the target tissues upon which pineal hormone(s) act(s).[1] To rule out the possibility that the failure of the testes to respond to melatonin after spontaneous recrudescence results from a protective action of the pineal, we pinealectomized newly recrudesced hamsters. They remained insensitive to exogenous melatonin (25 µg 3 times daily). Further, testing the target insensitivity hypothesis, we exposed hamsters whose testes had spontaneously recrudesced after 20 weeks of LD 14 : 10 or left them in LD 10 : 14 for an equivalent period. Upon pinealectomy and melatonin administration only those golden hamsters that had experienced the photoperiodic treatment sufficient to break refractoriness were found to have regained sensitivity to melatonin.

These findings of close correspondence between sensitivity to short days and responsiveness to melatonin allowed us to determine whether the pineal mediates the ability of long days to resensitize the neuroendocrine axis. We pinealectomized hamsters after 30 weeks of LD 10 : 14 and exposed them to either short or long days for an additional 20 weeks. Neither group was rendered responsive to melatonin by these treatments. These results indicate (*a*) that the pineal gland must be intact for long days to terminate refractoriness and (*b*) that mere withdrawal of pineal influences is not equivalent to long-day exposure. To my mind, these data compel the conclusion that the pineal does something other than secrete a hormone which causes the testes to regress.

You showed data indicating that photoperiods of LD 20 : 4 or greater cause less testicular growth in *Phodopus* than do other daylengths greater than LD 13 : 11. 1. Are these findings consistently reproducible? 2. Do they correlate with alterations in circadian entrainment of locomotor activity? 3. Can they be reconciled with the external coincidence model without adding assumptions not now included in that hypothesis?

Hoffmann: 1. The finding that a photoperiod of LD 20 : 4 is less inductive has been repeated twice. In all cases development after 45 days of exposure was significantly less than in LD 16 : 8. As figure 2 (p. 239) shows, this holds only for induction. 2. We have not recorded locomotor activity in these experiments. 3. I think these findings can only be reconciled with the model of Elliott with some difficulty and with additional assumptions. I want to point out, however, that in golden hamsters Elliott also found less induction in LD 18 : 6.[2] He states this 'might be due to some unknown complexity in the photoperiodic time measuring apparatus'.

REFERENCES

1. E. L. Bittman. *Science* **202**, 678 (1979).
2. J. A. Elliott. PhD Thesis, University of Texas at Austin, figure 32 (1974).

STEROID-DEPENDENT AND STEROID-INDEPENDENT ASPECTS OF THE PHOTOPERIODIC CONTROL OF SEASONAL REPRODUCTIVE CYCLES IN MALE HAMSTERS

Fred W. Turek and Gary B. Ellis

Department of Biological Sciences, Northwestern University,
Evanston, Illinois, USA

Summary

Photoperiodic control of the seasonal reproductive cycle in the male hamster involves an alteration in pituitary gonadotrophin release. Exposure to short days induces a reduction in serum levels of LH and FSH while these two hormones are elevated during exposure to long days. In this paper we present experimental evidence which suggests that two processes are involved in photic-induced changes in pituitary gonadotrophin release. One process appears to depend on steroid hormone feedback since it is now well documented that exposure to short days renders castrated male hamsters extremely sensitive to the negative feedback effects of steroid hormones on pituitary LH and FSH release. A second process does not appear to depend on steroid hormone feedback since the photoperiod can alter pituitary gonadotrophin release in the absence of the testes. The relationship between the steroid-dependent and steroid-independent processes is not clear at the present time, and it remains to be determined whether these processes are distinct mechanistically from one another or whether they are two manifestations of the same phenomenon.

INTRODUCTION

Over the last decade a great deal of progress has been made in elucidating the neuroendocrine events associated with the photoperiodic control of reproduction in both male and female mammals.[1] The male golden (Syrian) hamster (*Mesocricetus auratus*) has proved to be a particularly attractive species for photoperiodic studies. This is undoubtedly due to the clear effect that the length of the day has on testicular function in the hamster and the ease with which the animal can be raised and cared for under standard laboratory conditions. The testes of adult male hamsters maintained on long days (i.e. $\geq 12 \cdot 5$ h of light per 24 h) weigh about 3000 mg, contain mature spermatozoa and are capable of maintaining circulating levels of testosterone at about 3–5 ng/ml serum. In contrast, after a 10-week exposure to short days (i.e. <12 h of light per 24 h) there is a tenfold decrease in testicular weight, the seminiferous tubules are devoid of mature spermatozoa and circulating

Biological Clocks in Seasonal Reproductive Cycles 251–260 (1981) (ed. B. K. and D. E. Follett: Bristol, Wright).

levels of testosterone are reduced to less than 1.0 ng/ml serum.[2-5] Exposure of male hamsters to short days for a prolonged period of time (i.e. 20–25 weeks) results in a 'spontaneous' recrudescence of the testes.[6-9]

Associated with short-day induced testicular regression in the hamster is a decline in serum levels of two pituitary gonadotrophic hormones, luteinizing hormone (LH) and follicle-stimulating hormone (FSH).[3,7,10,11] In all seasonally breeding male mammals that have been examined, serum LH and FSH levels are reduced during the non-breeding season and elevated during the breeding season.[1] A reduction in serum prolactin levels is also associated with short-day induced testicular regression in the hamster,[12,13] but whether prolactin is involved in testicular regression in other seasonally breeding animals remains to be determined. Thus, it would appear that the photoperiodic regulation of pituitary LH and FSH synthesis and/or release is a major step in the photoperiodic regulation of seasonal reproductive cycles. The purpose of this review is to summarize what is known about how the length of the day regulates pituitary LH and FSH release in the male golden hamster.

The photoperiod is able to alter pituitary gonadotrophin release in the male hamster in at least two operationally distinct ways. First, the photoperiod can alter the negative feedback effect that gonadal steroid hormones have on pituitary LH and FSH release; we refer to this as a steroid-dependent response. Secondly, the light/dark cycle can alter pituitary gonadotrophin release in the apparent absence of circulating steroid hormone levels; we refer to this as a steroid-independent response. An examination of the experimental evidence which supports the hypothesis that the photoperiod regulates pituitary gonadotrophin release by both a steroid-dependent and a steroid-independent process is presented.

STEROID-DEPENDENT PHOTOPERIODIC REGULATION OF HYPOTHALAMIC–PITUITARY ACTIVITY

As is typical for mammalian species, the removal of the testes results in a rapid and sustained increase in serum LH and FSH levels in male hamsters.[14,15] This increase in pituitary gonadotrophin release is believed to be primarily due to a loss of the negative feedback effects of gonadal steroid hormones since the administration of testosterone to castrated hamsters results in a reduction of circulating levels of LH and FSH to levels at or below those found in intact hamsters.[10,16] The observation that an internal environmental factor (i.e. testicular steroid hormones) and an external environmental factor (i.e. the photoperiod) can both regulate pituitary gonadotrophin release raises the possibility that the photoperiod and steroid hormones may act in concert to regulate pituitary LH and FSH release.

An interaction between the photoperiod and steroid hormones is clearly demonstrated by the finding that castrated hamsters exposed to short days are extremely sensitive to the negative feedback effects of testosterone when compared with castrated hamsters maintained on long days.[10,16,17] The implantation of testosterone-filled silicone elastomer capsules of 2, 4, 8 or 20 mm in length into hamsters that have been exposed to LD 14 : 10 for 8–10 weeks results in a dose–response effect of testosterone with only the 20 mm long testosterone capsules causing a maximum suppression of circulating LH and FSH levels (figure 1). In contrast, all four doses of testosterone are able to suppress serum LH and FSH levels to minimal values if the castrated animals have been maintained on 8–10 weeks of

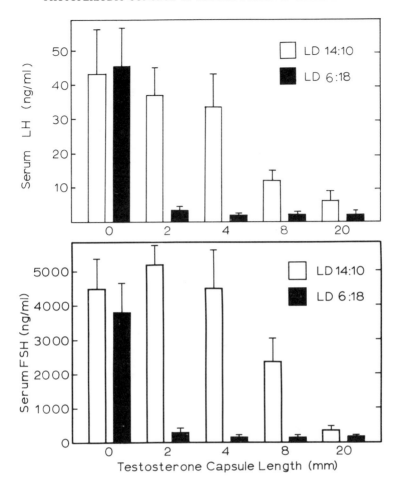

Figure 1. Mean (± s.e. mean) serum levels of LH (*top*) and FSH (*bottom*) in castrated male hamsters that had been implanted with either empty (0 mm) or testosterone-filled capsules of various sizes for 21 days. The animals were castrated during exposure to LD 14 : 10 and then either maintained on LD 14 : 10 (open bars) or transferred to LD 6 : 18 (solid bars) for 8 weeks prior to capsule implantation. (Redrawn from Turek.[17]).

short days before the start of testosterone treatment (figure 1). It is important to note that the photoperiod does not appear to alter the metabolic clearance rate of testosterone, since for a testosterone capsule of any given size, serum testosterone levels are similar in castrated hamsters maintained on LD 14 : 10 or LD 6 : 18.[16,17] Furthermore, the circulating titres of testosterone that are maintained continuously by testosterone capsules of 2–20 mm in length fall within the physiological range observed in intact untreated hamsters.[16] These results suggest that one way in which short days act to inhibit pituitary LH and FSH release in the intact hamster is to render the gonadotrophin control centre extremely sensitive to the negative feedback effects of gonadal steroid hormones.

The time course for photoperiod-induced changes in the responsiveness of the hypothalamic–pituitary axis to testosterone feedback has been defined for the male hamster.[18] It is quite clear that a change in the photoperiod does not induce an

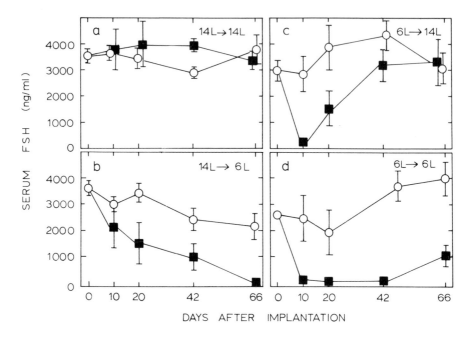

Figure 2. Mean (± s.e. mean) serum levels of FSH in castrated male hamsters that were implanted on day 0 with either empty (○) or 4 mm long (■) testosterone-filled capsules. The animals depicted in (*a*) and (*b*) were castrated and maintained on LD 14 : 10 for 60 days prior to capsule implantation. At this time they were either left (*a*) on LD 14 : 10 or transferred (*b*) to LD 6 : 18 for 66 days. The animals depicted in (*c*) and (*d*) were castrated and transferred to LD 6 : 18 for 60 days prior to capsule implantation. At this time they were either (*c*) transferred back to LD 14 : 10 or (*d*) left on LD 6 : 18 for an additional 66 days. (Redrawn from Ellis and Turek.[18])

immediate complete change in the sensitivity to steroid feedback. Instead, transfer of castrated hamsters from stimulatory long days to non-stimulatory short days induces a gradual increase in the responsiveness of the hypothalamic–pituitary axis to steroid feedback (figure 2*b*), while photostimulation induces a gradual decrease in the steroid-feedback sensitivity of the neuroendocrine axis (figure 2*c*). Photic-induced changes in responsiveness can take up to 42–66 days to reach completion, suggesting that a transfer from long to short or from short to long days results in a slow alteration in the functional relationship between testosterone and the constituent parts of the neuroendocrine system.

In figure 2*d* there is the suggestion that after a prolonged exposure to LD 6 : 18 there is a loss of the hypersensitivity to testosterone feedback that was originally induced by short days. Indeed, when castrated hamsters are exposed to a non-stimulatory photoperiod beyond 18 weeks, they become less sensitive to the negative feedback effect of exogenous testosterone on pituitary LH and FSH release as exposure to short days proceeds.[9,19] After a 26-week exposure to short days, the effect of testosterone on pituitary gonadotrophin release is similar to that observed in castrated hamsters maintained on photostimulatory long days.[18,19]

Photic-induced changes in steroid feedback sensitivity in the hamster are not specific to the testosterone molecule.[20] Silicone elastomer implants that are 4 or 8 mm in length and filled with either the 5α-reduced metabolite of testosterone, 5α-dihydrotestosterone (5α-DHT), or the aromatized derivative of testosterone, oestradiol-17β (OE$_2$), greatly suppress serum LH and FSH levels in castrates exposed to short days, but have little effect in castrates maintained on long days (figure 3). Higher doses of 5α-DHT and oestradiol (50 and 20 mm long capsules respectively) reduce serum gonadotrophin levels in the hamsters maintained on long days. These results suggest that testicular secretions besides testosterone, and/or extra-testicular conversion of testosterone to its metabolites, may be

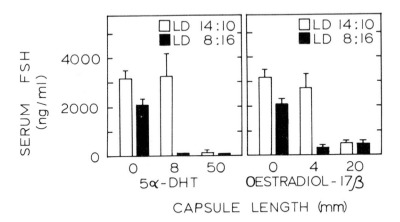

Figure 3. Mean (± s.e. mean) serum levels of FSH in castrated male hamsters that were implanted with silicone elastomer capsules containing either crystalline 5α-dihydrotestosterone or oestradiol-17β. Prior to hormone treatment the animals had been exposed to either LD 14 : 10 (open bars) or LD 8 : 16 (solid bars) for 65 days. Blood samples were collected prior to capsule implantation (0 mm) as well as 7 days after the steroid-filled capsules of various sizes were implanted. No standard error is presented if the hormone values were at the minimum amount detectable. (For a more complete dose–response curve for 5α-DHT and oestradiol, *see* Ellis and Turek.[20])

involved in the photoperiodic inhibition of the hamster reproductive system. Thus, 5α-DHT and oestradiol receptors, as well as testosterone receptors in the brain, may have functional significance in the endocrine dynamics underlying seasonal breeding.

Both the pineal gland and the suprachiasmatic nuclei (SCN) of the hypothalamus are clearly involved in the photoperiodic control of gonadal activity in the hamster. Removal of the pineal gland or ablation of the SCN both prevent short-day-induced testicular regression in the hamster.[6,21,22] In addition, pinealectomy results in testicular growth in hamsters whose testes have regressed due to exposure to short days.[23] While the precise mechanism by which the SCN and pineal gland regulate neuroendocrine–gonadal functions is not known, both appear to be involved in photic-induced changes in steroid feedback sensitivity. Short days do not result in a hypersensitivity of gonadotrophin control centres to steroid feedback if the pineal gland has been removed prior to exposure to short days,[17] and both pinealectomy and lesioning of the SCN reverse short-day-induced hypersensitivity to steroid feedback with a time course that is similar to that observed in hamsters transferred from short to long days.[23,24]

In addition to their role in the regulation of hypothalamic–pituitary gonadal function, the SCN are also involved in the circadian organization of a number of different mammalian species.[25,26] There is a clear relationship between the circadian system and light-mediated effects on the reproductive system in hamsters since photoperiodic time measurement involves the circadian system.[4] The role the SCN play in photic-induced changes in steroid feedback sensitivity may be due to their central role in the circadian organization or may simply be due to a disruption of pineal gland function once the SCN have been destroyed.[27]

In addition to altering the way gonadotrophin control centres are affected by circulating steroid hormone levels, the photoperiod can apparently also alter the response of neural centres which regulate male sexual behaviour.[28,29] In these studies, the administration of exogenous testosterone was more effective in restoring copulatory behaviour in castrated male hamsters that had been exposed to long days than in animals maintained on short days. In contrast, the response of peripheral androgen-dependent organs, such as the seminal vesicles and the flank glands, to exogenous testosterone was not affected by the photoperiod.[16,28] Taken together, the data suggest that exposure to short days results in specific changes in the responsiveness of central neural structures to testosterone; in the case of pituitary gonadotrophin release, there is an increase in sensitivity to the negative feedback effects of testosterone, while in the case of male sex behaviour, there is a decrease in the facilitatory effects of testosterone.

STEROID-INDEPENDENT PHOTOPERIODIC REGULATION OF HYPOTHALAMIC–PITUITARY ACTIVITY

While castration results in a dramatic rise in serum LH and FSH levels in male hamsters exposed to photostimulatory long days, the increase in serum gonadotrophin levels is greatly attenuated if castration occurs after short-day induced gonadal regression has taken place (figure 4).[10,11,14,15] These results suggest the possibility that short days inhibit pituitary gonadotrophin release independently of circulating steroid hormones since castration of those animals which are extremely sensitive to the negative feedback effect of steroid hormones could be expected to result in a rapid and sustained increase in serum gonadotrophin levels. An alternative interpretation is that the attenuated castration response observed in hamsters exposed to short days may be due to the hypersensitivity of the gonadotrophin-secreting system to steroid hormones of extragonadal origin. Evidence against this latter hypothesis is the observation that the attenuated castration response is also observed in castrated, adrenalectomized animals fed a steroid-free diet.[15]

The attenuated increase in serum LH and FSH levels observed in hamsters castrated after 10–12 weeks of exposure to short days is not maintained indefinitely even if the animals remain on short days.[7,9] Serum gonadotrophin titres eventually increase to normal castration levels. The time-course of this increase in serum gonadotrophin levels in castrated hamsters is similar to that for spontaneous testicular recrudescence and spontaneous changes in the negative feedback effects of testosterone which also occur after hamsters have been exposed to short days for a prolonged period of time.

Perhaps the most compelling evidence for the photoperiod being able to stimulate pituitary gonadotrophin release independently of steroid hormones is the observation that serum gonadotrophin levels increase rapidly when hamsters which have been castrated after a 10-week exposure to short days are transferred to photostimulatory long days (figure 4).[15,30] The photic-induced increase in pituitary

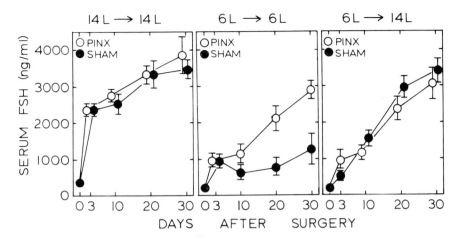

Figure 4. Mean (± s.e. mean) serum levels of FSH in male hamsters that were castrated and either pinealectomized (PINX) or sham-operated following blood sampling on day 0. The animals depicted in the left panel were maintained on LD 14 : 10 both before and after surgery. The animals represented in the middle and right panels had been maintained on LD 6 : 18 for 67 days prior to surgery. After surgery they either remained on LD 6 : 18 (*middle*) or were transferred to LD 14 : 10 (*right*). (Redrawn from Ellis and Turek.[30])

LH and FSH release occurs long before any spontaneous increase in hamsters maintained on short days after castration.

Surprisingly, while transfer of castrated hamsters from short to long days induces a dramatic rise in serum gonadotrophin levels, the transfer of castrated hamsters from long to short days results in only a slight or no decrease in serum levels of LH and FSH.[14,16] It is not clear why long days can cause a dramatic increase in serum gonadotrophin levels while short days have little effect in castrated hamsters.

The pineal gland appears also to be involved in the steroid-independent process. Pinealectomy is able to abolish the inhibitory effect of short days on the post-castration rise in serum FSH levels.[30] The time course in the increase in serum FSH levels following pinealectomy and castration of short day hamsters is similar to that observed in castrated hamsters transferred from short to long days (figure 4). While pinealectomy has an effect on pituitary FSH release in hamsters that are maintained on LD 6 : 18, it has no effect in hamsters that were castrated on long days or in animals transferred from short to long days at the time of castration (figure 4). The SCN also appear to be involved in the steroid-independent process since lesioning of the SCN abolishes the attenuated castration response (F. W. Turek, unpublished results).

DISCUSSION

In providing a detailed account of what is known about the steroid-dependent and steroid-independent processes for a single species, this review has focused entirely on the male golden hamster. However, it is now clear that the alteration of pituitary gonadotrophin release in other photoperiodic species appears to involve very similar mechanisms. The photoperiod has been shown to alter the negative feedback effect of steroid hormones on pituitary gonadotrophin release in a number of diverse species including quail, rats and sheep.[31–34] In addition, the photoperiod appears to be capable of altering pituitary gonadotrophin release independently of circulating steroid hormone levels in quail, grouse, turkeys, sparrows, pony mares, hares and rams.[35–42]

Experimentally, it appears that there are indeed two different ways in which the photoperiod is capable of altering pituitary gonadotrophin release. However, while the steroid-independent and steroid-dependent processes are operationally distinct from each other it is not known whether the two processes are simply manifestations of the same phenomenon. For example, during exposure to short days there may be a steroid-independent inhibition of pituitary LH and FSH release which automatically leads to a hypersensitivity to steroid feedback. The results of a recent study by Pickard and Silverman[11] enable us to formulate this hypothesis in a more specific manner. They found that hypothalamic content of LH-releasing hormone (LH-RH) was increased in both castrated and intact hamsters that had been maintained on short days when compared with hamsters maintained on long days. They interpreted their data to indicate that during exposure to short days there is a decreased secretion of LH-RH. Thus, exposure to short days may inhibit LH-RH release through a steroid-independent process and, once this has occurred, the LH-RH neurones may be very responsive to the inhibitory effects of low doses of testosterone.

One argument in favour of the hypothesis that the steroid-dependent and steroid-independent processes are simply two sides of the same coin is the observation that the two systems appear to respond identically to photostimulation, pinealectomy and prolonged exposure to non-stimulatory short days. When hamsters are castrated after testicular regression has occurred and transferred to long days or pinealectomized, there is a gradual loss of the attenuated castration response (figure 4).[15,30] Similarly, when castrated hamsters that have been implanted with testosterone-filled capsules are pinealectomized or transferred from a non-stimulatory to a stimulatory photoperiod, there is a gradual loss of the supersensitivity to the negative feedback effects of testosterone.[23] In addition, the attenuated castration response and the hypersensitivity to steroid feedback observed in castrated hamsters are both abolished if the animals remain on short days for 18–25 weeks.[7,19]

There is some experimental evidence which suggests that the two processes are indeed separate from one another. When sexually mature hamsters are castrated and transferred to short days it has been reported that there is a small but significant decrease in serum LH and FSH levels,[14,15] while in other studies no significant depression in circulating gonadotrophin was observed.[16,18] It would appear that the ability of short days to cause a decrease in the high post-castration titres of serum LH and FSH is marginal at best. Nevertheless, when a castrated hamster is transferred from long to short days the steroid-dependent process is clearly taking place. Whether the hamsters are implanted with testosterone at the time of transfer to short days, or after 60 short days, it appears that complete hypersensitivity to steroid feedback has occurred even though there is little evidence for any short-day induced, steroid-independent reduction in serum gonadotrophin levels.[16,18]

If the steroid-dependent and steroid-independent processes are indeed two distinct systems, the separation of the two systems could be at a number of different levels of organization. The two processes could involve different neural or endocrine pathways, although the observation that pinealectomy and lesioning of the SCN both abolish the attenuated castration response and short-day induced hypersensitivity to steroid feedback argues against this hypothesis. Alternatively, the distinction between the steroid-dependent and steroid-independent systems may be at the cellular level. That is, short days could have two different effects on the same cell. For example, short days may directly inhibit LH–RH release and at the same time render LH–RH neurones more sensitive to the inhibitory action of steroid hormones. If the differences in the steroid-dependent and steroid-independent processes are at the cellular level then the question of whether or not these two processes are indeed separate will be difficult to answer. Nevertheless, we believe that from an experimental point of view, the distinction between a steroid-dependent and steroid-independent process has been, and will continue to be, quite useful. Once we have elucidated the neural, endocrine and cellular events which underlie the steroid-dependent and steroid-independent processes involved in the photoperiodic response, then we shall undoubtedly have a more complete picture of how the length of the day can regulate seasonal reproductive cycles in mammals.

ACKNOWLEDGEMENTS

This work was supported by NIH Grants HD-09885, HD-12622, Research Career Development Award HD-00249 to F.W.T. and an NSF pre-doctoral fellowship to G.B.E.

REFERENCES

1. F. W. Turek and C. S. Campbell. *Biol. Reprod.* **20**, 32–50 (1979).
2. S. Gaston and M. Menaker. *Science* **167**, 925–8 (1967).
3. W. E. Berndtson and C. Desjardins. *Endocrinology* **95**, 195–205 (1974).
4. J. Elliott. *Fed. Proc.* **35**, 2339–46 (1976).
5. C. J. Gravis. *Cell Tiss. Res.* **185**, 303–13 (1977).
6. R. J. Reiter. *Anat. Rec.* **173**, 365–72 (1972).
7. F. W. Turek, J. A. Elliott, J. D. Alvis et al. *Biol. Reprod.* **13**, 475–81 (1975).
8. I. Zucker and L. P. Morin. *Biol. Reprod.* **17**, 493–8 (1977).
9. K. S. Matt and M. H. Stetson. *Biol. Reprod.* **20**, 739–46 (1979).
10. L. Tamarkin, J. S. Hutchison and B. D. Goldman. *Endocrinology* **99**, 1528-33 (1976).
11. G. E. Pickard and A. J. Silverman. *J. Endocrinol.* **83**, 421–8 (1979).
12. F. Bex, A. Bartke, B. D. Goldman et al. *Endocrinology* **103**, 2069–80 (1978).
13. A. Bartke, B. D. Goldman, F. J. Bex et al. *Endocrinology* **106**, 167–72 (1980).
14. F. W. Turek, J. A. Elliott, J. D. Alvis et al. *Endocrinology* **96**, 854–60 (1975).
15. G. B. Ellis and F. W. Turek. *Endocrinology* **106**, 1338–44 (1980).
16. F. W. Turek. *Endocrinology* **101**, 1210–15 (1977).
17. F. W. Turek. *Endocrinology* **104**, 636–40 (1979).
18. G. B. Ellis and F. W. Turek. *Endocrinology* **104**, 625–30 (1979).
19. G. B. Ellis, S. H. Losee and F. W. Turek. *Endocrinology* **104**, 631–5 (1979).
20. G. B. Ellis and F. W. Turek. *Neuroendocrinology* **31**, 205–9 (1980).
21. B. Rusak and L. P. Morin. *Biol. Reprod.* **15**, 366–74 (1976).
22. M. H. Stetson and M. Watson-Whitmyre. *Science* **191**, 197–9 (1976).
23. F. W. Turek and G. B. Ellis. In: A. Steinberger and E. Steinberger (eds.) *Testicular Development, Structure, and Function.* New York, Raven Press, pp. 389–93 (1980).
24. F. W. Turek, C. D. Jacobson and R. A. Gorski. International Congress of Endocrinology, Abstracts, p. 625 (1980).
25. R. Y. Moore. In: W. F. Ganong and L. Martini (eds.) *Frontiers in Neuroendocrinology*, vol. 5. New York, Raven Press, pp. 185–206 (1978).

26. B. Rusak and I. Zucker. *Physiol. Rev.* **59**, 449–526 (1979).
27. R. Y. Moore and D. C. Klein. *Brain Res.* **71**, 17–33 (1974).
28. C. S. Campbell, J. S. Finkelstein and F. W. Turek. *Physiol. Behav.* **21**, 409–15 (1978).
29. L. P. Morin and I. Zucker. *J. Endocrinol.* **77**, 249–58 (1978).
30. G. B. Ellis, S. H. Losee and F. W. Turek. *Acta Endocrinol.* (in the press) (1981).
31. J. Pelletier and R. Ortavant. *Acta Endocrinol.* **78**, 422–50 (1975).
32. D. T. Davies, L. P. Goulden, B. K. Follett et al. *Gen. Comp. Endocrinol.* **30**, 477–86 (1976).
33. J. C. Hoffmann and A.-M. Cullin. *Neuroendocrinol.* **23**, 285–96 (1977).
34. S. J. Legan, F. J. Karsch and D. L. Foster. *Endocrinology* **101**, 818–24 (1977).
35. G. J. Davis and R. K. Meyer. *Gen. Comp. Endocrinol.* **20**, 61–8 (1973).
36. F. E. Wilson and B. K. Follett. *Gen. Comp. Endocrinol.* **23**, 82–93 (1974).
37. J. Pelletier and R. Ortavant. *Acta Endocrinol.* **78**, 435–41 (1975).
38. W. R. Gibson, B. K. Follett and B. Gledhill. *J. Endocrinol.* **64**, 87–101 (1975).
39. M. Garcia and O. Ginther. *Endocrinology* **98**, 958–62 (1976).
40. P. W. Mattocks jun., D. S. Farner and B. K. Follett. *Gen. Comp. Endocrinol.* **30**, 156–61 (1976).
41. P. J. Sharp and R. Moss. *Gen. Comp. Endocrinol.* **32**, 289–93 (1977).
42. M. E. El Halawani, W. H. Burke, L. A. Ogren et al. *Gen Comp. Endocrinol.* **40**, 226–31 (1980).

DISCUSSION

Bittman: Two questions. First, do you mean to suggest that steroid-dependent and independent mechanisms are of differential importance at different stages of the reproductive cycle? Secondly, experiments in sheep indicate that progesterone is the most important steroidal suppressor of LH secretion in the breeding season while oestradiol assumes this role in anoestrus (*see* Goodman and Karsch's paper in this Symposium). Have you any evidence bearing on similar shifts in the effectiveness of gonadal signals relative to one another in the hamster's annual reproductive cycle (steroidal or otherwise)?

Turek: The suggestion that steroid-dependent and steroid-independent mechanisms may be of differential importance is based on our finding that transfer of castrated hamsters from short to long days results in an increase in serum gonadotrophin levels while transfer from long to short days has little effect on gonadotrophin release. This could indicate that the steroid-independent system is less important during the transition from the breeding to the non-breeding condition. However, there are other interpretations of these data which do not require that the steroid-independent system be differentially important during various stages of this reproductive cycle.

On the second question, we have no evidence that there is any change in the relative importance of various gonadal signals during different seasons. We do know that there are photically induced changes to oestradiol, 5α-dihydrotestosterone and testosterone, but whether there is a relative change in the importance of these signals under different photoperiods is not known.

Brown-Grant: To what extent can these results in male hamsters be duplicated in females? Also, have you checked whether pituitary responsiveness to GnRH has changed in the males? This could be important, as, for instance, inhibin appears to act at the pituitary level.

Turek: Whether or not the photoperiod can alter steroid feedback sensitivity in the female hamster (either positive or negative feedback) is not known. Pituitary responsiveness to GnRH in the long term castrated hamster is not altered by the photoperiod. Whether there is an interaction between the photoperiod and steroid hormones in altering pituitary responsiveness to GnRH remains to be determined.

THE PINEAL GLAND AND PHOTOPERIODIC CONTROL OF THE FERRET'S REPRODUCTIVE CYCLE

J. Herbert

Department of Anatomy, University of Cambridge

This paper sets out to discuss how much we currently understand of the mechanisms by which changing photoperiods regulate the female ferret's breeding season, with particular emphasis on the function of the pineal.

FEATURES OF THE REPRODUCTIVE CYCLE

The female ferret's reproductive cycle shows the characteristic features of an annual rhythm. Breeding begins in the spring (March–April in England) and continues until the autumn (July–September), a period which is quite constant for a given animal from year to year but much more variable between animals. Under natural lighting conditions, therefore, the onset of oestrus is better synchronized between different animals than its termination. Ovulation occurs only rarely unless the female is mated, a feature which allows uninterrupted study of the breeding season in unmated animals, in which oestrus is readily diagnosed by the appearance of enormous vulval oedema which disappears if the animal becomes pregnant (or pseudo-pregnant) or anoestrous. Early studies demonstrated the relationship between ovarian activity and the external sign of oestrus in this species, and established oestrogen as the principal hormone promoting oedema, an action which is antagonized by progesterone.[1]

THE EFFECT OF ARTIFICIAL LIGHT

It is important, in view of what follows, that the various effects of light on the female ferret's cycle are clearly distinguished. In general, these fall into three categories: (*a*) régimes which promote oestrus, (*b*) those that terminate it and (*c*) treatments that alter its frequency.

Advancing oestrus

It is well known that Bissonnette[2] was one of the first to show experimentally that artificial light could accelerate the onset of breeding seasons in mammals, using ferrets. The usual method is that anoestrous animals are exposed, during the autumn or winter, to additional illumination given either as an extra ration after the

Biological Clocks in Seasonal Reproductive Cycles 261–276 (1981) (ed. B. K. and D. E. Follett: Bristol, Wright).

end of daylight or as wholly artificial 'long' days, resembling those occurring naturally in summer. This reliably accelerates the season by up to several months, so that oestrous animals can readily be obtained in December or January (figure 1). Manipulations which disturb this response are, therefore, clearly evident, since the difference between animals so treated and their controls is usually tremendous. Exposing ferrets to photoperiods of different lengths has suggested that a critical duration (about 11–12 h) is necessary for oestrus to be advanced (compare the value of 12·5 h obtained by Gaston and Menaker[3] in hamsters), though whether the ferret's neuroendocrine system defines a 'long' photoperiod according to an absolute criterion, or on a comparative basis relating to the preceding one, is not yet clear.[4] Like some other mammals, the ferret's reproductive system is extremely sensitive to light, only 0·42 foot candles given as LD 14 : 10 being sufficient to induce premature oestrus.[5] A remarkable feature of the ferret's oestrous response to artificial photoperiods (as in the case of some other mammals) is that the effects take so long to appear; for example, about 6 weeks elapse between initiating artificial 'long' days and the onset of oestrus. One experiment showed that about 35–40 days of exposure to LD 14 : 10 was needed to advance oestrus.[5] This interval depends somewhat upon the time of the year, ferrets tending to respond more rapidly if treated during the latter part of the anoestrum.[6]

The technique of splitting the photoperiod into a 'skeleton' has not been used systematically in ferrets, though the results of Hammond[7]—who showed that LD 7 : 17 could not advance oestrus, whereas two blocks of 3½ h of daylight separated by 5 h of darkness was effective—indicate that rather similar findings to those obtained in other species (e.g. birds) might be obtained in the ferret. In summary, therefore, artificial 'long' photoperiods given to anoestrous ferrets can accelerate oestrus provided they are given to ferrets who are responsive, that is, anoestrous and not photorefractory.

This finding in the ferret contrasts clearly with that in the Syrian (golden) hamster in that 'long' photoperiods do not appear to have a stimulatory effect on reproduction in this species,[8] and with those on the Djungarian (dwarf) hamster in which stimulation of reproductive activity by long days, though evident, is clearly less significant quantitatively than the inhibiting effect of 'short' ones.[9,10]

Terminating oestrus

It is necessary to distinguish régimes which prevent the acceleration of oestrus from those that terminate it once it has begun.

There is now no doubt that exposing oestrous ferrets to 'short' photoperiods (e.g. LD 8 : 16) can prematurely terminate oestrus. Animals treated in this way (e.g. in May–June) go out of oestrus 4–6 weeks later (figure 1); similar results follow if animals are first brought into oestrus by being exposed to a long photoperiod (LD 14 : 10) and are then transferred to a short one (LD 8 : 16), whereas their respective controls (left in daylight or in 'long' photoperiods) remain in oestrus for many weeks.[11] It will be apparent from experiments to be described subsequently that it is the presence of short photoperiods, rather than the 'absence' of long ones, which is responsible for this effect. It is noteworthy that short photoperiods can terminate oestrus to a degree quantitatively similar to the activating effects of long ones on the initiation of breeding activity; that is, the response takes about 6 weeks to come about, and in both cases (if the lighting régime is applied at the appropriate time), the differences in the response between control (e.g. daylight-exposed) and

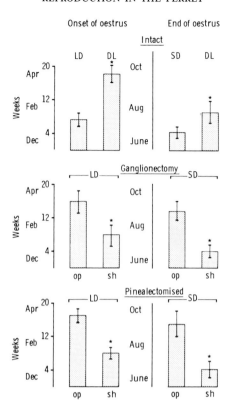

Figure 1. A comparison between the effects of various procedures on the onset of oestrus in anoestrous ferrets (*left*) and its termination in oestrous animals (*right*). *Upper panel: Left*, the effect of exposure to long days (LD) (beginning during late November) accelerating the onset of oestrus, compared with ferrets left in daylight (DL); *right*, premature termination induced by exposure to short days (SD), beginning in June, compared with controls receiving daylight (DL). *Middle panel:* The prevention of the effect of either LD (*left*) on activation of oestrus or of SD (*right*) on its termination by decentralizing the superior cervical ganglion. (C. Yates and J. Herbert, unpublished.) *Lower panel:* Similar effects on both activation of oestrus by LD or its termination by SD after pinealectomy. op, Operated; sh. sham-operated. Data shown are weeks from exposure to artificial light (means and standard errors); 6–12 animals per group. * $P<0.01$ (t test).

experimental groups (measured in weeks of oestrus elapsing after treatment starts) is also quantitatively similar.

Photorefractoriness has been described in several species of birds and in some mammals.[12,13] In the ferret it is most easily demonstrated by leaving animals in 'long' photoperiods until they go out of oestrus, or exposing them to long photoperiods shortly after the end of the natural breeding season, when they may fail to respond to further long photoperiods and not show the expected recurrence of oestrus within 6–8 weeks. Photorefractoriness in ferrets is not a very robust phenomenon; in a group so treated, whereas the majority stay anoestrous, some will show correspondingly persistent periods of continuous oestrus. However, it is clear that, in essence at least, a phenomenon similar to that described in 'long day' breeding birds or, more interestingly, in the short-day (inhibitory) response of hamsters[13] can be discerned in ferrets. Furthermore, to break refractoriness and restore sensitivity to 'long' days once again, requires the same treatment in ferrets as in other species, i.e. exposure for some weeks (it is not known whether there is a critical duration in ferrets) to the 'opposite' photoperiod—in this case 'short' days—which sensitizes the animal's neuroendocrine system to the stimulatory effects of subsequent 'long' days.[11,14]

Since short days can also induce anoestrus in oestrous ferrets (*see above*), is there a condition of refractoriness to short days in ferrets (as in hamsters) represented by failure of this kind of photoperiod to repress oestrus? The evidence is difficult to come by. Animals put into short days (LD 8 : 16) during November, and left there, come into oestrus at about the normal time in the following spring (J. Herbert, unpublished). Whilst this might be due to refractoriness developing to continuous short days, it could also reflect the operation of the animal's intrinsic timing mechanism in the absence of an external cue, as seems to be the case following pinealectomy (*see below*). Another experiment is to leave such animals in short days for several years. Adequate data are not available, but it seems that oestrus will recur for at least 2 years, although not at the 'expected' time during the second year (i.e., not 52 weeks after the onset of the first[15]). We do not know whether the ferret's reproductive system can free-run in the absence of varying light cues.

The question thus remains open, though of considerable interest. It might be expected, on theoretical grounds, that a species such as the ferret, in which both long and short photoperiods can change the reproductive cycle to apparent equivalent degrees (though in opposite directions), might be able to become refractory to either kind of photoperiod upon prolonged exposure. Alternatively, it can be suggested that a species starting its season in the spring (when day lengths are increasing rapidly) and continuing it throughout the summer to the autumn, when they are decreasing, would have little biological use for a well-developed photorefractoriness mechanism to either long or short photoperiods compared, for example, with the postulated role of refractoriness in hibernating species such as the hamsters, or in migrating species such as some birds.[8,12,13] Finally, it may be that a comparatively long-living predator (such as the ferret) is not so dependent upon maximizing fecundity as are hamsters (short-living prey species), and thus does not need a mechanism for regulating the breeding season so precisely, an argument that is considered again later in this paper.

Determining the periodicity of oestrus

Like other light-sensitive rhythms, the ferret's annual cycle can be 'driven' by imposing an alternating light cycle of 'short' and 'long' photoperiods with frequencies of less than 12 months.[5] Comparable findings have been made in birds, on the antler cycles of deer (*see* Pengelley,[16]) and in the seasonal rhythms in prolactin[17] of rams. The change in phase angle between the imposed and dependent rhythms as the frequency of the former changes is also that expected from the work on similar interactions involving circadian rhythms. Thus, in the ferret, the response lags progressively behind the phase of the entraining rhythm as the latter's frequency becomes shorter[5] (*see* comparable findings in sheep: Thibault et al.[18]).

There is a limit, which in the ferret is about 4 weeks, below which periodicities cannot entrain the reproductive rhythm, which then escapes and/or may show a demultiplying response.

To summarize: changes in photoperiods can advance, delay, prevent or synchronize the rhythm in the breeding season of the ferret.

IS THE PHOTOPERIOD NECESSARY FOR OESTRUS TO OCCUR?

Although alterations in photoperiods can entrain the ferret's breeding rhythm in the way described above, this does not prove that they are necessary for oestrus to occur. It is important to decide whether the changing photoperiod is required to initiate the mechanism inducing oestrus or anoestrus, or serves only to entrain a process which is energized independently from those factors that entrain it. The evidence strongly suggests the latter to be the case. Blind ferrets, who cannot respond to either long or short photoperiods, nevertheless show recurrent periods of oestrus even if followed for more than 3 years[19]—an interval which effectively rules out 'carry-over' effects from photoperiods experienced before operation (figure 2). Furthermore, the operations had been carried out neonatally, so that the animals had experienced postnatal light for only 4 days. Onset of the first oestrous period occurred at about the same time (though slightly earlier) than in controls kept in normal daylight, suggesting that some maturational factor, unconnected with day length, was responsible. Thereafter, periods of oestrus occurred apparently randomly, in contrast to the same controls, who showed recurrent regular breeding seasons starting in the spring of each year. Besides suggesting that changing photoperiods are not responsible for initiating oestrus (a contribution by extra-retinal receptors is unlikely since blind animals were not synchronized with each other), this experiment failed to demonstrate free-running circannual rhythms in individual animals. It might be that these ferrets were not followed long enough, but extreme irregularity of oestrus in both duration and periodicity seemed to be the rule. What, then, was deciding when oestrus would occur in these animals; or when it should terminate? Perhaps is is significant that, over the 3 years, blind ferrets spent about the same time in oestrus as their sighted controls, suggesting perhaps that some mechanism measures the total duration of oestrus, independently of its timing.

The photoperiod seems essential for the timing of oestrus, but not for its initiation or termination in the ferret. How is this related to the function of the pineal?

THE EFFECTS OF SUPERIOR CERVICAL GANGLIONECTOMY

During the 1950s there was a debate over the existence of an autonomic secretory motor nerve supply to the pituitary. In pursuit of this argument, Abrams, Marshall and Thomson[20] removed from anoestrous ferrets the superior cervical ganglia which were presumed to provide the putative nerve supply to the pituitary. Their expectations were apparently fulfilled when subsequent exposure of these ganglionectomized animals to long days failed to stimulate oestrus. However, inability to demonstrate convincingly the existence of intrahypophysial nerve fibres focused attention on the possibility that the bilateral ptosis, consequent upon this operation (though it tends to recover in a few weeks), might be responsible by limiting the amount of light entering the eye. This was also disproved by a series of control experiments.[21] Quite recently, it has been demonstrated that ganglionectomy (or severing the pre-ganglionic supply to the ganglion) also prevents short photoperiods prematurely terminating oestrus (C. Yates and J. Herbert, unpublished). Ganglionectomized ferrets, therefore, seem unresponsive to either the 'stimulatory' effects of long photoperiods or the 'inhibitory' effects of short ones (figure 1).

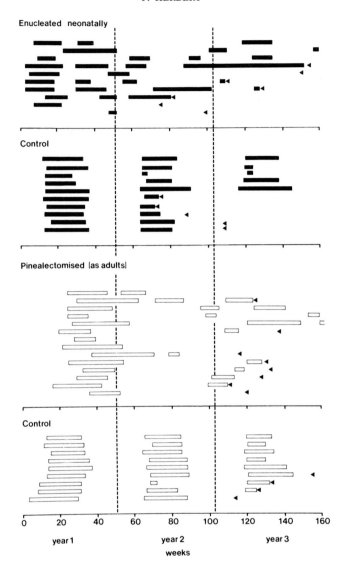

Figure 2. Recurrent periods of oestrus occurring in ferrets kept in daylight for more than 3 years after either being pinealectomized as adults or having the optic nerve sectioned neonatally. Their respective controls are also shown. Each bar represents a period of oestrus for an individual animal. Year 1 starts on 1 January of the year after operation. ◀ indicates that the animal died.

The pineal gland, it is well known, receives a generous autonomic supply from the superior cervical ganglion.[22] The afferent fibres run between the pinealocytes and, in the ferret, terminate principally in the large perivascular space surrounding the non-fenestrated capillaries of the pineal gland.[23] The club-like endings of the pinealocytes themselves are also found in the perivascular space. Ultrastructural studies have shown that noradrenergic 'false' transmitters are localized in the perivascular space in autonomic nerve endings, and that drugs which reduce the synthesis of noradrenaline (e.g., α-methyl-p-tyrosine) produce marked morphological changes in the autonomic and pinealocyte endings.[24] There are thus anatomical grounds, at least, for supposing that some kind of interaction between autonomic terminals and pinealocyte endings is possible, and that this occurs in the perivascular space of the ferret's gland (figure 3).

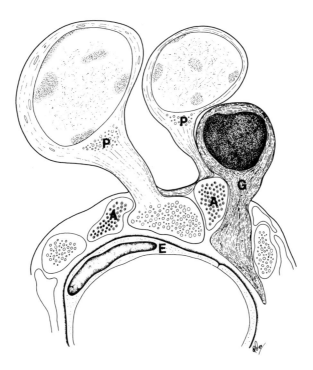

Figure 3. A diagram of the ultrastructural relationships in the perivascular space of the ferret's pineal gland. P, pinealocyte; G, glial cell; E, endothelial cell; A, autonomic (sympathetic) nerve ending.

THE EFFECTS OF PINEALECTOMY

Ferrets pinealectomized during the winter completely fail to respond to artificial long photoperiods.[25,26] No pinealectomized animal treated in this way has ever been found to show premature onset of oestrus, a statement made possible by the wide separation between the animals that *do* respond (and thus come into oestrus 6–8 weeks following exposure to long photoperiods in December and January) from those who *do not* but come into oestrus at the normal time in March–April (*see above*) (figure 1). It is important to note that these pinealectomized animals come into oestrus at the same time as normal ferrets would have done, if the latter are not exposed to artificial long photoperiods, i.e. in the spring.

Pinealectomy also prevents the early termination of the breeding season induced by 'short' photoperiods.[11] Ferrets pinealectomized during oestrus nevertheless go out of breeding condition at about the same time as unoperated animals kept under normal daylight conditions (figure 1).

These experiments lead us to conclude that (a) the regulatory effects of either long or short photoperiods on, respectively, the onset or termination of breeding are transmitted through the pineal gland, and (b) the initiation of the period of oestrus, or its termination once it has started, is not necessarily dependent upon the pineal and thus not upon the prevailing photoperiod.

IS THE PINEAL GLAND 'ANTI'- OR 'PRO'-GONADOTROPHIC?

It thus appears that, in the ferret, long and short artificial photoperiods, if administered at appropriate stages of the breeding cycle, can produce equivalently powerful effects on the cycle itself. Pinealectomy, or superior ganglionectomy, prevents both these effects, and thus the gland can be described as being both 'anti'- and 'pro'-gonadotrophic. However, the consistent finding that, in the absence of the pineal gland, both onset and termination of oestrus can still occur, argues against this description being a sufficient or a satisfactory one. Further experiments on the occurrence of breeding in pinealectomized ferrets kept in natural daylight confirm this reservation, and will now be considered. At this stage, however, we may reject the general conclusion, derived mainly from work on the rat and golden hamster, that the mammalian pineal is 'anti'-gonadotrophic.[8]

BREEDING SEASONS IN PINEALECTOMIZED FERRETS KEPT IN DAYLIGHT

Removing the pineal gland during anoestrus (i.e. the autumn) from animals kept in natural daylight has no effect on the timing of the next oestrus (that is, in the following spring). However, if the pineal is taken out at times increasingly earlier during the preceding oestrus (that is, during the summer) then the onset of the next breeding season occurs progressively later (figure 4), indicating that a timing mechanism has been disturbed by that operation and that 'escape' occurs in pinealectomized ferrets. Pinealectomized ferrets followed for 3–5 years showed clearly that the gland is essential for the breeding season to be timed even under natural conditions. Operated animals continued to come into breeding condition, and to go out of oestrus, but at irregular intervals and with no apparent reference to the time of the year (figure 2). Correlating these results with experimental ones described on p. 262 above, indicates that the timing of oestrus in any given year is not due to a process which occurs during the preceding anoestrum or, alternatively, that this timing mechanism shows a high degree of inertia such that a considerable length of time is needed to elapse between pinealectomy and subsequent oestrus for an effect to be evident. This must be greater than the 4–5 months elapsing between pinealectomy in the autumn and the expected time of the oestrus in the next spring. Secondly, pinealectomized animals left in natural conditions need to be followed for more than 1 year (i.e. until the onset of the second breeding season) following operation before valid conclusions can be drawn about the participation of the pineal gland in the timing of oestrus under natural conditions. Thirdly, our results suggest that the mechanisms which determine that oestrus will occur, or, having occurred, will come to an end, are not dependent on the pineal nor (since essentially similar effects follow if the optic nerve is sectioned in neonatal ferrets) upon the ability of the animal's neuroendocrine system to detect changes in

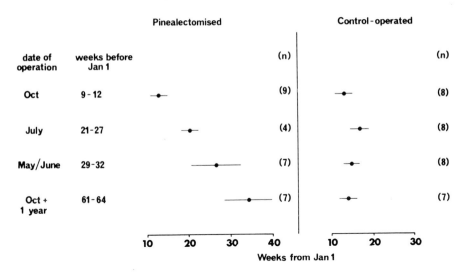

Figure 4. The onset of the oestrous period following pinealectomy related to the date of operation (*left*). There is a proportional delay in the timing of oestrus compared with controls (*right*) the earlier the pineal is removed in the preceding year.

photoperiod. However, since pinealectomized ferrets cannot adjust their breeding in response to either long or short photoperiods, or maintain annual periodicity for longer than a year if left in natural daylight, the photoperiodic factors that regulate the timing of the breeding season are absolutely dependent for their effect upon the pineal. We do not know what causes the apparently randomly occurring oestrus in pinealectomized (and blinded) ferrets, though we have calculated that, over several years, pinealectomized, blinded and intact animals all spend about the same proportion of time in oestrus,[19] suggesting that some other mechanism determines the amount of oestrous activity, but not its distribution over time.

THE EFFECT OF MELATONIN

Our work on the ferret has allowed us to address four questions concerning the action of melatonin. 1. Can it replicate the action of the pineal gland? 2. Does it act on the gland itself or on the brain? 3. Do its effects depend upon the time of administration relative to the phase of the photoperiod? 4. Is melatonin 'anti'-gonadotrophic?

1. Injecting melatonin (0·5 mg thrice-weekly or 1 mg per day per ferret) at 8 h into a long photoperiod (LD 14 : 10) induces anoestrus[11] in oestrous ferrets (figure 5). This occurs, at the highest doses, after about 6–8 weeks, a time comparable with that observed in intact ferrets exposed to short photoperiods. Ferrets kept in long photoperiods will become photorefractory (*see above*) but can be made to respond again to the stimulating effects of long days by being given injections of melatonin for about 7 weeks whilst being left in long days, an effect comparable to the results of being exposed to short photoperiods for a similar period.[11] These experiments suggest that at least some short day effects of light, transmitted to the pineal gland,

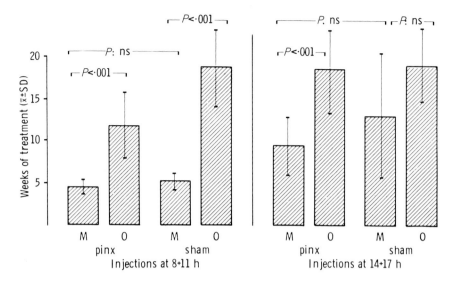

Figure 5. The termination of oestrus by melatonin (M) (0·5 mg given twice-daily 3 hours apart) in either pinealectomized (pinx) or sham-operated ferrets, compared with animals given oil vehicle alone (O). *Left*, the effect of melatonin given at 8 and 11 h after the start of the photoperiod. *Right*, the effect at 14 and 17 h. All ferrets were exposed to LD 14 : 10 throughout. (D. S. Carter and J. Herbert, unpublished.)

can be replicated by giving melatonin. However, other experiments in which pinealectomized ferrets (kept in long photoperiods) were given melatonin at 8 h (for 7–8 weeks) and then at 14 h (for a further 8 weeks) failed to show premature oestrus (D. S. Carter and J. Herbert, unpublished), and we have thus been unable, so far, to reproduce 'long day' effects of the pineal by giving melatonin.

2. Preliminary experiments suggest that melatonin is as effective in pinealectomized ferrets as in intact ones (figure 5, D. S. Carter and J. Herbert, unpublished). Ferrets pinealectomized during oestrus, and then given melatonin 8 h after the onset of light, go out of oestrus as rapidly as intact animals also given melatonin, and long before either intact or pinealectomized animals given oil. At least as far as this effect is concerned, the action of melatonin does not appear to require the presence of the pineal itself. Furthermore, additional evidence (*see below*) suggests that melatonin may act on the 5HT-containing system of the brain.

3. Another set of observations indicates that in both intact and pinealectomized oestrous ferrets, melatonin (1 mg per day per ferret) is more effective in inducing anoestrus if it is given 8 h after the start of the photoperiod (LD 14 : 10) than if the injection is delayed until 14 h (figure 5, D. S. Carter and J. Herbert, unpublished). Thus, it seems that the greater efficacy of melatonin given at 8 h is not simply that it summates with the animal's own melatonin, but, rather, that there may be a true circadian change in sensitivity to melatonin which is independent of the presence of the pineal. More data are required, however, before this point can be established fully.

4. Melatonin (1 mg per day per ferret) injected 8 h into a long photoperiod (LD 14 : 10) into anoestrous animals effectively antagonizes the expected advance of oestrus (J. Herbert, unpublished). Such ferrets, however, despite continued treatment, do not remain anoestrous. They come into breeding condition during

March–April, the time at which they would have done had they been left in daylight (or exposed, throughout the same period, to 'short' photoperiods).

Melatonin, it seems, has prevented the breeding season from being re-timed by artificial long days, but a mechanism persists which determines that oestrus will occur in the spring—an event which, as we have seen, is independent of both lighting cues and the pineal. Ferrets kept in short days (LD 8 : 16) throughout the same period and given melatonin 8 h after its onset also come into oestrus about the same time. Although refractoriness to melatonin has not yet been ruled out as a contributory factor in these experiments, it seems unlikely that melatonin is simply an 'anti-gonadotrophic substance' in the sense that it inhibits the secretion of gonadotrophins in a way analogous, say, to the steroid hormones. Rather, we think that melatonin may act on the timing mechanism itself.

THE PINEAL, MELATONIN AND THE BRAIN

If it is true that the pineal gland, and melatonin, act on a timing mechanism, where is it located, and what is its nature?

There must, it seems, be a neural mechanism interposed between the pineal gland and the pituitary, which responds to the photoperiodic time-measuring mechanism factors described above. Such a system clearly must be light-sensitive, innervate the areas which are known to be concerned in the light response and there ought to be a prima facie case supporting its involvement in the control of endocrine rhythms. For these reasons, we have chosen the 5HT-containing system as a possible candidate.

A number of other experiments have implicated this system in regulating the circadian rhythms of both adrenocortical and luteinizing hormone release.[27,28] Furthermore, it is well known that the suprachiasmatic nucleus (SCN) and other parts of the brain concerned with optic input (including the ventral nucleus of the lateral geniculate body, which projects to the SCN), receive a large input from the 5HT-containing system.[29]

There has not, so far, been any direct experimental demonstration that modulating 5HT can mimic or prevent photoperiodic effects on the breeding season of any species, though the well-known ability of the 5HT system to sprout, if part of it is damaged or removed, or for receptors to change their numbers in denervated areas, should warn us against simple interpretation of either negative or positive effects of such procedures.

Nevertheless, if the 5HT system did represent one mechanism by which the central nervous system became aware that the animal was either in long or short photoperiods, then it might show different biochemical parameters in the two conditions. Recently, we have found that the circadian rhythm in 5HT content of the pineal gland and the hypothalamus, two areas which are implicated in the photoperiodic response, showed differential shapes in animals put into either long (LD 14 : 10) or short (LD 8 : 16) days.[30] In short photoperiods, rhythms in the pineal and the hypothalamus resemble those in the rest of the brain, a prominent nadir occurring at 6 h after the lights come on (figure 6). However, in long photoperiods the 5HT levels in the pineal become arrhythmic, whereas those in the hypothalamus were at their highest at 6 h; that is, a direction exactly opposite to that in the same area of the brain in animals kept in short photoperiods (figure 6). These different effects of the two photoperiods were not observed if animals were

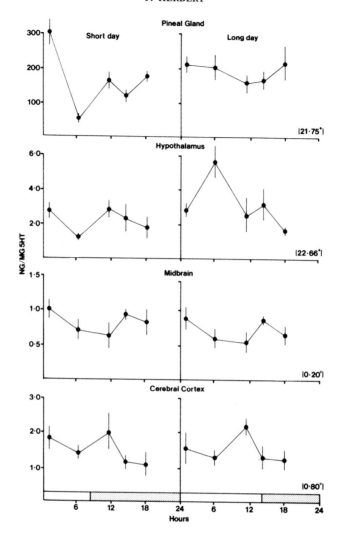

Figure 6. Circadian variations in 5HT content of the pineal, hypothalamus, midbrain and cerebral cortex in ferrets kept in either short days LD 8 : 16 (*left*) or long days LD 14 : 10 (*right*) and killed at five time points. Means ± s.e. from 6 animals per sample point.

either pinealectomized or ganglionectomized—procedures which prevent the phys-iological response to them. Measuring 5HT levels gives only a rather superficial idea of what changes may be occurring in this neural system. In preliminary experiments, therefore, we have investigated the dynamic aspects of the 5HT system in animals kept in either long or short photoperiods (Powiesnik, Herbert, Christiansen and de la Riva, unpublished). It is known that one of the most important ways by which such a system regulates itself is by re-uptake of transmitter released from the pre-synaptic terminal. We therefore wondered whether a change in this parameter might underlie the changes in 5HT concentra-tion observed in our earlier experiments. Preliminary results indicate that this might be the case. Calculations of the *Km* (that is, the affinity of the uptake mechanism) and its *V* (that is, the maximal rate at which 5HT is taken up into the system) demonstrate that, whereas the former remains unchanged, the latter is greatly increased in the hypothalamus of ferrets kept in long photoperiods, when the tissue is examined 6 h after lights come on (i.e., the time at which, in our previous experiments, 5HT levels were observed to be at their highest). This effect has been seen in the hypothalamus but not in the cerebral cortex, indicating that there may be regional specificity similar to that observed in assays of 5HT content.

Since melatonin has been observed to replicate some of the short day effects of the pineal gland, it was of interest to see whether or not it could also restore the neurochemical effects of short days. Experiments showed that the circadian hypothalamic 5HT rhythm in ferrets kept in long photoperiods could be modified to resemble that of those kept in short photoperiods, if the former were given a daily injection of melatonin 8 h after lights came on.[30]

Since we wanted to see whether there was a similar correlation between the lesser effectiveness of melatonin given at 14 h into the photoperiod (*see* p. 270 *above*), we were interested to find that melatonin given 14 h into the photoperiod in animals kept in LD 14 : 10 was unable to produce the typical short day pattern in the hypothalamus, but, rather, a circadian variation which more resembled animals left in 'long' day.

COMPARISON WITH OTHER SPECIES

It is now pertinent briefly to consider the results discussed in this paper on the ferret with some of the findings described in other species, notably the hamster and the vole. In the latter two species, it is clear that, with the possible exception of some results in *Phodopus*,[10] the breeding season is regulated principally by the inhibition exerted by short days, and that this acts through the pineal gland.[8] It is difficult to show a stimulatory effect of long days, and the physiological effect of this kind of photoperiod seems to be limited to dispelling photorefractoriness to short days, which develops after 25 weeks or so.[8] It is important to note that it is these results which have given rise to the now widespread notion that the pineal gland is 'anti'-gonadotrophic. In the hamster and the vole this is very likely to be the case, since the short lifespan of the animals and their high mortality rate dictate that nearly all members of these species will experience only one breeding season. As Reiter[8] has pointed out, it is therefore of utmost importance that the animal should emerge from hibernation in maximum reproductive condition, which can be assured by the development of photorefractoriness to short days during hibernation. Furthermore, since animals are born during the summer, it is important that they either be driven, or kept, out of oestrus by the shortening photoperiods of autumn, signalling the onset of an environment unpropitious to breeding. It is therefore unnecessary for these animals to either develop or retain a 'long' day mechanism which times the onset of breeding. The situation in the ferret is quite different. Not only is this animal a carnivore, and therefore a predator rather than a prey, it is also much longer-lived. Successive breeding seasons will thus need to be timed on an annual basis, in a way that is not required for a species such as the hamster. Thus, the development or retention of photorefractoriness may not be as essential in the ferret. Not only can the onset of its breeding season be timed by long photoperiods, but its end can also be determined by the shortening day lengths after the summer solstice, a mechanism which obviates the necessity for photo-refractoriness. This explanation would only apply to species (such as the ferret) in which the breeding season lasts for most of the summer. In others, such as the mink, in which the breeding season is very short (and terminates whilst the days are still lengthening), photorefractoriness may play a much more important role.

Though information on pineal function is incomplete in sheep, this species is interesting in that short days induce oestrus.[4] It seems likely that the pineal gland will also be found to be essential for this effect, since ganglionectomy prevents the endocrine effects of exposure to artificial long or short photoperiods.[17] We do not yet know whether injections of melatonin at 8 h into a 'long' photoperiod will induce oestrus in sheep; if this proves to be the case, it would strengthen the arguments presented above that melatonin cannot be considered simply as an 'anti'-gonadotrophic substance. The principal conclusion, therefore, is that the effects of the pineal gland on the reproductive system depend upon the ecology and reproductive biology of the species in question. In general, whatever effect increasing or decreasing photoperiods have upon reproduction, it will be transmitted by the pineal; however, what that effect may be depends not upon the pineal gland, but upon the way the central nervous system (or perhaps the pituitary gland itself) responds to changes in pineal function. Thus, it can be suggested that the adaptive mechanisms which determine the way in which photoperiods regulate the breeding season in different species reside in the neural response to the pineal gland.

ACKNOWLEDGEMENTS

This work was supported by grants from the ARC. I thank my colleagues D. S. Carter, J. Powiesnik, P. A. Thorpe and C. A. Laud (née Yates) for their collaboration and P. M. Stacey, L. Christiansen and C. de la Riva for their technical assistance.

REFERENCES

1. J. R. Hammond and F. H. A. Marshall *Proc. R. Soc. B.* **105,** 607–20 (1930).
2. T. H. Bissonnette *Proc. R. Soc. B.* **110,** 322–36 (1932).
3. S. Gaston and M. Menaker *Science* **167,** 925–8 (1967).
4. N. T. M. Yeates *J. Agric. Sci.* **39,** 1–42 (1949).
5. J. Herbert and D. S. Vincent *Proceedings of the Fourth International Congress of Endocrinology* **273,** 875–9 (1972).
6. W. A. Marshall *J. Endocrinol.* **24,** 315–23 (1967).
7. J. R. Hammond *Nature (Lond.)* **167,** 150–1 (1952).
8. R. J. Reiter *Ann. Rev. Physiol.* **35,** 305 (1973).
9. K. Hoffmann and I. Kuderling *Experientia* **31,** 122–3 (1975).
10. K. Hoffmann *Experientia* **35,** 1529–30 (1979).
11. P. A. Thorpe and J. Herbert *J. Endocrinol.* **70,** 255–62 (1976).
12. B. Lofts and R. K. Murton *J. Zool.* **155,** 327–94 (1968).
13. R. J. Reiter *Anat. Rec.* **173,** 365–72 (1972).
14. B. T. Donovan *J. Endocrinol.* **39,** 105–13 (1967).
15. D. S. Vincent *J. Endocrinol.* **48,** iii (1970).
16. E. I. Pengelley (ed.) *Circannual Clocks.* Academic Press, New York (1974).
17. G. A. Lincoln *J. Endocrinol.* **82,** 135–47 (1979).
18. C. Thibault, M. Courot, L. Martinet et al. *J. Anim. Sci.* **25,** Suppl., 119–142 (1966).
19. J. Herbert, P. M. Stacey and D. H. Thorpe *J. Endocrinol.* **78,** 389–97 (1978).

20. M. E. Abrams, W. A. Marshall and A. P. D. Thomson *Nature* (*Lond.*) **174**, 311 (1954).
21. W. A. Marshall *J. Physiol. Lond.* **165**, 27–28P (1962).
22. P. Rio del Hortega In: W. Penfield (ed.) *Cytology and Cellular Pathology of the Nervous System*, vol. 2. Hoeber, New York, 635–703 (1930).
23. J. Herbert In: G. E. W. Wolstenholme and J. Knight (ed.) *The Pineal Gland:* Ciba Foundation Symposium. Edinburgh, Churchill Livingstone, 303–27(1971).
24. J. R. Johnson, P. A. R. Mayer, D. A. Westaby et al. *J. Anat.* **118**, 491–506 (1974).
25. J. Herbert *Acta Endocrinol.* (*Cobenh.*) [Suppl.] **119**, 46 (1967).
26. J. Herbert *J. Endocrinol.* **43**, 625–36 (1969).
27. U. Scapagnini, I Gerendai and G. Clementi et al. In: I. Assenmacher and D. S. Farner (ed.) *Environmental Endocrinology.* Berlin, Springer Verlag, 137–43 (1978).
28. M. Héry, E. Laplante and C. Kordon *Endocrinology* **49**, 496–503 (1976).
29. K. Fuxe *Z. Zellforsch.* **65**, 573–96 (1965).
30. C. M. Yates and J. Herbert *Brain Res.* **176**, 311–26 (1979).

DISCUSSION

Gwinner: To evaluate your data on the oestrous periodicity of blinded and pinealectomized ferrets under natural daylengths it would be important to know how the pattern of oestrous activity is maintained in a constant photoperiod in intact ferrets. Have you done such experiments?

Herbert: There are some data of Diana Vincent's which indicate that ferrets kept for a period of 2 years in long days show highly irregular breeding seasons but I think that the occurrence of photorefractoriness (an irregular phenomenon in ferrets) complicates the picture. Ferrets kept in short days showed mutual synchrony, but these experiments did not last long enough to be fully informative in the context in which you raise them.

Turek (comment): In support of Dr Herbert's and Dr Hoffmann's suggestion that melatonin may interfere with the ability of the animal to respond appropriately to photoperiodic signals, we have observed that melatonin will prevent long days from terminating the refractory condition in the hamster.

Rusak (comment): It is interesting that you report higher hypothalamic serotonin levels in ferrets kept in long days than in those kept in short days. Arendash and Gallo have described an inhibition of episodic luteinizing hormone (LH) release by serotonin of midbrain origin reaching the hypothalamus in rats. If a similar mechanism regulated LH release in a short-day breeder like the ferret, one would hope to find the seasonal pattern of hypothalamic serotonin levels that you have described.

Arendt and Symons (comment): In collaboration with Dr G. A. Lincoln (MRC Reproductive Biology Unit, Edinburgh) we have shown that circulating immunoreactive melatonin is greatly reduced in ganglionectomized Soay rams whose ability to respond to photoperiod changes is impaired. This observation provides indirect evidence for the involvement of melatonin in photoperiodic responses in sheep. We have also investigated the circulating levels of melatonin in Suffolk cross-bred ewes kept in natural light. Melatonin was assayed by radioimmunossay (RIA) at 3 h intervals throughout a 24 h period at different times of the year. The most important variations found were the frequent presence of double peaks, in the main from May to September, a light phase at 1200 h in October and a shift in the time of maximum secretion to the late dark phase in early spring and summer (figure 1).

A partial explanation for both quantitative and qualitative aspects of the

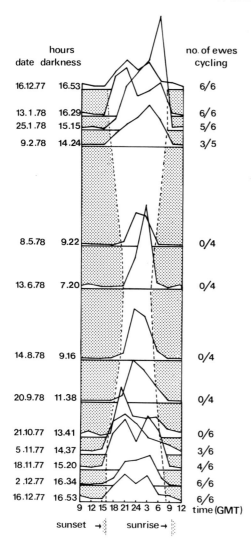

Figure 1. Variations in circulating melatonin (mean pg/ml) in 4–6 Suffolk cross-bred ewes at 3 h intervals during 24 h at different times of the year. Standard deviations are left out for reasons of clarity: analysis of variance indicates that light/dark, number of hours after sunset, number of hours after sunrise and time of year, all have significant effects ($P<0.01$–0.001) in the observed values.

observed patterns might be provided by the 'signals' for pineal secretion operating at spaced intervals from sunset and sunrise: where close coincidence of signals occurs, then secretion is augmented. The amount of circulating melatonin at different times (maximum in January and November) was not consistent with an 'anti'-gonadotrophic function for melatonin. We suggest that if melatonin is the pineal principle concerned with the timing of seasonal breeding in sheep, then pattern variations may supply biochemical time cues for triggering the appropriate seasonal response.

Pittendrigh (comment): Having seen these data, I wonder whether melatonin acts on the pacemaker?

TWENTY YEARS ON*
THE ANNUAL COLSTON LECTURE

Professor Jürgen Aschoff

Max-Planck-Institut für Verhaltensphysiologie, Andechs,
Federal Republic of Germany

After I had accepted the illustrious task of delivering the Annual Colston Lecture, my first plan was to discuss the development of our field over the past 20 years, from the Symposium at Cold Spring Harbor in 1960[1] until the present time. Follett agreed with this proposal and suggested as a title 'Twenty Years On'. In preparing the lecture, however, I was forced for reasons which become obvious later, to go back a further 20 years. I then thought that it would not matter too much if another trifling 280 years were added. So the final title of my talk might be more accurately paraphrased as '300 years before, and 20 years after, with Cold Spring Harbor as a milestone between these two time spans'.

Before entering upon my subject I should introduce the topic of our present symposium to those guests who are not biologists. The title is 'Biological Clocks in Seasonal Reproductive Cycles'. We are actually considering two kinds of biological clocks, expressed in plants and animals as daily and seasonal rhythms. These rhythms are based on endogenous, inherited, periodic processes which persist in constant conditions like self-sustaining oscillations. They 'free-run', as we say, with a period slightly different from that of the environmental periodicity which they mimic. This is indicated by the prefix 'circa' and thus we speak of them as being 'circadian' or 'circannual' clocks. These clocks can be used by organisms to measure time of day and time of the year. In two-thirds of my talk I shall concentrate on problems associated with circadian clocks, in the last third on annual rhythms in man.

Let me begin by outlining briefly the history of the search for the circadian clock. We learned from Bünning[2] that it was the French astronomer de Mairan who, in 1729,[3] first noticed the endogenous nature of circadian rhythms. He was curious to find out why the leaves of plants move up and down during the course of each day, and he observed that these movements continued when the plants were kept in continuous darkness. de Mairan must have been one of those colleagues who never prepare abstracts when they present a paper at a meeting! The short note on his observations which appeared (anonymously) in the Proceedings of the French Academy, has as its last sentence the following: 'La marche de la véritable phisique, qui est l'epérimentale, ne peut être que fort lente.' de Mairan probably knew that progress in science is slow at the beginning but accelerates later on.

---

* The reader should be aware that this after-dinner speech was given on the evening of April Fool's Day 1980! Roman numbers in the text (e.g. [I]) refer to notes added at the end for clarification of a few details.

Biological Clocks in Seasonal Reproductive Cycles 277–288 (1981) (ed. B. K. and D. E. Follett: Bristol, Wright).

Before giving you evidence for such a growth of knowledge in our field, I should like to ponder briefly upon de Mairan's botanical interests. His age was famous for richly ornamented floral sundials. True floral horloges make use of plants which open their flowers at different times of day: Linné[4] is said to be their inventor. One wonders, however, whether Andrew Marvell[I] had such a clock in mind when he wrote, about 80 years prior to de Mairan's experiment, the last verse of his poem 'Thoughts in a garden':

> How well the skilful Gardner drew
> Of flow'rs and herbes this Dial new;
> Where from above the milder Sun
> Does through a fragrant Zodiak run;
> And, as it works, th'industrious Bee
> Computes its time as well as we.
> How could such sweet and wholesome Hours
> Be reckon'd but with herbs and Flow'rs!

Whatever the dial may have been to which Marvell referred[II], it is astounding for a biologist to learn that a time sense was attributed to bees about 280 years before the bee's 'zeitgedächtnis' was demonstrated by Beling.[5]

In mentioning Beling, I have passed over many members of our genealogy: I should mention a few. It was the French medical student, Virey,[6] who in 1814 first coined the expression 'l'horloge vivante' to describe human circadian rhythms. Early in this century Semon[7] published the first records of leaf movements persisting in continuous light. Bünning, in 1932,[8] then emphasized that the deviation of the period from 24 hours was a crucial argument for the endogenous nature of this rhythmicity in plants. Finally, Johnson[9] demonstrated the first free-running activity rhythm in a rodent. Two of his records, together with the period analysis which we have made from his data, are shown in figure 1. In dim light, there is a clear rhythm of activity (black horizontal bars) which has a mean period of 24·6 hours. In bright light, activity seems to be arrhythmic, but mathematical analysis reveals a rhythmic component still occurring with a period of 25·6 hours. Johnson concluded: 'This animal has an exceptionally substantial and durable self-winding and self-regulating clock, the mechanism of which remains to be worked out.' He also pointed out that his results 'show an exponential relationship between light intensity and rate of shift in time of daily active period': an observation substantiated later by many workers (Aschoff[10]).

The second figure (figure 2) lists all the major discoveries which have led from de Mairan's initial observations to Johnson's self-winding clock. I hardly know of any better example of exponential growth in science[III]. At the very top of this curve, I could also have added the name of Palmgren[11] who published the first record of a free-running rhythm in a brambling. He called this 'derailment' of the rhythm.

Twenty years after the publication of Johnson's and Palmgren's records, the Cold Spring Harbor Symposium took place. This was a milestone. Its centre of gravity was the twenty-five page paper by Colin Pittendrigh.[23] So many facts had been accumulated by this time that he was able to list sixteen major generalizations about circadian clocks on the first two pages alone. All of these 'rules' are still relevant today. Pittendrigh's next four and a half pages deal with a model of the 'pacemaker'—a kind of masterclock—which consists of an A- and a B-oscillator. Earlier this week Pittendrigh demonstrated to what degree of sophistication such a model can be developed (pp. 1–35). I must, however, admit to being

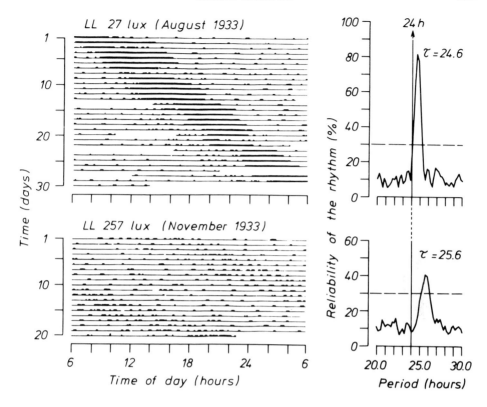

Figure 1. Rhythms of locomotor activity in two white-footed mice (*Peromyscus leucopus noveboracensis*), kept in continuous light (LL) with two intensities of illumination. Original records taken from Johnson.[9] The two right-hand diagrams show period analyses of these activity records (τ = mean circadian period).

somewhat confused about the present state of the B-oscillator. Is it still part of the 'pacemaker' (as in 1960), or are overt rhythms and B-slave oscillators the same thing? Possibly it is only my ignorance, or naivety, that sees a problem here. Whatever the answer may be, the description of a self-winding clock by Johnson, and Pittendrigh's pacemaker story of 1960 have resulted in the search for, and the discovery of, localized units which may turn out to be true masterclocks (e.g. pp. 113; 129; 137; 171; 219). Funnily enough, this line of research runs counter, at least in part, to one of Pittendrigh's own statements at Cold Spring Harbor: 'We are forced to abandon the common current view that our problem is to isolate and analyse '*the* internal clock', and are faced with the conclusion that the organism comprises a population of quasi-autonomous oscillatory systems' (Pittendrigh[23] p. 165). As a result of these considerations, the final twenty pages of Pittendrigh's contribution[23] deal with what he calls 'the circadian organization: a multi-oscillator system'. The emphasis on a multi-oscillatory structure is, of course, not contradictory to the concept of one or several pacemakers; it is rather a complementary aspect that deserves as much attention as the physiology of the masterclock itself.

Let me contemplate this complex system for a while. When I visited Pittendrigh this February, he reminded me of a picture which he had presented in Cold Spring Harbor and which he called 'Pittendrigh's nightmare'. You can see it—the activity

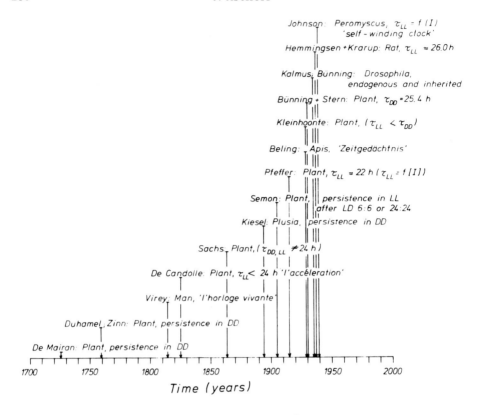

Figure 2. 'Milestones' in clock research up to the early forties. Symbols: LL, continuous light; DD, continuous darkness; τ, mean circadian period; I, intensity of illumination. (For sources see references[3,5–9,12–22].)

record of a Tundra vole—in figure 3. I have added to it 'Aschoff's nightmare', the activity rhythm of a rat, recorded last December in collaboration with Kenichi Honma. The vole's record shows the rhythm after a release from continuous light into continuous darkness. The initial long period of the free-running rhythm continues even after a second activity component with a shorter period has emerged at about day 25. The rat's record shows a rhythm free-running in continuous dim light, with food available all the time during the first and third section of the experiment. When food is only offered for 4 hours daily, a second component of activity appears whose phase leads the feeding time, and which persists as a 24 h band for several days after the feeding restriction is terminated. Both these records demonstrate that there are distinct and dissociable components within one and the same overt rhythm, one of which can be entrained by a zeitgeber (*see* the rat's record) while the other continues to free-run. I am not convinced that these two components represent the 'dawn and dusk oscillators' which have been discussed in relation to other splitting phenomena,[24] but I would like to know to what degree they represent separate pacemakers or slave oscillators. To quote Pittendrigh[23] once more: 'The autonomy and potential dissociability of distinct oscillatory components in the circadian system is the crux of the viewpoint, and its implications, I am going to develop' (Pittendrigh[23] p. 170) [IV].

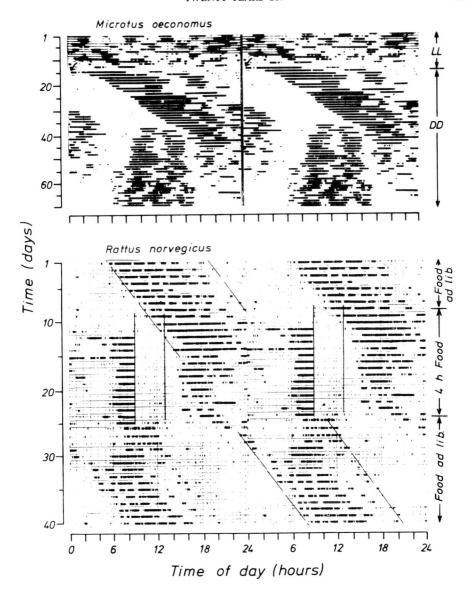

Figure 3. Circadian rhythms of locomotor activity in a Tundra vole (*above*) and in a rat (*below*). Constant conditions for the vole were continuous light (LL) and continuous darkness (DD), and for the rat (kept in dim LL) restriction of food availability, as indicated on the right margin. Vole data from Pittendrigh;[23] rat data from K. Honma and J. Aschoff (unpublished).

Research into pacemakers and central masterclocks *is* of prime importance, as we have learned in this meeting, but I consider it of as much interest to study the interaction between components (central as well as more peripheral) of the circadian system. To illustrate this further, I will make use of data from human subjects kept for weeks in an isolation unit. In the first of three diagrams (figure 4), the rhythm of wakefulness and sleep of such a subject is shown, together with the

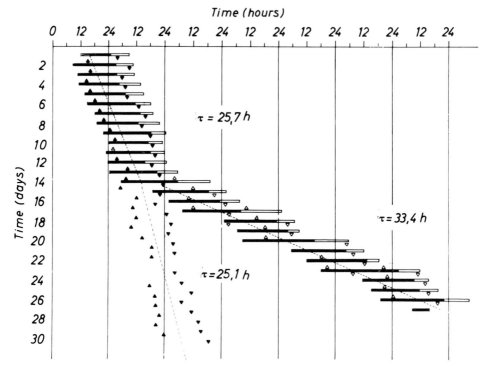

Figure 4. Circadian rhythm of wakefulness and sleep (black and white bars) and of rectal temperature (triangles for maxima and minima) in a human subject living in isolation without time cues. Spontaneous internal desynchronization occurred on day 14. τ = mean circadian period. (Taken from Wever.[25])

rhythm of rectal temperature. During the first two weeks of the experiment, both rhythms free-ran with the same mean period of 25·7 hours. On day 14 the activity/rest cycle suddenly lengthened, for still unknown reasons, to a period of 33·4 hours. This was, obviously, too long a period for the rhythm of rectal temperature and it separated from the activity/rest cycle and continued to free-run with a period close to 25 h. During such a state of 'internal desynchronization' the two rhythms are sometimes in phase with each other, sometimes out of phase. All this time the two rhythms, or the pacemakers to which they are coupled, are still interacting, the amount of the interaction depending on the mutual phase relationships. This is illustrated by the fact that the duration of sleep depends on the phase of the temperature rhythm (figure 5): sleep is longest when it coincides with decreasing temperatures (hour prior to zero on the abscissa), and is shortest when it coincides with low and increasing temperatures. Also noteworthy is the bimodal distribution of sleep onsets as a function of the phase of the circadian temperature cycle (upper diagram in figure 5). The probability of falling asleep depends on circadian clock time as given by an autonomic function. Lastly, I would like to draw your attention to a phenomenon which we call 'partial entrainment'. In the isolation unit, subjects sometimes develop a free-running rhythm even in the presence of a light/dark cycle. If in such a case internal desynchronization occurs, the rhythm of rectal temperature does not continue to free-run (as in constant conditions, c.f. figure 4), but locks on to the light cycle (figure 6). This observation

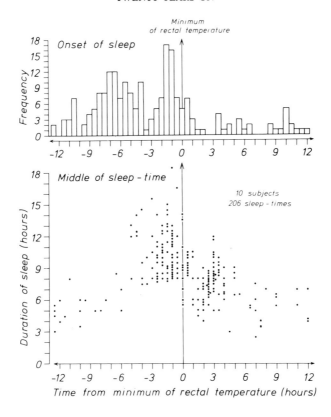

Figure 5. Onset of sleep (*above*) and duration of sleep time (*below*) as a function of the phase of the temperature rhythm. Data from subjects with internal desynchronization only (c.f. figure 4). (From Aschoff.[26])

not only supports strongly the concept of a multi-oscillatory structure for the human circadian system, but also demonstrates that parts of the system can be entrained by light while other rhythms free-run.

How many circadian clocks, then, are there? For some organisms we have the answer, given yesterday by Page (pp. 113–124): 'It is frightening to see so many oscillators in insects!' I wonder what the answer may be for vertebrates after another 20 years have passed.

I now turn to the last third of my talk. When I read the programme of this symposium, I was disappointed that reproduction in man—presently a most serious problem—was not included, and yesterday I was asked by one of the mammalian experts whether there was seasonality in human reproduction at all. Therefore, I think it appropriate to point out, at least in an after-dinner speech, that seasonal rhythms can be found in man and that some of the human data may be of direct relevance for current problems in zoology.

Even a young girl at the age of 12, or a boy of 15 years, can notice changes in his mood and temper when spring approaches. Such a young boy was Thomas Chatterton who, with Colston, is among the more famous citizens of Bristol. When 15 years old, Chatterton wrote the following song, addressed to a Miss C of Bristol[V]:

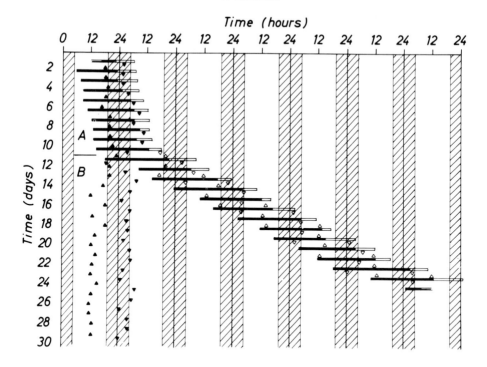

Figure 6. Circadian rhythm of wakefulness and sleep and of rectal temperature (c.f. figure 4) in a subject living in an isolation unit under the influence of a light/dark cycle (shaded area: darkness), with additional reading lamps available. (From Wever.[25])

As spring now approaches with all his gay train,
And scatters his beauties around the green plain,
Come then, my dear charmer, all scruples remove,
Accept of my passion, allow me to love.

Without the soft transports, which love must inspire,
Without the sweet torment of fear and desire,
Our thoughts and ideas, are never refin'd,
And nothing but winter can reign in the mind.

But love is the blossom, the spring of the soul,
The frosts of our judgments may check, not controul:
In spite of each hindrance, the spring will return,
And nature with transports refining will burn.

Before discussing with you the consequences of a 'burning nature', I cannot resist the temptation to give you another little song of Chatterton written in anticipation of what we all know so well to be true of Colin Pittendrigh[VI]:

Young Colin has a comely face
And cudgels with an active grace,
    In everything complete;
And watch how he can dance divine,
Gods! how his manly beauties shine,
    When jigging with his feet.

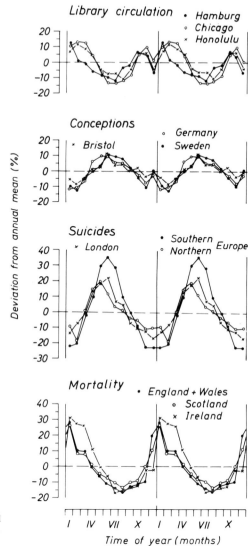

Figure 7. Annual rhythms in man. Means from several years, drawn twice to emphasize the rhythmic pattern. (From various vital statistics.)

Let us now return to spring and to love! As you can see in the second diagram from the top of figure 7 (based on data from Germany, Sweden and Bristol) monthly conception rates reach a maximum in May or June with a secondary (minor) peak in December. Mirror images of these curves are those representing the circulation of books taken out of the central libraries of three cities[VII]. It is up to you to draw whatever conclusion you wish from the negative correlation between reading and making love. In the lower two diagrams showing seasonal rhythms in suicides and in general mortality I have included suicide data from London because it was there that Chatterton ended his life, in August 1770, when only 18 years old.

Data on such seasonally timed events are available for many countries from all over the world. I have fitted sine functions to the original raw data, and derived

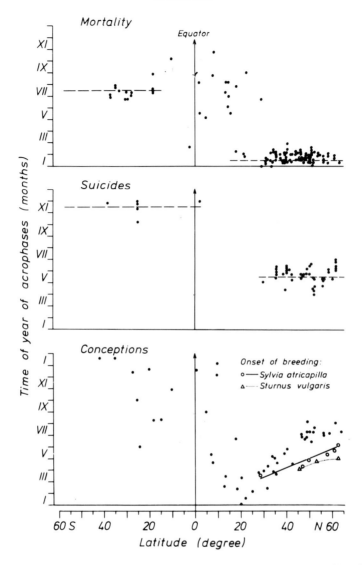

Figure 8. Dependence of the acrophase of human annual rhythms on latitude. Data on onset of breeding seasons in birds from P. Berthold and E. Gwinner (personal communication).

therefrom measures of the acrophase (i.e. the month of maximal values) and of the relative amplitude (i.e. the difference between the maximum and the annual mean, expressed as a percentage of the mean). One of my interests was to see how these two parameters depend on latitude. The amplitude is, as you may expect, minimal in areas close to the equator, and increases towards southern and northern latitudes. This increase, however, does not continue up to the Arctic or Antarctic, respectively. At least in the northern hemisphere where most of my data come from the amplitude of the seasonal rhythm in conception rates reaches maximal values at 20 ° to 25 °N and becomes smaller again further north. On the other hand, the amplitudes of the suicide and mortality rhythms are maximal at about 40 °N (c.f. figure 10 in Aschoff[27]). So far, no explanation can be given for these latitudinal variations in amplitude.

Of even greater interest, at least for biologists, are the changes in acrophase with latitude (figure 8). In most parts of the northern hemisphere, general mortality is maximal in January: south of the equator, the maximum occurs in July. Surprisingly, the 6-monthly shift in acrophase does not seem to occur close to the equator but is already apparent 10 ° to 20 ° north of it. Suicides have their maxima in May (north) and November (south), respectively. In strong contrast to these two rhythms whose acrophases are, within one hemisphere, largely independent from latitude, the acrophase of the rhythm in conception rate depends on latitude drastically. In the high North (50 ° to 60 °N), conception rates are maximal in July, but as we move southward the acrophase occurs progressively earlier, eventually approaching January at 20 °N, i.e. at about the latitude where the amplitude is maximal. South of 20 °N, the acrophase seems to move back again, but the scarcity of data from these areas does not allow strong conclusions.

In summary, the analysis of these data has convinced me that there are seasonal rhythms in man which have a biological basis, that there must be periodic factors in the environment other than the photoperiod which control these rhythms, and that suicides and mortality on the one hand, and conception rates on the other, are controlled by different environmental factors. I am further convinced that there is a 'biological equator' north of the geographical equator which may be related to the 'meteorological equator'[28] and which needs to be investigated with regard to animal breeding cycles. I can offer no explanation for the effects of latitude on the acrophase of conception, i.e. on why the biological spring in man occurs about two months earlier in Lisbon than in Bristol. It is a trend similar to that seen in the breeding seasons of birds (figure 8). The hypothesis that a common environmental factor influences all these reproductive cycles cannot be ruled out *a priori*.

I hope I have been successful in convincing you that daily and annual clocks are at work in man as in animals, and that it is worth while (as well as entertaining) to study them. However, there is always the danger that we take our studies too seriously. Science, including clock science, tells us a lot, but will it have the last word in our world so beset with problems? Pittendrigh added a motto from Walter de la Mare to the last chapter of his milestone paper: 'Why this absurd concern with clocks, my friend?'. In the same sense, I finish my talk with another little song from Thomas Chatterton[VIII]:

> What are clocks and all their joys?
> Useless mischief, empty noise.
> What are data-trophies won?
> Spangles glittering in the sun.
> Rosy Bacchus, give me wine:
> Timeless happiness is thine!

NOTES

I. The poems and letters of Andrew Marvell (1621–78) were edited by H. M. Morgilouth. The third edition, revised by P. Legouis and E. C. Duncan-Jones, is published by Oxford University Press (1971).

II. According to K. J. Höltgen (*Notes and Queries*, **214**, 381, 1969), it is unlikely that Marvell who, during his visit to Rome, probably saw a floral sundial with a gnomon in the garden of the Villa Aldobrandini, refers to a true (Linnaean) floral horloge.

III. The doubling time of about 50 years which can be derived from figure 2 agrees with estimates of B. Glass (*Q. Rev. Biol.* **54**, 31, 1979) for the growth of biology from the seventeenth to the twentieth century, and of genetics in the twentieth century.

IV. Note that the quasi-autonomous oscillators of p. 165 have become autonomous on p. 170! Anyone familiar with German literature is reminded of the famous essay of Heinrich von Kleist 'Uber die allmähliche Verfertigung der Gedanken beim Reden' which may be translated here as 'On the progressive development of ideas during writing'.

V. From *The Complete Works of Thomas Chatterton*, edited by D. S. Taylor in association with V. H. Hoover and published by the Clarendon Press, Oxford (1971) (vol. 1, p. 559). Chatterton lived from 1752 to 1770.

VI. Second verse of 'The Virgin's Choice' (*Complete Works*, vol. 1, p. 633). Friends of Colin may also read 'Colin instructed' (vol. 1, p. 681).

VII. Reliable statistics on fiction and non-fiction literature—not April Fool's Day data!

VIII. Modified (in a 'Rowleyan' manner) from 'A Bacchanalian' (vol. 1, p. 632).

REFERENCES

1. A. Chovnick (ed.) Biological Clocks. *Cold Spring Harbor Symp. Quant. Biol.* **25** (1960).
2. E. Bünning *Cold Spring Harbor Symp. Quant. Biol.* **25,** 1–9 (1960).
3. J. J. D. de Mairan *Histoire de l'Academie Royal des Sciences, Paris.* p. 35 (1729).
4. C. von Linnaeus *Philosophia Botanica* 273 and 274, Stockholm (1751).
5. I. Beling *Z. Vergl. Physiol.* **9,** 259–328 (1929).
6. J. J. Virey Thesis, University of Paris (1814).
7. R. Semon *Biol. Zentralbl.* **25,** 241–52 (1905).
8. E. Bünning *Jb. Wiss. Bot.* **27,** 439–80 (1932).
9. M. S. Johnson *J. Exp. Zool.* **82,** 315–28 (1939).
10. J. Aschoff *Z. Tierpsychol.* **49,** 225–49 (1979).
11. P. Palmgren *J. Ornithol.* **89,** 103–23 (1941).
12. H. L. Duhamel du Monceau *La Physique des Arbres,* vol. 2. Paris (1758).
13. J. G. Zinn *Hamburgisches Magazin 22,* 40–50 (1759).
14. A. P. de Candolle *Physiologie Végétable,* vol. IV. Paris (1832).
15. J. Sachs *Flora 46,* 465–72 (1863).
16. A. Kiesel *Sber. Akad. Wiss. Wien [Math.-nat. Klasse]* **103,** 97–139 (1894).
17. W. Pfeffer *Abh. Sächs. Akad. Wiss.* **34,** 1–154 (1915).
18. A. Kleinhoonte *Archs Néerl Sci.* **V,** 1–110 (1929).
19. E. Bünning and K. Stern. *Berl. Bot. Ges.* **48,** 227–52 (1930).
20. H. Kalmus *Biol. Gen.* **11,** 93–114 (1935).
21. E. Bünning *Berl. Dtsch. Bot. Ges.* **53,** 594–623 (1935).
22. A. M. Hemmingson and N. B. Krarup. *Biol. Meddr* **13,** (1937).
23. C. S. Pittendrigh *Cold Spring Harbor Symp. Quant. Biol.* **25,** 159–84 (1960).
24. C. S. Pittendrigh In: F. O. Schmidt and F. G. Warden (eds.). *The Neurosciences: Third study program.* Cambridge, Mass., MIT Press, pp. 437–58 (1974).
25. R. Wever (ed.). *The Circadian System of Man.* Berlin, Springer-Verlag (1979).
26. J. Aschoff In: L. C. Johnson, D. I. Tepas, W. P. Colqhoun et al. (eds.). *The 24-hour Workday.* Washington, National Institute for Occupational Safety and Health (1981).
27. J. Aschoff In: J. Aschoff (ed.) *Handbook of Behavioral Neurobiology,* vol. 4: *Biological Rhythms.* New York, Plenum Press (1981).
28. H. Flohn *Geogr. Rdsch.* **12,** 129–42, 189–95 (1960).

SPECIES INDEX

Acyrthosiphon pisum, 127, 128
Aëdes atropalus, 93, 95
Anolis carolinensis, 27, 28, 137, 139, 143–52, 203
Antheraea peryni, 95, 113, 115
Aphis fabae, 125–7, 129, 131–4

Blaberus fuscus, 122
Blackcap, 169
Blaps gigas, 113, 122, 123
Brambling, 278

Callisaurus draconoides, 143
Calotes versicolor, 143
Chenopodium rubrum, 47, 187
Chicken, 171, 172, 174–8, 180, 181
Cockroach, 2, 27, 113, 116, 119, 122–4
Coleonyx variegatus, 139
Cow, 224
Cricket (*Teleogryllus commodus*), 2, 113, 115–16, 122–4
Culex quinquefasciatus, 81

Deer, 197, 224, 264
Desmodium gyrans, 40
Diatraea grandiosella, 84, 91, 101
Drosophila melanogaster, 37, 113, 115
Drosophila pseudoobscura, 2–7, 9, 10, 13, 15, 17, 21, 24, 26, 29, 33, 68, 69, 76, 80, 102

Epilobium adenocaulon, 47
Euonymus europaeus, 126

Ferret, 215, 237, 240, 241, 261–76
Flying squirrel (*Glaucomys volans*), 217

Greenfinch (*Chloris chloris*), 198
Grouse, 258

Hamster (golden), 27, 29, 139, 142, 188, 203, 206, 217, 219, 237–40, 244, 249, 251–9, 262, 263, 267
Hamster (*Phodopus sungorus*), 236–47, 262, 273
Hare, 258
Hemidactylus turcicus, 139
House finch (*Carpodacus mexicanus*), 29, 185, 188
House sparrow (*Passer domesticus*), 2, 27, 171–5, 180–1, 195, 212, 258
Hyalophora cecropia, 113, 115

Java sparrow (*Padda oryzivora*), 183
Junco (*Junco oreganus*), 195

Kalanchoe, 38, 47–54

Lacerta muralis, 142
Lacerta sicula, 139
Lamphropholas quichenoti, 142
Lathyrus pratensis, 126
Lemna gibba, 38
Leucophaea maderae, 102, 113, 115–18, 122

Man, 37, 224, 277, 281–3
Mare, 258
Medaka (*Oryzias latipes*), 150
Megachile, 37
Megoura viciae, 77, 79–80, 111, 125–34, 146, 149, 150
Mink, 274
Monkey, 224
Myzus persicae, 127, 131, 132

Nasonia vitripennis, 31, 84, 90, 97, 101

Ostrinia nubalis, 30, 31

Pectinophora gossypiella, 16, 26, 33, 34, 95
Penicillium claviforme, 39
Periplaneta americana, 115, 116
Pharbitis nil, 42
Pieris brassicae, 83, 84, 86, 89–93, 95–7, 101
Plodia interpunctella, 84, 97, 101–2
Podospora anserina, 37, 39
Potato, 54
Pterostichus nigra, 93

Quail (*Coturnix coturnix japonica*), 171–2, 179, 181, 185–201, 224, 258

Rabbit, 224
Roach (*see* Cockroach)

Sarcophaga argyrostoma, 29, 31, 33, 67–81, 93, 95–6, 149–50, 188, 191–2, 212
Sceloporus clarkii, 139
Sceloporus magister, 139
Sceloporus occidentalis, 140
Sceloporus olivaceus, 139, 140, 141
Sheep, 193, 197, 223, 224, 258, 260, 264, 274, 275–6
Soy bean, 29
Starling (*Sturnus vulgaris*), 139, 153, 156–67, 171, 172, 174
Stickleback (*Gasterosteus aculeatus*), 150

Tobacco, 47
Trachydosaurus rugosus, 142
Tree shrew, 139
Trout, 169
Turkey, 258

Vole, 205, 237, 273, 280

Warbler (*Sylvia borin*), 153, 155, 155–9, 161–2
White-footed mouse (*Peromyscus leucopus noveboracensis*), 279
White-crowned sparrow (*Zonotrichia leucophrys*), 187, 188, 195
White-throated sparrow, 197, 198
Woodchuck (*Marmota monax*), 166

Xanthium strumarium, 51, 57–63, 66
Xantusia vigilis, 139

GENERAL INDEX

Abscisin, 52
Action spectra, 130
Adrenal, 256
After-effects, 215
Alate morphs, 125–6, 128, 131–2, 134
Anthesin, 46
Apterization, 125–6, 128–9, 131–4
Asymmetrical skeleton (*see* Skeleton photoperiod)
Azide ions, 41

Behavioural rhythms, 113
Bistability, 67, 73, 75
Body weight cycle, 238, 240, 243, 247
Bünsow-type experiments, 83, 85

Calcium ions, 37, 41
Carbachol, 38, 43
Carbon dioxide, 45, 49–50
Carry-over, 193–5, 265
Castration (*see* Gonadectomy)
Cerebral lobes, 115
Chloroplasts, 49
Circadian 'surfaces', 30–3
Circannual rhythms, 1, 153–4, 158–9, 163–8, 223, 261, 264–5, 268–9, 277
Cold Spring Harbor Symposium, 2, 277–9
Conception rate (in man), 285–7
Critical daylength (photoperiod, day length, night length), 32–3, 51, 54, 55, 57, 68, 69, 73, 77, 80, 88, 97, 101–2, 106–7, 130, 144, 151–2, 185, 195–8, 204, 238–9, 262
Critical thermoperiod, 103, 107
Cuticle deposition rhythms, 113, 122, 123

Diapause, 30, 67–8, 72, 74–9, 83, 85–6, 90, 92, 99, 101, 110, 126
5α-Dihydrotestosterone, 254, 255
Dormancy, 54

Eclosion hormone, 115
  rhythm, 2–5, 7, 12, 14, 21–3, 26, 37, 67–71, 76, 78, 115, 189
EDTA, 38
Electro-retinogram rhythms, 113, 122, 123
Embryogenesis (in aphids), 127
Emergence (of insects) (*see also* Eclosion), 26
Entrainment, 1, 9, 23–5, 28, 37, 67, 72–4, 76–8, 84, 111, 114, 116, 119, 124, 137, 139, 154, 156, 172, 174, 175, 189, 198, 203, 206, 245, 265, 280
Epidermis (leaf), 48
External coincidence, 28, 30, 33, 67, 68, 69, 72, 73, 74, 76, 77, 79, 95, 97, 101–3, 105, 143, 148, 149, 189, 191, 192, 212, 250
Extra-retinal photoreceptor, 3, 41, 129, 137, 139, 171, 172, 265
Eye, 39, 116, 118, 121, 137, 139–41, 220, 265

Flight activity rhythm, 116

Floral sundial, 277, 287
Florigen, 46, 54
Flowering hormones, 45–8, 55
  inhibitors, 45, 47, 52, 53
  responses/induction, 45–6, 49, 52–3, 57, 67
Follicle stimulating hormone (FSH), 185, 189, 190, 192, 193, 198, 234, 239, 251–9
Food, 220, 280
Free-running period, 3, 68–9, 154, 205, 277, 278, 280
Frequency demultiplication, 153, 154, 155, 158, 162, 165–7
  following/dividing, 24–5

Gallic acid, 52
Gibberellin, 52
Gonadectomy, 190, 192, 196, 197, 224, 228, 230–1, 240, 252–4, 256, 257
Gonadotrophin (*see* Luteinizing hormone and Follicle stimulating hormone)
Gonadotrophin-releasing hormone (GN–RH, LH–RH), 185, 189, 191, 192, 193, 194, 196, 225, 258, 259, 260
Grafts, 47
Gynoparous morph, 126, 129, 132, 134

Harderian gland, 138, 179
Hibernation, 273
History dependence, 215
Hour-glass, 28, 67, 79, 96, 111, 130, 137, 143, 144, 146, 147, 152, 185, 203
5HT (serotonin), 271–3, 275
Hyamine 1622, 39
Hydroxyindole-O-methyltransferase, 142, 179

Indole-3yl-acetic acid, 52
Internal coincidence, 27–9, 33–4, 40, 67, 76, 78, 97, 148, 149, 189, 191, 192, 212
Interval timer (*see also* Hour-glass)
Ionophores, 39

Juvenile hormone (JH), 125, 131–4

Kinoprene (ZR–777), 125, 131, 132

Leaf, 45, 48, 59, 61
  rhythms, 39, 40, 42, 277, 278
Lithium, 221
Locomotor rhythm, 113, 115, 116, 117, 118, 120, 122, 123, 124, 137, 148, 155–6, 171, 172, 173, 174, 188, 204, 205, 219, 222, 250
Luteinizing hormone (LH), 185–98, 223, 235, 239, 251–9, 271, 275

Melatonin, 1, 137, 138, 142, 152, 171, 174–80, 215–16, 222, 237, 242–7, 249, 269–73, 274, 275–6
Membranes, 61–5
Mortality rate (in man), 285–7
Moult, 113, 153, 155, 158–60, 168, 169, 240

291

Multi-oscillator systems, 2, 11, 21, 26–7, 29, 34, 37, 41, 68, 74, 78, 79, 113, 114, 116, 122, 123, 140, 215, 220–1, 279–80, 282

N-acetyltransferase, 38, 43, 138, 142, 171, 175–9, 237, 245–7
Nanda–Hamner (resonance) experiment, 28, 29, 31, 33, 66, 70, 83, 84, 86, 93, 111, 137, 144, 145, 149, 150, 185, 186, 205
Night-break experiments (see Skeleton photoperiods)
Noradrenaline, 175, 177, 267

Ocelli, 116
Optic lobe, 113, 115, 116, 118, 120, 122, 124
Oviparae, 125–8, 146
Oviposition, 26, 188
Ovulatory (oestrous) cycles, 197, 200, 201, 223, 225–7, 228–9, 261–76

Pacemaker, 9–10, 12, 22, 26, 77, 113, 119, 171, 173, 174, 214–15, 276, 278
Pacemaker–slave oscillators, 10, 11, 14, 16, 20, 22, 23, 25–6, 29, 31, 114
Parthenogenesis, 126 et seq.
Pelage cycle, 238, 240, 243, 247
Phase relationship, 6, 13, 14, 15, 19, 22, 26, 31, 41, 105, 118, 122, 173, 174, 205
Phase-response curve, 3–5, 7, 21, 37, 39, 67, 69, 71–2, 76–7, 96, 119, 188, 206, 207
Photo-inducible rhythm (phase), 33, 67–9, 72–5, 77, 80–1, 83, 99, 111, 137, 143, 144, 148, 150, 185, 187, 188, 189, 191, 192, 198, 203, 204, 207, 209, 212, 215
Photomimetic, 38, 43
Photoperiodic time-measurement, 1–2, 28–34, 39, 41–2, 57, 60, 64, 68, 73, 76, 83, 84, 86, 93, 95, 96, 101, 102, 111, 114, 126, 137, 138, 143, 144, 149, 150, 152, 154, 185, 186, 190, 195, 196, 203, 205, 212–15, 250, 271, 277
Photoreceptor, 3, 37, 39–40, 57, 63–4, 80, 106, 107, 114, 116, 118, 121, 123, 125, 126, 129, 130, 137, 139, 172, 178, 219–20, 265
Photorefractoriness, 168, 204, 205, 244–5, 249–50, 263–4, 269–70, 273, 275
Photosynthesis, 48, 50–1, 60
Phytochrome, 45, 48, 49, 54, 57, 63–5
Pineal gland, 2, 38, 43, 137–42, 152, 171–83, 215, 222, 237–47, 249–50, 255, 257–9, 261–76
Polymorphism (in aphids), 125 et seq.
Prolactin, 234, 239, 252, 264
Propanolol, 38
Protocerebrum, 113, 116, 129
Pulsatile (episodic) secretion, 192, 193, 223–35, 275
Pulse generator, 223–35

Refractoriness (see Photorefractoriness)

Retina, 138, 171, 179, 180
Retinohypothalamic tract, 219–20
Retinol binding protein (RBP), 198–200

Salicylic acid, 38
Selection (on rhythms), 25, 26
Sexual behaviour, 256
Single pulse experiments, 185, 187, 188, 210–12
Singularity, 39, 71
Skeleton photoperiod, 38, 42, 47, 67, 72–5, 77–8, 80, 83, 85, 86, 88, 90, 93, 95–6, 99, 101, 106–7, 111, 146–7, 150, 191–2, 198, 212, 217, 246–7, 262
  thermoperiodic experiments, 93, 99, 104, 106–7
Sleep–wakefulness rhythm, 1, 2, 27, 40, 281–3
Spermatophore production rhythm, 116
Splitting (of circadian rhythms), 137, 139, 140, 215, 280
Steroid feedback, 189, 190, 196, 197, 225–35, 271
Steroid-dependent changes, 231–5, 239, 251–9, 260
Steroid-independent changes, 230, 231, 251–9, 260
Stomata, 48–9
Stridulatory activity rhythms, 113, 116, 122
Suicide rate (in man), 285–7
Superior cervical ganglion, 172, 239, 246, 265, 275
Suprachiasmatic nucleus (SCN), 1, 2, 43, 141–2, 171–2, 180–3, 213, 215, 219–22, 239–40, 245, 255–9, 271
Supra-oesophageal ganglion, 113
Symmetrical skeleton (see Skeleton photoperiod)

T-experiment, 9, 25, 28, 30, 31, 67, 76, 77, 137, 144, 149, 150, 156, 157–68, 185, 188, 207–9, 212–15, 217
Temperature effects, 2, 4, 24, 27, 32, 33, 37, 40, 61, 83, 101, 104, 143, 149, 152
  pulses, 118, 119, 120, 121, 124
  rhythm, 44, 172, 281–3
Testicular cycle, 155–69, 203, 238, 242–5, 251–9
Testosterone feedback, 251–9
Thermo-inducible phase, 83, 96, 97, 99, 105, 111
Thermoperiodic experiments, 108, 109
Thermoperiodism, 78, 83–4, 88–9, 92, 97, 101–3, 104, 111
Thyroxine, 199, 200
Thyroxine-binding prealbumin (TBPA), 198–200
Tocopherol, 58
Tocopherol oxidase, 57–66
Transients, 16, 17, 19, 20, 25, 172
Translocation, 53
Transplantation, 114, 115

Vanadate ions, 41
Virginoparae, 125–34, 146
Vitamin E, 57

Xanthoxin, 52

Zeitgedächtnis, 1, 2, 78